WATER WORKS

Water gazes. Photos: Henriëtte Waal & Coltrane McDowell

WATER WORKS
Ecosocial Design

Henriëtte Waal &
Clemens Driessen
(eds.)

Contributors

Jon Ardern
Mari Bastashevski
John van Boxel
Janna Bystrykh
Center for Genomic Gastronomy
Yi-Fei Chen
Natsai Audrey Chieza
Ștefan Constantinescu
Ahmed Salem Dabah
Carolina Domínguez-Guzmán
Clemens Driessen
Brecht Duijf
Faber Futures
Julien Fargetton
Sylvain Hartenberg
Shahnoor Hasan
Clara Sika Helbo
Ils Huygens
Anab Jain
Zairah Khan
Mahmoud Khattab
Studio Klarenbeek & Dros
Sjoerd Knibbeler
Sophie Krier
Rebecca Lave
Atelier LUMA
Mary Maggic
Francesca Masoero
Coltrane McDowell
Kent O. Miller
mirko nikolić
OOZE Architects
Roric Paulman
Julia Pelka

Fernando Felipe Pérez Riojas
Joana Pestana
Mariana Pestana
Eva Pfannes
Benedetta Pompili
Marjetica Potrč
QANAT Collective
Rimini Protokoll
Sandrine Rozier
Rutger Schrijer
Susan Schuppli
Moreno Schweikle
Veronika Sedlmair
Islam Shabana
Esha Shah
Brynjar Sigurðarson
Superflux
Mihnea Tănăsescu
Serina Tarkhanian
John Thackara
Billy Tiller
Daniela Tokashiki
Moniek van de Vall
Jennifer Veilleux
Heidi Vogels
Kees van der Vooren
Henriëtte Waal
Ria Waal
Daniel Wetzel
Thijs de Zeeuw

Valiz, Amsterdam

CONTENTS

p. 148

p. 205

p. 255

p. 309

p. 354–355

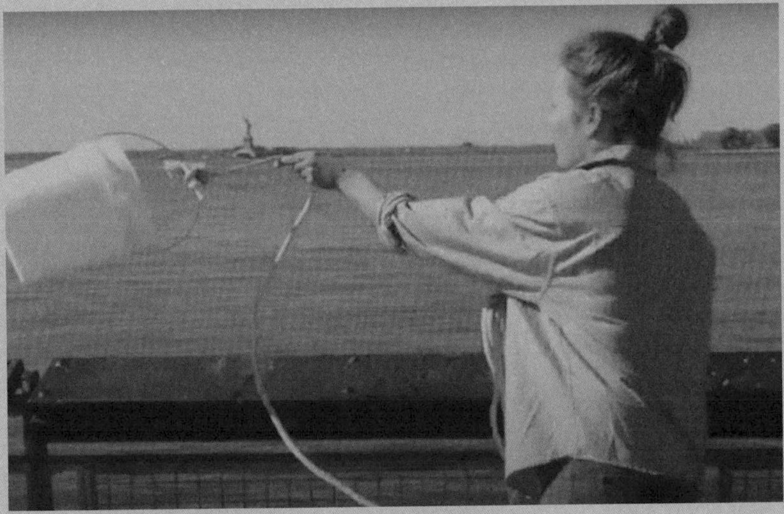

A Taste of the River

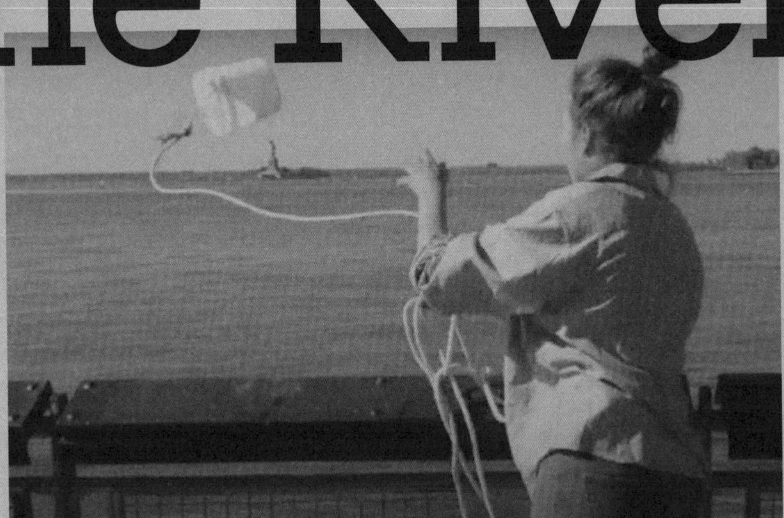

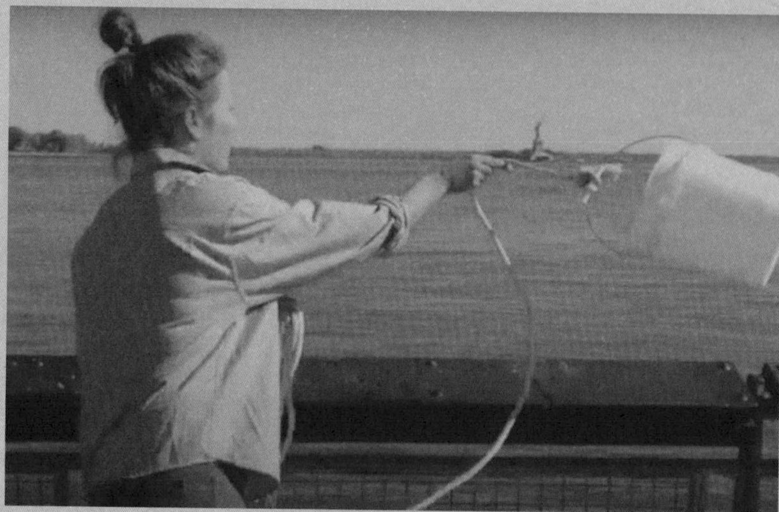

A Taste of the River

Prologue

Brewing *Sloterplas Blond*, Amsterdam, 2010. Photo: Henriëtte Waal

A Taste of the River

BY HENRIËTTE WAAL & CLEMENS DRIESSEN

We are standing on the bank of the Amstel river, in the south of Amsterdam. The atmosphere is festive. A group of people has gathered holding fresh beers. We clink our glasses and take a sip, with half an eye on the water flowing by, slowly, murkily?

Henriëtte Waal, *Buitenbrouwerij*, gathering wild plants for the brew, 2010. Photo: Jorn van Eck

Two months ago, Henriëtte was at this same spot with her mini-trailer. It has an open frame in the shape of a little house, holding large stainless-steel pots at different heights to allow brewing without heavy lifting. Various filters, a large stirring spoon, pH testers, thermometers and refractometers to measure density and sugar content hang on the frame and can be used when needed. With a small built-in ladder you can reach the top pot. Henriëtte has branded this her Outdoor Brewery—*Buitenbrouwerij*.

She hoisted water from the river and then started brewing on the spot. Because the river water is slightly brackish, she chose something similar to a Gose, a forgotten salted sour wheat beer style using wild fermentation.

But now, tasting the beer, one cannot help but wonder what is upstream. What runoffs flow into this river, are there pesticides, hormones in there? Beer brewing was a common water purification method in the Middle Ages. But does this actually purify the water to take out modern contaminants? Just swimming in this river doesn't look very appealing, let alone drinking from it. But here today a new collective is coming together, bodily invested into relating to the watershed. Drinking beer becomes an act of

Opening images: Henriëtte Waal harvesting water in New York Harbour, 2014 solidarity with fish and other water-based life; an invitation to enter into new ecosocial relations. People are surprisingly willing to join the party and engage. Nobody (else...) seems to have any hesitations about drinking the beer. Ready for another glass even.

With her mobile beer brewery—a sober but elegant assemblage of steel, pots and pans, and years of skill in working with (wild) yeasts and more or less helpful local brewing enthusiasts—we get a fresh taste of the river and its landscape, including some of its complexities and interdependencies. We become aware of the vulnerability we share with other organisms, with whom we are all filtering each other's medium of life.[1] All brought together into something uplifting and easily digestible.

1. Astrida Neimanis, *Bodies of Water: Posthuman Feminist Phenomenology*, Bloomsbury Academic, 2017.

OUTSIDE BREWERY BACKGROUND

Buitenbrouwerij with Anomaly and Village Pump, *Gracht Unreasonable Pale Ale*, 2016

Waal uses beer brewing to make 'dirty' waters drinkable: crafting a fresh way of engaging with, tasting, the landscape. Water purification now mostly takes place out of sight in closed-off treatment plants, while purifying water and brewing beer once was something people did themselves. In 2008, when beer was still mostly a product of large corporations, Waal designed an autarkic mobile water-cleaning system annex beer brewery on a trailer to be able to brew unique and local beer.

Several special beers were developed with local/seasonal ingredients such as rainwater, snow, blossoms, pioneer species, coffee grind, mushrooms, and microbial cultures. With rivers including the Waal, the Rhine, the Meuse, the Gulp, the Zenne, the Emscher, in a carwash and an allotment garden and—as a provocation—at the world's leading trade fair for horticulture in Berlin. On an island in the Wadden Sea, in wetlands and tidal flats, the Outside Brewery developed Zeebier, Rampzalig, Eilander and Parelduiker, beers with sea water and mineral-rich dune water from the freshwater bubble under the island.

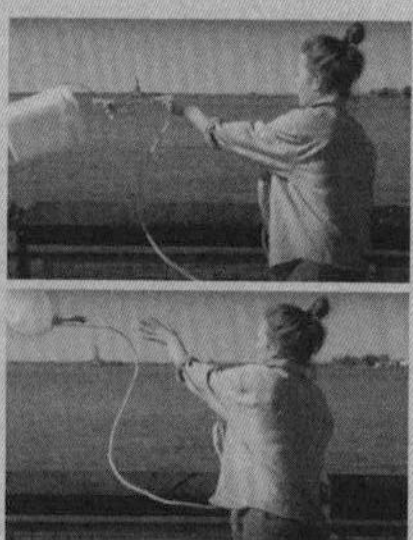

Henriëtte Waal harvesting water in New York Harbour, 2014

Traveling through landscapes in the Netherlands, Belgium and Germany, the mobile brewery allowed local brewers, beer enthusiasts and those passing by to taste their local waters, leaving a trail of recipes and microbreweries behind. It also led to a platform with monthly meetings in Amsterdam about beer, tasting, microorganisms, and fermentation, that tests and discusses the latest beer design, but also the failures and crossovers.

Jetze de Beer and Henriëtte Waal, *Buitenbrouwerij/Sea Beer*, 2012.
Photo: Geert Snoeijer

Henriëtte Waal, tapping dune water from the fresh water bubble of Vlieland island, 2015

Henriëtte Waal, *Buitenbrouwerij*, 2010, Sloterplas, Art at the pool festival. Photo: Jorn van Eck

Henriëtte Waal, *Buitenbrouwerij*, pumping and filtering ditch water with charcoal and ceramic filters, The Hague, 2010

Henriëtte Waal, *Buitenbrouwerij/Amstel Beer*, 2018, Amstel river, exh. 'Going Nomad', Zone2Source testing ground for art & ecology, Amsterdam. Photo: Coco van Duijvenvoorde

Henriëtte Waal, *Bottle Beach*, 2014, rain water brew in wild harvested old bottles from Bottle Beach/Dead Horse Bay, New York (US), Photos: Henriëtte Waal

In the years subsequent to inventing her mobile brewery, Henriëtte has sought to embed ecosocial design as an approach at regional scales of the Camargue/Rhone Delta and across the Mediterranean (Atelier LUMA), and more recently in the Frisian peat meadows (Veenweide Atelier). This book is a reflection on—and of—these efforts.

BEER MAKING

You first mix crushed grains (or another carbohydrate source) into water and heat at different temperatures to break the starch into simple sugar molecules that can ferment. Then the grain husk is used as the first filter and sparging step, which filters larger particles or little animals out of the brew. The second step is cooking, with herbs for bitterness. The third step happens after cooling down and allowing yeasts to eat the sugars and 'pee' alcohol, which enhances the taste and also preserves your brew.

Introduction

BY HENRIËTTE WAAL & CLEMENS DRIESSEN

WATER JUST BEING WATER

'But water is just water, right?' is a response Henriëtte gets a lot when she says she works with water. This book can be read as one long attempt to move away from this assumption. And to get a sense of the incredible infrastructure and amount of work needed to produce this idea, this experience of a universal substance that is mundane, often seemingly tamed, present in the modern home.

Forgotten, ignored, paved and pumped—we have inherited particular water systems which also fix infrastructures of the mind: a certain shared imagination, of who is an expert (real experts), how to think of quality (uniform), and who is in a position to do something (not us). As a result, many types of conversations around water fail to see ecosocial connections, even while mainstream awareness of water issues appears to grow.

Water is a process, not an object. It flows by, evaporates, rains, circulates through bodies. Water is a medium of ecological and social relating. Water *works*, by suddenly appearing in unexpected ways, by filling in cracks, seeping through, condensing, leaking out of containers. Water *makes*, and is also a *tool for making*—boiling, cooling, washing, acting as a solvent, a vehicle for enzymes.

In this book, divided into seventy odd chapters, we learn from a range of efforts to make, fix, and rework water. To in new ways understand, to care, to fight for water, to learn to love water. Thereby we unpack the abstraction of 'modern water' to get to know the hydrosocial relations it contains, or mediates.[1] We treat these efforts as so many approaches to 'ecosocial design'. How to understand that in the context of this book?

1. Jamie Linton, 'Modern Water and its Discontents: A History of Hydrosocial Renewal', *Wiley Interdisciplinary Reviews: Water*, *1*(1), 2014, pp. 111–120.

ECO

Ecologies are not neutral nor natural. Ecological thought can help envision the irreducibly complex ways in which designs may turn out into the world, in contrast to linear, mechanical forms of imagining causal impacts.[2] But

2. Viveka Hocking, 'An Ecology for Design: From the Natural, through the Artificial, to the Un-Natural', *Sustainability Research by Designers: An Anthology*, Common Ground Publishing, 2011.

the particular way the natural world is imagined, what it is made up of and how elements are seen to relate is not obvious, and deeply political.[3] What are relevant ways to learn about the incredible interdependencies that we live in? Anna Tsing contrasts natural histories of place-specific relations between deeply entangled organisms in environments, with a formalized science of ecology. A science that emerged together with neoclassical economics in which the still dominant fantasy of isolated individuals competing against each other are seen as the drivers in functional systems.[4] Nature then is imagined as a collection of species, positioned on a linear scale between extinct, threatened, vulnerable, or 'of least concern'. Biodiversity in this approach is understood in terms of genetics of individual organisms, functionally related in certain compositions that produce ecosystem services. This simplified nature, reduced to urgent if limited human interests, actively ignores all other ways in which ecologies could be understood and valued.[5] It easily glosses over the ever-changing variation of organisms learning to live in particular places, which demand but also maintain water.

3. Mihnea Tănăsescu, *Ecocene Politics*, Open Book Publishers, 2022.

4. Anna Tsing, *The Mushroom at the End of the World: On the Possibility of Life in Capitalist Ruins*, Princeton University Press, 2015, p. 28.

5. Esther Turnhout & Casey Lynch, 'Raising the Carbonized Forest: Science and Technologies of Singularization', *Environment and Planning*, 2024; Irene J. Klaver, 'Radical Water', *Hydrohumanities: Water Discourse and Environmental Futures*, 2021, pp. 64–88.

There is now no place on Earth where the climate is not a looming prospect that evaporates all plans, threatening to destroy life. There is the global problem of fossil fuel-intensive agricultural and energy production. But this focus on CO_2 emissions ignores how the climate is always also local. Vegetation does not just capture carbon and consume water, but also regionally influences rain and evaporation, shapes temperatures while retaining water in place, keeping soils moist and alive. Getting to know ecologies in particular places is more than mere noting presence or absence of species, but tuning into processes of adapting and relating, learning about stories produced by and meaningful to all kinds of creatures. And water always plays a key part in these processes of learning to live in places.

SOCIAL

We can do something. But assuming an obvious 'we' carries the risk of erasing differences, inequalities, and levels of exposure and despair. While also all too easily disconnecting an exclusive 'human' isolated from all other organisms and substances. Starting from a social world that is never independent of ecological landscapes, we can, with Astrida Neimanis, also enter into a different sense of self, a different phenomenology, another way of experiencing ourselves in environments.[6] When our bodies are always permeable and overflowing bodies of water, it is impossible to isolate an 'us' from intimate relations and ecosocial processes. Social structures divide the world into those without the ability to defend their own water and those who in the name of immediate return on investment end up destroying the capacity of water bodies to restore themselves.

6. Astrida Neimanis, *Bodies of Water: Posthuman Feminist Phenomenology*, Bloomsbury Academic, 2017.

How to envision collectives that learn to move water differently? All the terms representing social groupings should not be taken at face value: when we hear 'stakeholders', that usually means there will be limited forms of involvement; 'the community' suggests more unity than is actually present, while we all know community and participation are riddled with power and interests, and awkward personal relations.[7] When seeing design as not just oriented to users, consumers, or maybe citizens, we can go beyond individualizing the affective relations that make up the social, while embracing the joy of coming together to do something.

7. Mihnea Tănăsescu, *Ecocene Politics*, Open Book Publishers, 2022.

As a designer you can either help competitive neoliberal consumers by seeking to optimize smooth and carefree devices and infrastructures that satisfy their preferences, or you can generate forms of collective experiences that open minds to storying, and allow for the creation of meaning and solidarity across unlikely partners, human or otherwise.[8] While sometimes these options—designing for individualizing or communing—are hard to keep apart.

8. Agnieszka Wołodźko, *Affect as Contamination: Embodiment in Bioart and Biotechnology*, Bloomsbury Publishing, 2023.

DESIGN

Implicated in histories of industrial destruction, haunted by the wastefulness of inventing and selling ever-more stuff, (industrial) design is struggling to find its role in a time of combined ecological and social crises.[9]

The time for ironically keeping the world at a distance with a knowing smile seems over. But artistic design can bring other forms of excitement, desire and enthusiasm, plus fresh dynamics to break through scenarios of only doom.

9. Arturo Escobar, *Designs for the Pluriverse: Radical Interdependence, Autonomy, and the Making of Worlds*, Duke University Press, 2018.

In product design, the interest of companies remains central. The promise of ecosocial design is a layered approach that draws in an endless range of figures, methods, and roles: an ecosocial designer can be a performance artist, a healer or an engineer—and will probably find out they need to all three, in order not just to inspire some people but also establish alternative networks of production. There is enormous potential to engage in systemic design, integrating critical systems thinking and design practices, bringing together a deep understanding of interconnectedness, of ecological and political processes with technological and aesthetical expertise and deep insights into production systems and socio-cultural behaviour of all kinds of users of their environment.

Design thereby has to take into account a multidimensional set of factors, relations and networks of existence. It is about carefully working through these, rather than about superficial aesthetics and immediate profit.

ECOSOCIAL DESIGN

All design is ecosocial design,[10] but it matters what kinds of naturecultures are made and who gets to shape and narrate these. Ecosocial design then is about creating conditions under which the difference between better and worse can be worked through in the most interesting, critical, and creative way.

Ecosocial design starts from the presumption that 'human nature is an interspecies relationship'.[11] This means that shaping and staging the relations between humans and all the lifeforms and materials we are entangled with is key to intervening in our world. The ways in which our environments are ordered and formed may promote or diminish this sense of connectivity; it may erase relations or make us see them with new eyes.[12]

Ecosocial design is about taking up the challenge of actively crafting ecologies and societies as embodied relations, as ways of knowing and telling stories. And at the same time water makes very particular kinds of material connections, the sense of wetness that muddles what may seem solid features of land.[13] Integral to designing then is actively engaging with the ethical and political nature of these relations. This type of work can best be understood by seeing 'objects' of design as ever-changing multilayered relations.

10. Jan Boelen and Michael Kaethler (eds.), *Social Matter, Social Design*, Valiz, 2020.

11. Anna Tsing, 'Unruly Edges: Mushrooms as Companion Species', *Environmental humanities*, *1*(1), 2012, pp. 141–154.

12. Henriëtte Waal & Clemens Driessen, 'Walk like a Designer: Following the Haagse Beek, Artist Krijn Giezen, and the Water System of The Hague', in Christian Ernsten & Nick Shepherd (ed.) *Walking as Embodied Research; Drift, Pause, Indirection*, Routledge, 2024, p. 224.

13. Anuradha Mathur & Dilip da Cunha, *Design in the Terrain of Water*, Applied Research+Design Publishing, 2014.

WATER

Does it make ecosocial sense to focus on water at all, separating it from its ecologies, muddy sediments, spots where water whirls, speeds up or slows down? What good does it do, to gather around this 'single' universal substance? This is a book about freshwater. The oceans may evoke an image of a blue planet with abundant water; but freshwater scarcity, pollution, and violence generate distinct types of ecological and social questions. Three per cent of water is freshwater, two thirds of which is polar and glacier ice, the rest is ground water, leaving only 0.3 percent of freshwater 'readily available'. This type of data emphasizes water as a single homogenous global resource: 'virtual water' that can be ascribed to certain products and thereby subject to trade, encapsulated in avocados and mangos, making for a logistic flow of water from economically marginal to rich places.[14] Looking at water more closely, all the issues it is related to come to the surface: sanitation and gender, geopolitics and hydropower, fish and fishing. What does it do then to focus on water, rather than talk about access to land, food, how to conserve biodiversity, how to care for heritage? Water changes things; everything.[15] Tracing water offers a way to understand landscapes as interconnected in often surprising ways. In many places, water is already being depleted and polluted to the point of ecological and social destruction. But with rising temperatures, water as a multi-sided, multi-sited mega-problem in the making absorbs everything and leaks into everything—with water, things are never only local, but water connects, disperses, and maybe offers new directions to think and create.

14. John Anthony Allan, 'Virtual water: a strategic resource', *Ground water*, (36:4), 1998, pp. 545–547.

15. Arianna Tozzi & Irene Leonardelli, 'How Water Changes (Every) Things: A Feminist Study of how 'Water Worlds' Shape Processes of Rural Agrarian Transformations in Maharashtra, India', *Routledge Handbook of Gender and Water Governance*, 2024, pp. 357–371.

WORKS

For this book we sought to treat a large range of ways of making, crossing vastly different kinds of conditions and sets of relations, making for very different water bodies and water works. Besides a set of thematic essays by relevant experts, this book contains interviews with makers; reports of site visits, and we delve into images we were captivated by. There are unresolved frictions between chapters, between solutions and despair, commerce and violence, making and letting go.

Water Works moves across and between the arts, design, film, the sciences, technology and engineering—seeing these as working alongside but also crossing each other, not evidently different in their approaches or motives. We draw on river scientists and engineers—fish-lovers, wastewater treatment plant operators, and artists, filmmakers, architects, farmers, and anthropologists. We approached them symmetrically, basically asking the same questions regarding their experiences of seeking to intervene in both the material infrastructures and cultural imagination of water.

This is not the definitive overview of key solutions to all water problems, nor do we provide a universal blueprint for changing relations to water for the better. Instead, these chapters are a series of portals into different ways of living with water; invitations via water to learn new things about ecologies and possible ways of being human—or more-than-human, or at least more than isolated individuals—in relation to water.

We see how societies and ecologies have been designed, by designers, engineers, or informally, by experts—some of whom may not be human. We then see more than only extractivist water grabbing as a form of land grabbing: water as the ultimate liquidity and vehicle for cheap nature that can be used and depleted unseen. How do actual people—and beavers, eels, and microbes—employees and users, designers and other makers, navigate, critique, work around, or work with these powerful institutions and dreadful developments?

These works offer counterpoints to a tendency to think in terms of technological design as if these solutions work without complicating ecological or social considerations. Moving away from large technological solutions and associated systems thinking as the only way to understand and intervene meaningfully, whereas infrastructure, product design, architecture, all tend to be built around the ready availability and the immediate discharge of water.

We believe it is important to understand these efforts by learning from people closely involved in making, the trouble you get into—and stick with—when seeking to realize change. Key here is sensitivity to experiential worlds of others; to cultivate curiosity, empathy. To realize that what seemed like a single obvious universe in fact is a pluriverse.[16]

16. Arturo Escobar, *Designs for the Pluriverse: Radical Interdependence, Autonomy, and the Making of Worlds*, Duke University Press, 2018.

None of these projects claim to be perfect. In writing, especially of the critical academic kind—there may be a tendency to search for moral clarity and political purity. To divide the world into good and bad, to seek clear divisions between friends and enemies, extractivist and indigenous, essentialist and open ended, commerce and commons, small and big, citizens and consumers, engineering and art. Oppositions that in practice mostly turn out way more muddled, with weird middle grounds or imperfect blends.[17]

17 Anthony Dunne & Fiona Raby, *Speculative Everything, With a New Preface by the Authors: Design, Fiction, and Social Dreaming*, MIT press, 2024.

Ecosocial design is then about muddling through. This means reflecting on ecosocial design is not all about form and material, but seeing relations themselves as what is to be shaped. Evoking the process of this type of making is the aim of this book, which orients the writing, interviews and images towards the sense of a world made of relations.

SEVEN THEMES

The chapters are organized under seven themes, each of which highlights a particular friction for ecosocial design in relation to water: *Purity*, *Wild*, *Scale*, *Representation*, *Violence*, *Infrastructure*, and *Commerce*. Every theme can be seen as 'material' to creatively repurpose, to engage with actively—reinventing what counts as infrastructure, how to think of scaling up, or out, or unscaling; how to get paid without selling out; how to resist aggression and make space to play; how to love water when it's permanently polluted?

Let's dive in.

Purity

Purity

1
Purity

BY HENRIËTTE WAAL & CLEMENS DRIESSEN

There's something in the water. A sense of unseen agents affecting everyone. Or some more than others.[1]

Pollution is often defined as the condition under which a water body is no longer able to naturally purify itself. In this perception of water as a resource to make optimal use of, purity is not the ideal; the maximization of 'safe' pollution is.[2] Threshold values of individual harmful substances define a free pollution space.[3] A built-in economics of dilution. This is how, in times of drought, rivers dwindling into meagre streams suddenly become too toxic. Whereas water abundance reveals toxicity too: flooding events tend to generate dangerous cocktails of sewage and toxic chemicals. Sandbags used to protect against floods are to be treated as chemical waste after the water has subsided.

Coming together to protest against pollution is a powerful form of activist gathering. Especially in the name of pristine bodies of water. But the fight against pollution can also turn into an elusive ideal, implicitly isolating humans, disconnecting them from a harmful outside world, ignoring the impurity of life. 'Beware of purity, it is poison to the soul'.[4]

Imagined as an unpolluted state, purity is hard to work with in a 'permanently polluted world'.[5] Especially when toxicity is hard to establish, think of the countless toxins that are undetectable to our human senses: clear rather than murky water may be holding profound dangers.

The ideal of purity runs the risk of promoting a substance, a form of water that is dead, static, separate from connections and processes of change; an essentialist, timeless dream of isolated bodies. Forever tainted by the insertion of any alien molecule whatsoever.[6] Pushing cleaning to the limit in the name of pure water, nano filtration is proposed as the new standard, promising to remove 'forever chemicals' and heavy metals, but also completely drains the life from the water.[7]

1. Ingrid R.G. Waldron, *There's Something in the Water: Environmental Racism in Indigenous & Black Communities*, Fernwood Publishing, 2021.

2. Max Liboiron, *Pollution is Colonialism*, Duke University Press, 2021.

3. Ronnie D. Lipschutz & Doreen Stabinsky, *Environmental Politics for a Changing World: Power, Perspectives, and Practice*, Rowman & Littlefield, 2018.

4. Michel Tournier, 'Gardez-vous de la pureté, c'est le vitriol de l'âme', *Vendredi ou les limbes du Pacifique* (Vol. 373), Gallimard, 1972.

5. Max Liboiron, Manuel Tironi, & Nerea Calvillo, 'Toxic Politics: Acting in a Permanently Polluted World', *Social Studies of Science*, (48:3), 2018, pp. 331–349.

6. Ibid.

7. David. L. Chandler, 'New Filtration Material Could Remove Long-Lasting Chemicals from Water', *MIT News*, 6 September 2024. www.news.mit.edu/2024/new-filtration-material-could-remove-long-lasting-water-chemicals-0906

Opening image: Brecht Duijf, *Duik* (*Dive*), 2023, photograph, Loosdrechtse plassen

The ideal of purity comes with an approach of knowledge that isolates, and is expensive, while neutralizing every potential life form in the name of protecting pure states. In purity, modernity appears as a double-edged operation—both defining purity and then inexorably polluting—whereby purity is then sold back to us in plastic bottles.

And purity can be unhealthy, for instance when pure H_2O in the body of an organism becomes an agent of extraction, as the molecule's electric charge attracts mineral particles that living bodies need for muscle and nerve control.

How can we design for a world that is inherently messy, diverse, and irreversibly polluted, without enforcing a sanitized, uniform echo chamber that stifles diversity? Ecosocial design suggests we start in 'dirty labs': grounded, real-world spaces where communities can find ways to experiment and work within the messy situation they find themselves in, seeking not so much purity but justice, through slow and intimate forms of activism, which can also be a way to speed things up.[8]

8. Max Liboiron, Manuel Tironi, & Nerea Calvillo, op. cit., pp. 331–349.

Rather than promising cruelly optimistic solutions, ecosocial design provides a shared space to address the psychological and emotional impacts of living with constant, often invisible, dangers.[9] It acknowledges feelings of loss and grief while offering a guide to navigate the anxiety and uncertainty that come with our messy, impure bodies and environments.

9. Lauren Berlant, *Cruel Optimism*, Duke University Press, 2020.

To be meaningful, however, ecosocial design must also confront the broader forces that are intent on dumping even more harmful stuff into water bodies. As long as we keep producing huge amounts of single-use plastic full of endocrine disruptors, no effort of cleaning will ever attain anything.[10]

10. Max Liboiron, *Pollution is Colonialism*, Duke University Press, 2021.

In this state of impurity, ecosocial design must find ways to grieve, and sometimes to simply dwell in denial, allowing us to coexist with ever-present pollution without becoming toxic ourselves. This means embracing the diversity and generative potential of mutating impurities and learning to work with these agents of surprise when control is no longer viable; to collect blooming micro-organisms that remediate toxic metals from water and integrate them in objects as colourants; to capture and embed toxic clay in our everyday environments as cultural objects, and to open up impure bodies to each other.

Ils Huygens is a curator, writer and researcher currently connected to the University of Antwerp. She was a curator at Z33—House for Contemporary Arts, Design & Architecture for 15 years where she developed numerous exhibitions focusing on art, ecology and extractivism ('On-Trade-Off: Charging Myths', 'The Sustainist Gaze', 'Nuclear Culture', 'Seasonal Neighbours').

1.1 Chemical Forever

BY ILS HUYGENS

If we are going to live so intimately with these chemicals, eating and drinking them, taking them into the very marrow of our bones—we had better know something about their nature and their power.[1]

1. Rachel Carson, *Silent Spring*, Houghton Mifflin Harcourt, 1962.

FOREVER CHEMICALS

PFAS (Per- and polyfluorinated alkyl substances) are a group of chemical compounds that are used to make products water and grease repellent. PFOS and PFOA are currently two of the most discussed types of PFAS, as they are considered the most toxic of a chemical family of over 12,000 types. 'Mapping the universe of PFAS, and trying to understand which are bad and which aren't, is an important activity but that is going to take decades and by then it could be too late to do anything about it', says PFAS expert Ian Cousins.

After PFAS came under scrutiny, the companies worked out new types of compounds with shorter chemical chains (and as the industry claims, less hazardous effects), some of which also were rebranded as GenX.[2] Currently companies are continuously changing the chemical structure with small details, chasing tails with new regulation and legislation in a game of extremely unequal trade-off between profit and future human and earthly wellbeing.

2. GenX, introduced in 2009 by DuPont to replace PFOA is part of a newer generation of short-chain PFAS designed to have fewer carbon molecules than the original long-chain PFAS. These were initially believed to be less toxic and more quickly excreted from the body. But some new studies suggest that these relatives may pose many of the same risks as their predecessors.

The origin of the white, waxy substance goes back to the 1930s and 1940s, like many modern technological innovations, it was initially developed in a military context. Scientists working on the 'Manhattan Project' first discovered how fluorine gas bonds with carbon molecules, producing the most highly resistant coolants ever known, crucial for the development of the Trinity—the first atomic bomb that was detonated in 1945. It extinguishes fires and repels grease, dirt and water better than any substance ever created. Chemical companies like DuPont and 3M began manufacturing it on a commercial scale to produce waterproof and stain-resistant qualities for many modern consumer products from Scotchgard tape to waterproof fabrics and Teflon coatings for kitchenware. Over the next decades, thousands of different PFAS chemicals emerged and found their way into a dizzying array of products that we encounter, wear and eat daily: from dental floss and toilet paper to food packaging, paints, firefighting foams, to waterproof clothes and shoes. A 2024 study found that it even appears in vegetables all over Europe, with the highest rates in the Netherlands and Belgium, thanks to PFAS mixed into pesticides.

What gives PFAS its magical capacities are the carbon-fluorine bonds which form an unbreakable chemical chain. This simply means it does not degrade over time, is highly persistent and bio-accumulative, so once it is present, it stays in bodies and builds up in the food-chain. From the start it was clear to the producers that this substance, which they cynically dubbed 'forever chemicals', was hence highly toxic and should never be released in nature, let alone in the earth's water systems.

LEAKY BUSINESS

Before 2021 most people in Belgium had probably never heard of PFOS—a kind of PFAS, when a huge scandal broke out involving the 3M factory near Antwerp. Huge areas of PFOS-contaminated soil were discovered on a construction site and were reported to the government already in 2017. Despite the seriousness of the situation, the discussion was kept behind closed doors, no real action against 3M was taken and it took another four years and persistent activist struggling to get broader public attention for the impact of the contamination. Moreover, once journalists, and later government committees, started digging deeper, it became clear that massive PFOS/PFAS contamination

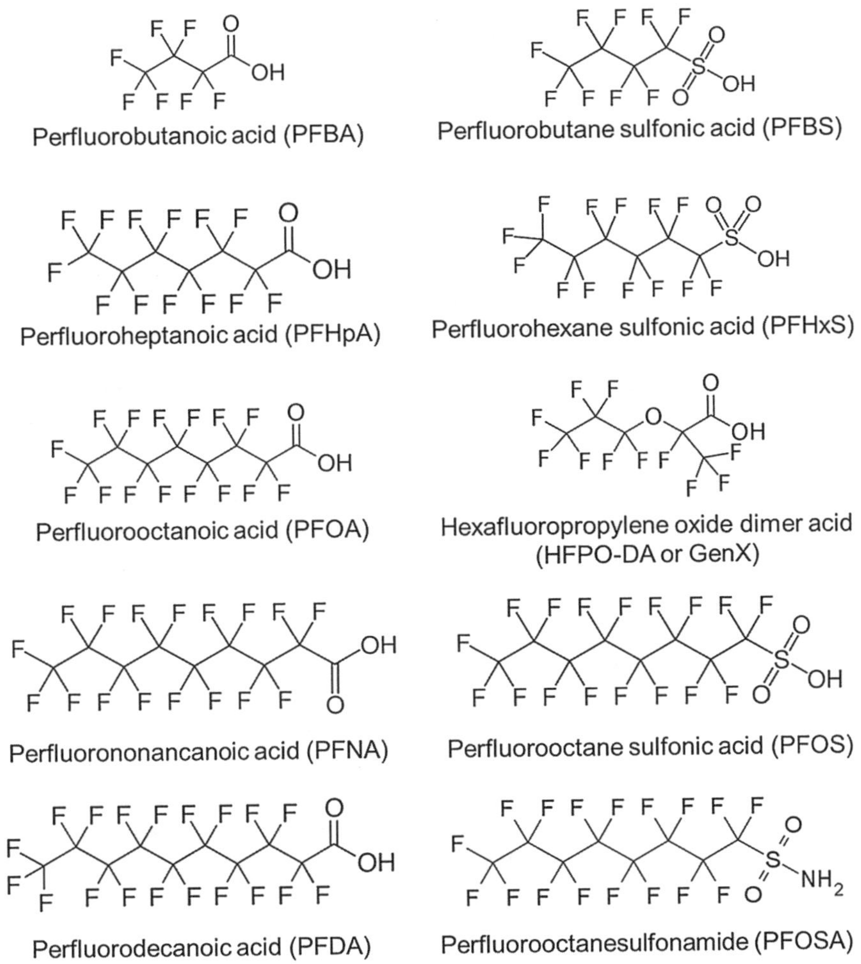

Bevin Blake and Suzanne Fenton, the structure of common PFAS molecules, 2020

had been taking place for decades through air and by dumping wastewater into the Schelde river. In 2021, first blood samples were taken from people living in a range of five kilometres from the 3M factory to test for PFOS and thirteen other dangerous PFAS-types. The relatively small study revealed only nine per cent of the people were considered 'safe' in terms of health regulations, some people even showing values that are over a thousand times above HBM-I or HBM-II values (human biomonitoring). People in the direct vicinity of the plant were advised to stop eating eggs and vegetables from their garden, to wash off sand and dust and stop using groundwater. Not everyone was keen on sending their blood samples in for testing.

Meanwhile in the Netherlands a similar scandal emerged around Dordrecht where Chemours (formerly DuPont) has been based since 1962. Journalists revealed a trail of reports, showing how both DuPont and 3M were aware of the health dangers of PFAS as early as 1978 and had even been exchanging information. Internal documents emerged from the end of the 1970s up until the end of the 1990s showing how American corporate scientists were sharing research results linking the substance to cancer, immune system deficiencies, diabetes and many other health risks: clearly showing they were well aware of the environmental threats, and the water pollution near Dordrecht. It wasn't until 1999, when American lawyer Rob Bilott started to dig deeper and acquired access to these files, that the toxicity of PFAS and its omnipresence was beginning to dawn on the public. Bilott's research led to the world's first PFAS-related lawsuit in Ohio and West-Virginia—a story that became famous via the Hollywood film *Dark Waters* in 2019.

Today, PFAS is linked to major health threats such as various types of cancer, fertility problems, changes in metabolism, increased risk of obesity and much more. However, it remains hard to obtain proof to link effects directly to the contamination, let alone to make a claim on the long-term and intergenerational impact this man-made chemical compound will have. This makes it difficult to file legal claims against PFAS-producing companies. In the early 2000s, both PFOS and PFOA (considered the more dangerous PFAS types, produced near Antwerp and Dordrecht) have been phased out, following new directives of the mother company in the US, where PFOS was banned in the wake of the West-Virginia case. But this is a far cry from stopping the PFAS pollution. Meanwhile, new types of PFAS are being released on the market on a daily basis, and due to a lack of regulation for each specific type, still being dumped in the environment.

Since PFAS is more actively monitored in the wake of the recent scandals, it is simply popping up everywhere. New contaminated sites are discovered every day, all over the world, not only via direct producers like 3M or DuPont, but also through secondary chains that work with PFAS (textile, paper and plastics factories, fire stations, construction sites). We are only just beginning to realize the massive scale of the problem. Precisely because of its water-repellent magic it has managed to seep into every leaky substance around the globe, moving through oceans (up till now especially in the northern hemisphere) as well as the clouds in air droplets and rain, contaminating our soil, surface and drinking water, moving up the food chain, flowing into our blood and breast milk, being passed on to the bodies of future lives.[3]

3. Studies have shown it to be found in the blood of 97% of Americans: www.niehs.nih.gov/health/topics/agents/pfc/index.cfm#footnote1

Globally, wherever measurements are done, it is found, from the ice layers in the Arctic to the Mount Everest. Because of their persistence and unbreakable chemical chain, these 'forever chemicals' will remain present for centuries even after production and emissions have ended. And although research and experiments on decontamination are being developed, there is no viable solution yet to break them down, let alone to remove them from the earth's water systems.

PFAS AND THE CULTURAL IMAGINARY

For the larger public it is extremely hard to come to terms with the scale of PFAS' global impact. The discussion of PFAS in the media focuses its attention to evermore discoveries of new contamination sites and on tracing corporate (and sometimes also political) cover-ups. Scientists are busy monitoring and debating health impacts with the industry, while governmental committees try to work out who was responsible for what, and judges analyse claims from affected citizens. Each of them seeks to break the PFAS issue down into manageable problems that can be addressed and tackled through legal claims, financial compensation or regulation, all of which is of course good, but also carry the potential of normalizing and naturalizing the danger.

The linear model of disaster time makes the fading of disaster to normalcy both the condition and the ongoing goal of expert and policy interventions, with the aim of packaging perpetual uncertainties into acceptable risks. Yet, as Ulrich Beck has observed, risks in late industrial environments cannot be contained 'in terms of time and space', nor can their delayed effects be predicted 'according to established rules of causality, blame, and liability'.[4]

By focusing on the risk and mitigation narrative, the extreme leakiness and the scale of space and time slips from our attention. PFAS cannot be addressed only by focusing on localized events, it is a form of 'slow violence' almost invisibly stretching out over the surface of the earth and into the depths of time.[5] Slow violence may affect the lives of people in unequal ways, meaning low-wage workers and communities living near toxic areas are often more affected.

4. Mankei Tam, 'Skin of the Earth: On Soil, Collaboration, and Temporality after Fukushima', *Environmental Humanities*, (15:2), 2023, pp. 39–61.

5. Rob Nixon, *Slow Violence and the Environmentalism of the Poor*, Harvard University Press, 2011.

PFAS however is becoming so ubiquitous it cuts across any social or (post)colonial inequalities. Like nuclear waste, the plastic in the ocean or global warming, it has become the next toxic phenomenon that is so vast in scale, spread out in space and time that it exceeds the human capacity to grasp, visualize or understand, let alone to tackle in a holistic manner. Like nuclear waste, PFAS is a phenomenon that is a real constituent of the physical world yet at the same time, we cannot access its materiality via our senses, we do not smell, see or taste these substances, yet they can permeate and affect us, like radiation. Due to their overflow of materiality, we can only register them via abstract data and technoscientific instruments. Nevertheless, these globally dispersed *hyperobjects* are real material forces that directly affect human and other bodies as well as the earth's systems.[6] And it is precisely their typical characteristics of vastness and opacity that make it difficult to objectively assess what immediate, scientifically provable effects they have on the environment and human health.

6. Timothy Morton, *Hyperobjects: Philosophy and Ecology after the End of the World*, University of Minnesota Press, 2013.

One of their other characteristics of hyperobjects is viscosity, meaning they are fluid and at the same time sticky and elastic. 'The more you get rid of them, the more you realize you can't get rid of them. They seriously undermine the notion of "away". Out of sight is no longer out of mind.'[7] This gives them their spectral and otherworldly nature, making it hard to fit them into our modes of interpretation and representation of the world, and hence difficult to respond to. This takes us back to the situation in Belgium, where the discovery of the massive PFAS contamination led to a surge in public attention, which after a few years has come to the same static point as climate change: we know it's there, we know it's a massive problem, we don't know how scared we ought to be and nobody knows how to deal with it, so we turn to either living in anxiety or denial.

7. Ibid. p. 36.

QUEERING CONTAMINATION

Like other toxic materials, PFAS confronts us with the darker side of the new relational and multispecies understanding of the world, our entanglement not only with plants and animals but also with the active and agential forces of dangerous chemical materials and other 'Anthropocene products' such as plastic waste. When acknowledging the fundamental interrelatedness of beings and matter, we can rethink toxicity and contamination as an example of more-than-human relationality: seeing it as the contingent encounter of bodies and toxic elements; instead of only as a characteristic of a chemical substance, like the way it figures in risk analysis. Furthermore, these thinkers use contamination as a concept that urges us to rethink how to survive in the ruins, and to reshape our ethical relation with the world around us.

In *The Mushroom at the End of the World,* Anna Tsing writes: 'we are contaminated by our encounters', not only referring to encounters with the more-than-human presences around us but also how they are shaped in histories of socio-economic entanglements. 'Contaminated diversity implicates survivors in histories of greed, violence, and environmental destruction'.[8] Ecocritical feminist scholars Stacy Alaimo and Mel Chen similarly see how the permeability and porosity of our bodies to toxic substances opens up to post-reproductive hybrid beings, queering earthly bodies into new narratives that do away with the old paradigm of closed systems, notions of purity, and strictly distinguished dualities such as human/nature. Queer theories allow and welcome the idea of toxicity since the permeability of bodies makes it resist precisely the categorization and dualist thinking that lie at the foundation of our toxic earth.

8. Anna L. Tsing, *The Mushroom at the End of the World: On the Possibility of Life in Capitalist Ruins,* Princeton University Press, 2015, p. 33.

Contemporary artists develop these notions by speculating on future hybrid beings, as in Sissel Marie Tonn's *Plastic Hypersea*. The work was inspired by an event in 2019, when 350 containers spilled overboard near the Dutch island of Schiermonnikoog. Millions of kilograms of plastic and other toxic materials were released into the sea. In her work, which developed as artistic research into a sci-fi documentary and sound walk, the artist explores what Tsing's idea of 'shared contamination' could mean in the future, presenting a speculative scenario where the presence of microplastics in the bloodstream triggers unexpected physiological changes. Toxic elements are not defined as external intruders entering individual and closed bodies but as material entanglements that cause mutations that are not only harmful but are carriers of transformative potentialities, new material becomings. Like in Mary Maggic's work on hormone biopolitics and Pinar Yoldas' projects on bio-engineering and genetics, these works question Western humanist ideal notions of 'the human'. More importantly they show how change and mutation can also bring new stories of hope and resilience. Like Tsing, Alaimo or Mel

Chen, these artists seem to embrace material mutations caused by toxicity in the Anthropocene, speculating on new hybrid life forms beyond Western heteronormativity, opening up spaces for multispecies becoming. Reaching an alternative understanding of what constitutes 'the human' enables us to enter into a relationship of care for the world; or as Tsing has shown through her pursuit of the Matsutake mushroom, contaminated diversities can teach us how to survive, live and counteract in the ruins of the Anthropocene.

New conceptualizations of contamination in current ecofeminism and posthuman anthropology as well as artistic practices hence draw on the emancipatory qualities of toxicity pushing forward new multispecies epistemologies, based on the fundamental relationality between beings and their surrounding material worlds.[9] However, as David Bond, an American activist, artist and anthropologist who has worked a lot on PFAS, critically notes, this poses serious ethical questions: 'While perhaps philosophically stimulating, such a move begins with a concession: the political struggle for the present is lost', which can result in an 'inadvertent allegiance between certain forms of queerness and the petrochemical industry'.[10] Both share the conviction and acceptance of the toxic earth, that total contamination is the starting point of the contemporary Anthropocene era. A danger emerges that theory becomes detached from 'lived realities', the suffering and unequal realities these toxins inflict. Presenting toxicity as the proof of the failure of binary modes of thinking, is not of much help to people who are directly affected and who are caught in a web of disinformation and uncertainties about the safety of their own body, their home, their garden. So how can such narratives speak to people living close to a PFAS contaminated site, to those overcome with fear of this viscous goo flowing through blood, bodies, placentas.

Taking a step back to Anna Tsing, contamination should always be seen as a double bind, on the one hand acknowledging the damage we are living in, but on the other hand considering contamination as a 'transformative encounter'. By this, Tsing stresses the importance of collaborative approaches to address complex social and environmental challenges. She sees contamination as a transformative narrative, by defining it as a call for learning to live with the damage through 'collaborative adaptation to human-disturbed ecosystems'.[11] In his own engagement with PFAS, which has affected his own

9. Giovanna Di Chiro, 'Polluted politics? Confronting toxic discourse, sex panic, and eco-normativity', *Queer Ecologies: Sex, Nature, Politics, Desire*, 2010, pp. 199–230; Antke Engel & Renate Lorenz, 'Toxic assemblages, queer socialities: A dialogue of mutual poisoning', 2013. e-flux (44:4), 2013, pp. 80–89.

10. David Bond, 'Contamination in Theory and Protest', *American Ethnologist*, 2021, pp. 1–18; Kim Fortun, *Advocacy after Bhopal: Environmentalism, Disaster, New Global Orders*, University of Chicago Press, 2009.

11. Anna Tsing, 'Contaminated Diversity in 'Slow Disturbance': Potential Collaborators for a Liveable Earth', in: 'Why Do We Value Diversity? Biocultural Diversity in a Global Context', edited by Gary Martin, Diana Mincyte & Ursula Münster, *RCC Perspectives*, (9), 2012, pp. 95–97.

community, Bond proposes a combination of anthropological and scientific research, with education and information, creative practice, and activism. It follows the line of the 'radical pedagogy' proposed by Kim Fortun, a cultural form of research that involves artists, citizens and scientists in a collaborative way with the goal of providing the public with a combined mix of tools for visualization, education, citizen-based monitoring. They advocate for an approach that operates bottom-up and not merely top-down, and although perhaps equally unsure, it involves at least local public discussion on the topic, providing citizens with tools to access or interpret information better, overcoming the stillness caused by fear and denial.

From this perspective, artists and designers can visualize imperceptible toxic presences in completely different ways than how we are accustomed to see them through data visualization, graphs and numbers. They can equally produce tools for community-based education, prototype open-source instruments for citizen monitoring and testing, offering different pathways to information, adding critical perspectives that disentangle social, political and economic blind spots. Artistic and cultural approaches to contamination offer a more holistic approach than the common risk-based discourse that runs amok today in the mediatized debate producing short-term policy reactions, such as testing and regulation. And finally, it also creates a social space to address the psychological and emotional effects of living with yet another invisible danger. It acknowledges feelings of mourning and loss while at the same time offering us a compass to navigate through the anxiety and uncertainty, and to continue practicing hope.

acne treatments
artificial grass
artificial turf
apricots

building materials (...long list)

camping gear
carpets
cleaning material
cosmetics (nail polish, eye makeup)
cucumbers

dental floss

See for instance: www.time.com/6281242/pfas-forever-chemicals-home-beauty-body-products/

eggs
electric vehicle batteries
electrical circuit sheathing in airplanes
endives

fabric softeners
fast food wrappers
firefighting foam
freshwater fish
fry wrappers

Gore-Tex shoes and clothes
grease-resistant paper
guitar strings

hip prostheses

kebab wrappers
kitchenware

leather shoes
microwave popcorn bags
milk

paints (as a binding agent and to give a smooth finish)
paprika
pizza boxes
Post-its
pesticides

nonstick pans and cookware

rain clothing

sealants
shampoo
shoe spray
ski wax
solar panels
sports gear
strawberries
sunscreen

tampons
tap water
toilet paper

upholstery

wood lacquers
...

1.2
Pastel Dyeing

INTERVIEW WITH SANDRINE ROZIER
BY HENRIËTTE WAAL & CLEMENS DRIESSEN

People want me to show it—well I don't want to sell a machine—I just want to change the world.

For the last twenty years, Sandrine Rozier has been experimenting with natural dyes for textile colouring. In an old villa in Montpellier she has dedicated her ground floor to fermentation dyeing. In her garden stands a large vessel that harvests rainwater for her dyes. Pieces of fabric in various shades of blue hang on lines of rope to dry. Indoors, you can smell the fermentation process in her vats somewhere between putrid and sweet. Her largest earthenware pot is 600 litres; bigger than herself. In baskets and on drying racks, there are black pulp balls, handmade from the woad plant (Isatis tinctoria), also known as 'dyers weed'. The Clostridium bacteria, indigo pigment and cellulose in these naturally magic plant balls can be incubated with some water. Mixed with wood ash and hot water, it starts a fermentation process that produces a range of astonishing blue colours known as pastel in French and woad in English.

When we speak with Sandrine, she radiates with excitement—a company has developed a machine for her to radically reduce the need for water in the process. It is slightly bigger than a washing machine. She is conducting tests to produce statistics on how fabrics and colours turn out.

SR Big industry is so incredibly wasteful and polluting! Especially in an environment where water is scarce, it feels very strange.

CD How do you know how to operate this device? And how do you select your results?

SR It's experience, I have to, we all have to, experience much more. I can't answer your question with a handbook. When I started twenty years ago, I didn't want to only research on a lab scale. I started to work on costume design as it was a good opportunity to test everything on a normal scale. To gain some experience on how the colour lives on stage and on the body, how you wash the colour, and how it ages. When I dye, I respond to what I get; I don't have total control. It's embracing diversity and the process. You can always put it back in the vat if you're not satisfied.

Sandrine Rozier stirring the woad vat in her atelier, 2022. Photo: Coltrane McDowell

Recently, Sandrine also started to develop alternative waterproof coatings from plant protein. And she has made colours for a new performance of the Opéra-Comique in Paris. In film and theatre, colours do not need to last forever. She shows an image on her phone, with vividly coloured costumes, the brightest red, yellow and blue.

HW Making woad balls and pastel dye became rare in France. Where did you experience it?

SR I have travelled to about fifteen countries to meet the dyers and to make natural dyes in professional places that I could not find here. In 2003, I made my first journey into indigo, to the Mong communities in Vietnam. And I have been to Japan. In 2019, I activated a pastel vat with indigo in my house with tap water, after that trip. I was so moved by the range of blues that you can achieve with this dyeing method that I wanted to do it myself.

HW What made you decide to use rainwater instead of tap water for your vats?

SR Covid was a period that changed my way of working. Stress emerged in society. People were very afraid of something in the water. I had problems with the vats during that period; I could not understand it. Every time I lifted the cloth from the vat or after I rinsed, something was wrong. I thought it was my fault; that I was off. It took me two to three months—I wasn't aware of the global water situation. Later, I figured out that they doubled the levels of chlorine during the summer of 2020. I went up and down to a Roman source nearby with my car and brought the water for my vats. But now I just work with rainwater. Although tap water seems okay again, I don't want to take the risk; I don't have a chlorine meter. With the rainwater, the 600-litre vat is very stable. The bacteria are very well established there. I advised the studios of Bry-sur-Marne, one of the largest film studios in France, to install a rainwater collector during Covid as well. We continue to make dyes with it.

CD Would you describe your approach as reinventing colours by reinventing the systems of making them?

SR I really don't invent anything. Together with Dr Dominique Cardon and by practice-based research, I try to understand the knowledge of the dyers of the eighteenth century; how they used to work with natural dyes on a large scale. Now, this kind of dyeing is reserved for the domestic scale, but they were farming woad as a crop and dyeing on a large scale, near the river. Wool was heavy; they didn't rinse between each step; they also didn't have a situation with water like today. Dominique and I initiate workshops with designers who are involved in natural dyes from across Europe. We explore ingredients, quantities, volumes. We work very concentrated and go back and forth: from old manuscripts, observations of old samples, to our experiments. You get a very different result, especially with yellow, if you change the pH of the water. And woad must stay at a constant temperature and pH for a certain time. Former dyers had empirical techniques to achieve that. [pH could not be formally measured until the invention of the pH scale by Søren Sørensen at the Carlsberg

Sandrine Rozier, *Indigo Soie*, 2020. Photo: Sandrine Rozier

Atelier LUMA in collaboration with Sophia Guggenberger, *Camargue Sandal,* coloured with Isatis tinctoria (woad) by Sandrine Rozier. 4 parts (insole, outsole, upper and strap) sewn together. Made from Mediterranean wetland materials hemp, wool and natural rubber. Courtesy LUMA Arles. Photo: Coltrane McDowell

Compared scales of blues, Regional Archives in Montpellier, France. Photo: Sandrine Rozier

research laboratory in 1909.] The dissolved pastel is greenish in colour; the immersed cloth turns blue when exposed to air. Colour is a living thing. We try to find four shades of blue: Blue de lait, blue mignon, blue azur, and blue a roi. Like that we have a range of possibilities together with the other primary colours.

HW In contrast to the eighteenth-century masters who kept their recipes a mystery to protect their market position, you create moments to share your empirical knowledge with an international community. What is your aim?

SR It's like making an acrobatic position between history and the potential of today's techniques and possibilities. Natural dyes are a world of interactions. I'd like to inspire the younger generation. I expect that we will see an explosion of experiments in the coming generation and shifting aesthetics. We are fighters for life, for diversity, for abundance. Society is now afraid. Fear, control and predation are accumulating. We need to fight this.

CD You seem to scale up by creating a community. What should happen next?

SR Big corporations find it difficult to switch. They wash in big machines using toxic chemicals; you cannot go inside those machines, so there is no other way to clean them. Maybe a middle scale is better, using slightly larger vats and inventing sustainable machines, taking more time, and spreading knowledge. By slowing down processes, fermentation at colder water temperatures, you can achieve colours that last a long time, very energy- and water-efficient. Small is beautiful and also good for the diversity of plants.

There is also the medicinal side to dyeing with plants that Sandrine is interested in. In her current project with the hospital in Grenoble, she is making a scientific protocol for five different natural dye products, for people who are in the hospital for long periods. Together with scientists at the hospital, she researches whether wearing a certain red can help people with skin disease or insomnia. From the communities built around dyeing, a world is opening up in which colour is connected to health, knowledge, and new ways of experiencing bodies.

1.3 DIY Estrogen & Yogyakarta

INTERVIEW WITH MARY MAGGIC
BY HENRIËTTE WAAL & CLEMENS DRIESSEN

The workshops progress from demystifying sciences, allowing more amateur exploration of molecules, and resisting neoliberal trashy X to embracing Xeno feminist perspectives.

OPEN SOURCE ESTROGEN

Open Source Estrogen began in 2015 as a collaboration between artist Mary Maggic and professor Byron Rich, exploring the idea of an open source birth-control pill. Then it evolved to address political and access issues. And that led to an investigation into the presence of estrogen that's all around us as a form of environmental 'toxicity'. By 2017, the project expanded its focus, formulating a six-point plan for hormone queering resistance. The six-point plan outlines the project's research goals and is still used today as a guideline during group workshops. The workshops progress from demystifying sciences, allowing more amateur exploration of molecules, and resisting neoliberal trashy X to embracing Xeno feminist perspectives. Points four to six delve into questioning normativity, combating purity politics through queering, and emphasizing the micro-performativity of hormones as collaboration, not just colonization.

MM I think number six is the hardest for participants to grasp. Because we are taught in this very toxic discourse that we have to other certain molecules because they're considered toxic. We have to consider them as other, as invasive, as harmful, as something we have to purge from our bodies. And this is coming from the purity politics of fixed bodies that should not change or deviate outside of the norm. I'm actually learning a lot by giving these workshops. It's almost like a research methodology to give these workshops and to learn from the participants. I'm a big advocate of artists also being workshop-ologists.

HW Yes, with a group you have multiple sensors/senses. When I explore a new landscape or situations with my students, I feel so enriched doing that in a learning group. It's a more diverse knowledge that comes up. I'm interested in what 'aiming to neutralize fear' looks like in your workshops. What is needed to make it happen apart from talking, what are the exercises?

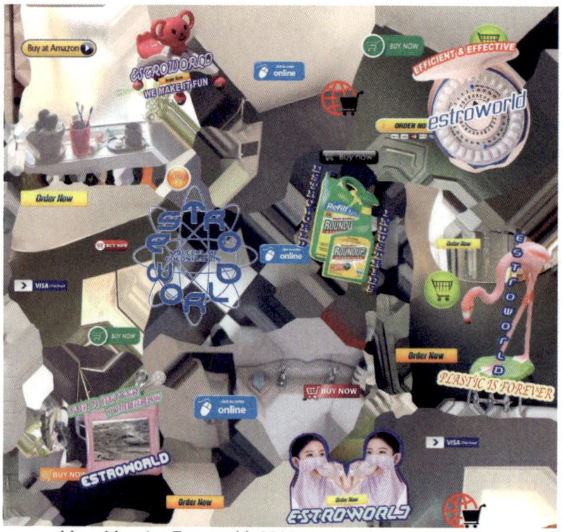

Mary Maggic, *Estroworld*, 2015

MM I've been experimenting with these durational workshops that go over two days, three days, sometimes four days, as it is a slow process. One workshop that I've been doing is arriving at this site where toxicities have been leaking across bodies across space and time and then scavenging for these materials in silence. And we call them alien materials because they have transformed over time. Like plastic waste hybridized with organic materials. The blind sensing exercise, it's about one hour long and can be very exhausting. Afterwards, we do a scenography building exercise. So, we take all the materials, and we start to construct our new world. And this is where we get into collective world making, which is one way to find that connectivity and belonging in this very alienated world. So, it's a very embodied exercise. And again, no speaking.

CD How does the isolation of estrogen work? What do people have to learn or do?

MM The process involves a hardware part, constructing a column using recycled glass bottles, silica gel, and cigarette filters for column chromatography. These DIY alternatives can trap hormones, because they create a chemical environment that the hormones are attracted to. Methanol is used to elute molecules, leaving hormones dissolved. Evaporation of methanol then isolates the hormones, achieving basic separation without traditional glassware.

HW I like how you describe hormones not as a substance, but as something that is attracted to something else, to the cigarette butt. It is something relational.

CD Does the science then come in as a sort of tangible practice? Or do you have to reshape the practice of science to some extent to better suit this kind of embodied approach?

MM I want to move away from 'is it science or is it not' because I think it all flows into each other. I think that the way that we're blind sensing is also a form of science, it's just not the traditional way of producing knowledge. So, the extraction is one way of seeing something that's previously invisible, which is the hormones in our urine.

CD Can you give some examples of what people come up with through the workshops?

MM They have their own ways of fixing, or making or producing, or even just playing.

When we're doing this in silence, they get to be in their own world, and they can just be themselves without any explanation or verbalization. When we talk, we talk about what it means to be an alien, what it means to be othered, how it feels. Or we talk about anxiety, having panic attacks, you know,

Mary Maggic, *Milik Bersama Rekombinan Mandala*, 2019

all of these things become associated with toxicities, with becoming alien. So, you start to realize how interconnected everything is. And that's why it doesn't make sense to me to only do a science workshop, because it's all connected to the body, to the environment, to emotions, to the immaterial, everything. To lived experiences.

CD I find it really fascinating also because it plays on this tension between, on the one hand, this idea of toxicity as something that is completely toxic while assuming an ideal of purity and impermeability. But at the same time, you also want to maybe resist a certain form of purification or promote certain ideas of wholeness. Is it also a critique on science?

Mary Maggic, *DIY Estrogen workshop*, 2015

Or do you see it as another version of science or working with it in a more grounded way?

MM I think, to a certain degree science is promoting these purity politics, because of the way we study science, how we perform science is always in controlled conditions. So, we are already separating and isolating certain processes, biological processes, so we're always in the process of isolating. But it's mainly a critique on the dominant ideology of purity politics.

When I'm critiquing the purity politics, it's not that I'm advocating that we should all just start ingesting as much toxicities as possible, but it's about questioning our current strategies, which is to barricade, or to just scoop everything up and put it in a landfill or flush it down the toilet. It's about questioning why we do this and how it is based in some kind of fear or panic of our bodies changing. So, it's thinking about what other strategies we can come up with that are not separating oneself from the other. Because that's impossible. This is a fantasy. We can't separate because we're all connected. And we're all implicated.

YOGYAKARTA

During a ten-month residency with Lifepatch, an arts, science, and technology citizens' initiative, Maggic focused on the Code river in Yogyakarta, Central Java, Indonesia. Her close collaboration with river communities made it clear that introducing biohacking without understanding local cosmologies is ineffective. Residents perceive the river not as alive but as a highway, where spirits can cross between the kingdoms of the North (Volcano Merapi) and the South Sea. Despite industry contributing to pollution, domestic waste is the main source of trash ending up in the river. Maggic's initial efforts involved six months of interviews to comprehend residents' cosmologies and their relation to the river.

CD The spiritual highway sounds very new-materialist to me, kind of super-materialist. But what did you do there after the half year of understanding the river?

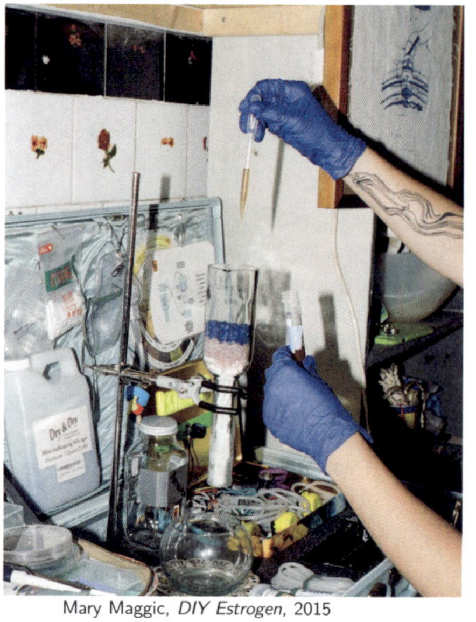

Mary Maggic, *DIY Estrogen*, 2015

MM I ended up collecting a lot of trash from the river and creating an artistic installation out of it at Jogja National Museum. It is a time-based sculpture with blue agar and mushroom that gets messy over time. The trash came together on a latex that resembled skin, kind of representing the view of the river from above. You see, like water and trash floating on top. Lastly, I manipulated all trash in a projection of a

Mary Maggic, *Milik Bersama Rekombinan*, installation in Jogja National Museum, Yogyakarta. Image: ARTJOG Team, Jogja National Museum

mandala that is always moving, and it has this kaleidoscopic effect.

HW And how did you work with local communities to re-envision the river?

MM We held several workshops just presenting a more educational approach to the way that the water is connected with the well-water, with the drinking, the cooking and cleaning water. So, we mapped these connections. We also did workshops on how to filter the water and how to make it safer for consumption. We tried to blend the practical with introducing a new cosmology about interconnectivity. We also tried to teach the workshop in the language of their mysticism because the context that we're working in is this Javanese mysticism about these spirits that cross the river. We also talked with people who were considered shamans of the village, so they have their own kind of spiritual wisdom because they're able to communicate to spirits. They're able to see them, to observe their presence. We had to really construct our strategies in a way that is part of, and built into, the local context.

CD What did the shamans think of the plastic? Or how did you work with it?

MM I think for the shamans their main concern was never about the trash but about the flooding because during the rainy season, there's always a really big risk of flooding. But at the same time, the floods wash away the trash. So somehow there's this cyclical cleansing of the river, that is also very dangerous and very risky for the people. Every year they pray that the floods are not too intense, in other words that the Queen of the South Sea will be most merciful. But it just seems like no one was really afraid of the trash. Even the children. They said: We're not afraid of the trash, we're afraid of the water snakes. So, you understand it's so normalized to find trash in the river, and to be part of the trash in a way.

CD Which resonates with this idea of openness, permeability, and, maybe, how can you protest against pollution not in the name of purity, but in the name of becoming a particular form of difference? Or how do you engage if we're not supposed to protest its toxicity in the name of purity?

MM Well, I think that any kind of protest has to be very specific; it has to be case-by-case because one approach to a certain situation with a certain company or certain whatever will be different in another location. In this case, what the people could protest for is, you know, better transparency with the hotels and the hospitals that are dumping their waste into the river.

They could advocate for this, or they could advocate for more sewers and water treatment facilities in the area, or they can advocate for more educational resources. So, this is very specific to that location. And it's not reinforcing purity. It's just about basic human rights or basic rights of nature.

HW Water and soil have little choice than to allow our new molecules.

MM I think the Earth, the planetary is in this cocreation process with all industrial molecules that we're creating. The Earth itself is trying to figure out or trying to adapt to all these molecules that have their own age and show power. In the scientific way of studying molecules, they're just objects with no agency, but actually, they have much creation power, regardless of human input. I guess I'm going into new materialisms and material agency. And that relates, of course, back to water, because water has its own material agency as well.

1.4 Atelier LUMA: The Happy Algae Lab

BY HENRIËTTE WAAL

Microalgae are 1.5-billion-year-old waterborne organisms living in colonies. They are considered the first complex life form on the planet.

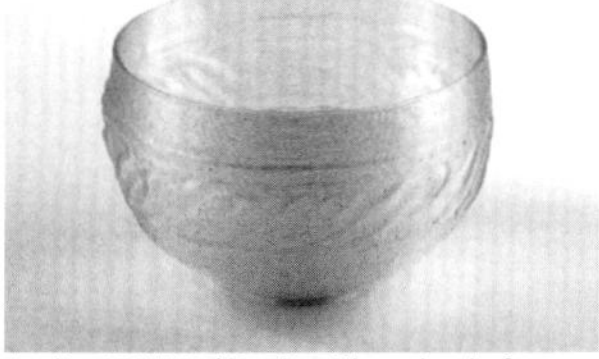

Atelier LUMA/Studio Klarenbeek & Dros, *Algae Glass*, 3D scanned roman ceramics translated to 3D printing with microalgae and sugar cane based biofilament, 2017, France. Courtesy LUMA Arles. Photo: Antoine Raab

Alongside herds of black bulls, white horses, and pink flamingos, microalgae are a key species of the Camargue wetland region. Some reveal themselves as strawberry-candy-coloured water bodies in summer, when there is a mix of high salt levels, high temperatures and sunlight. With their affective presence, the pink pools are an unmistakable part of Camargue identity. These pink waters are made by the Dunaliella microalgae protecting themselves by producing reddish beta-carotene. They are the ones nourishing and colouring the flamingos. Through their photosynthesis, microalgae actively absorb CO^2, so that algae-based materials may have a negative carbon footprint. There are hundreds of thousands of microalgae. Only a few hundred of them have been studied by scientists.

In 2017, drawing on their previous experience with seaweed (macroalgae) as a material, Klarenbeek & Dros moved their studio to Arles for a year to build—together with the local team—algae material from scratch and spearhead the Algae Lab as a biolaboratory within Atelier LUMA exploring algae from various water bodies in Camargue wetlands for design applications.[1] Algae Lab provides the conditions required for microscopic observations, control of key culture growth parameters, formulation of culture media, post-harvest biomass processing and bench-top chemical manipulations such as pigment extraction. Converted into filaments for biopolymers, algae can be used for 3D printing, allowing on-site production.

1. Atelier LUMA is a bio-design lab that is part of Maja Hoffmann's LUMA Arles, a unique cultural center situated in Arles at the edge of the Camargue wetlands in the south of France. The lab that was founded in 2016 aims to reshape our understanding of place-based innovation and the potential of bio-regional material cycles to inspire global change.

Atelier LUMA/Yasmine Elwani, *Shisha Forever*, 2019, 3D printed microalgae and sugarcane based biofilament, 22 cm and 15 cm. Part of Algae Lab Cairo. Courtesy LUMA Arles. Photo: Joanna Luz

Atelier LUMA in collaboration with Studio Klarenbeek & Dros, Algae Lab Cairo, 2017. Courtesy LUMA Arles. Photo: Antoine Raab

Atelier LUMA/Cirkel Sector Genk, Algae Lab field research. Photos: Kristof Vrancken

In 2018, the microalgae filament gained attention for its contributions to material innovation in the service of ecological and social sustainability.[2] I used the momentum to take the Algae project to other places in the Mediterranean region tapping into the unexplored connections between local and raw biomaterials, makers and production methods associated with them. Combining 3D scanning of cultural archives, 3D printing and biology, Algae Geographies explored new models for circular production through bio- and decentralized fabrication. Simultaneously, the project has become an aggregator of knowledge and expertise on algae, exploring innovative ideas and new types of collaborations.

2. New Material Award 2018.

Atelier LUMA/Henriëtte Waal & Mevce Çıracı icw Cirkel Sector Genk, *Ecover Bottle*, 2019. Photo: Kristof Vrancken

The promises around algae for making fuels, foodstuffs and materials tend to be based on generic organisms assumed to come without attachment to place.

Pictured here are results of workshops at Atelier LUMA in Arles, LUCA School of Arts & Design in Genk, Industrial Modernisation Centre in Cairo and Bilgi University in Istanbul, where locally sourced algae were used in biofilaments to 3D print prototypes related to water and waste issues. Drawing on particular cultural archives from each location, a hybrid traditional-contemporary craft-digital aesthetic emerged, encouraging us to unlearn the limiting preconceptions with which we value materials and production techniques.

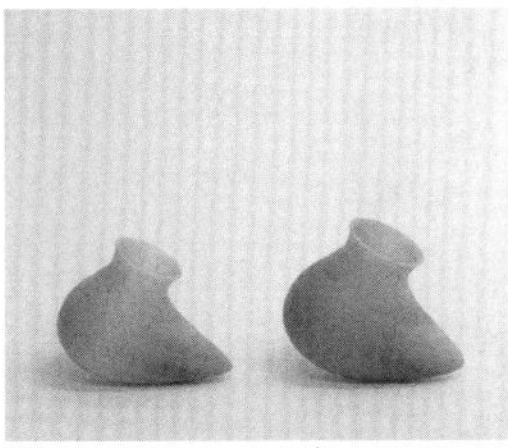

Atelier LUMA/Beste Tekdoğan, *Muhabbet (Tittle-Tattle)*, 2019, 3D printed microalgae and sugarcane based biofilament, 7.9 and 7.10 cm. Part of Algae Lab Istanbul, Atelier LUMA in collaboration with Bilgi University/Mevce Çıracı and Bogazici University/SO Architects, Turkey. Courtesy LUMA Arles. Photo: Joanna Luz

Algae Lab has since continued to build its library of strains and develop further research into applications of algae. The library contains the Spirulina microalgae named *Camarguaise*. It was very special to find this species, which was once discovered in Camargue itself, and to start working with it. Phorphyridium microalgae we call *Toujours jeune*—Forever young, as its pigments have anti-aging properties and are used to contribute to the elasticity of human skin. Scenedesmus microalgae has been deployed to absorb the CO^2 generated by the industrial processes of Fos-sur-Mer. We harvest it from the smoke treatment basins. We call it *Démon de midi*—Noontime Demon. The extremophile Dunaliella is green but turns reddish when stressed in a salty environment. Spirulina looks like a tiny spiral, has superpowers and is highly nutritious. Their colour is quite unruly, with shades of green and blue that can be beautifully revealed through a random cone dyeing process. We name it after a waterbird known by the name Cacharel in Camargue. As with the pink flamingos, colours from algae are often found in water animals as well. Biologist Adeline Weppe, who interacts with microalgae on a daily basis, speaks very lovingly about them: 'The more their spirals curl up, the happier they are'.

Atelier LUMA in collaboration with Studio Klarenbeek & Dros, *Algae Lab Arles*, 2017, showing Eric Klarenbeek & Maartje Dros harvesting wild algae, Camargue region, France. Courtesy LUMA Arles. Photo: Victor Picon

Pollution regenerating algae basin at Arcellor Mittal, steel plant in Fos-sur-Mer, France

Bogazici University, Turkey's first biorefinery making fuel out of algae, 2022, Istanbul. Photo: Bogazici University, Istanbul

1.5 Toxic River Pottery

BY BENEDETTA POMPILI

Water molecules attract clay-loving minerals. And the feeling is mutual—clay particles are hydrophilic—they love water.

Benedetta Pompili, throwing with wild clay from the Meuse river. Photo: Benedetta Pompili

When a little clay is added to moving water, a slurry forms because the clay distributes itself evenly throughout the water. Another important property of clay minerals is the ability to attract ions to the surface of its particles where they get taken up within the structure of these minerals. With this quality, clay can be an important vehicle for transporting and widely dispersing contaminants from one area to another.

In *Conversing with Matter,* Benedetta Pompili investigated the reclamation of potentially toxic clay from the river Meuse. River clay can be deployed in multiple ways, for instance as brick, roof tile or to seal dikes. A gradual harvesting method allows riverbanks to clean and regenerate, whereas the resulting clay preserves local colours, textures, and properties while promoting a refamiliarization with regional river bodies. River clay urges us to care for what goes into urban waters as clay absorbs chemical contamination.

The marbling of the river clay with a common studio clay, a white stoneware from Germany, allowed for an easier application of the wild clay body, which otherwise is hard to manage. On the potter's wheel the two clay bodies materialize into pots inspired by Jacoba jugs, reflecting the origins and history of the two clays. Jacobas were traded from Germany, the origin of the stoneware, to the Netherlands, where the Meuse clay was deposited. Some of the ships used for their transport have sunken, so many Jacobas are in shipwrecks on the riverbed. Inspired by the archaeological finds, the vessels share the stories of the materials and highlight their geographical origins while crafting a shape to archive the process, sealing the toxic elements under a layer of glazing.

Benedetta Pompili, ceramic bowls made with wild clay from the Meuse river (work in progress). Photo: Benedetta Pompili

(Possibly) toxic clay deposit of the Meuse river, the Netherlands.
Photo: Benedetta Pompili

Benedetta Pompili, *Conversing with Matter*, 2021, wild clay from the Meuse river and Sibelco stoneware, Germany.
Photo: Femke Rijnemans

A pottery wisdom is that you need to throw at least one hundred pots on the potter's wheel before you can begin to understand the clay you're working with. Hence, one hundred *conversations*, were thrown.

Clay may appear to be abundant, hiding its finite nature as a resource and the damages caused by its extraction. Starting from the conversation with the clay on the potter's wheel, the material research of working with the Meuse clay explored the economic, social, and technical aspects entangled with the mining and the making. Meanwhile, the clay pots propose an alternative journey to making while reducing the effects of extraction.

1.6 Microbial Bathhouse

By Serina Tarkhanian

AN EMBODIED COLLECTIVIZED MEDICAL EXPERIENCE FOR RESTORING HUMAN MICROBIAL IMBALANCE.

Co-Healing is an embodied, collectivized medical experience that uses the unique microbiome of participants to restore microbial imbalances of others. The work materializes a radical revisioning of Western healthcare, making everyone both a receiver and giver of care. These new rituals for health are designed to be experienced in a 'microbial bathhouse', structured around a series of implements for safely exchanging microbes between our bodies. One such device, the **Lung Microbiota Exchange Tool**, is designed in collaboration with a medical microbiologist, Maxime-Antoine Tremblay, and Ad Waterschoot, scientific glassblower at Eindhoven University of Technology, and allows visitors of the microbial bathhouse to gather and share microbes from their respiratory tract. By positioning the health of our bodies as a collective, shared and self-governing experience, **Co-Healing** aims to subvert the prevailing hierarchies of power within the dominant model of Western medical care, moving away from isolationist, depersonalized and paternalistic practices of healthcare.

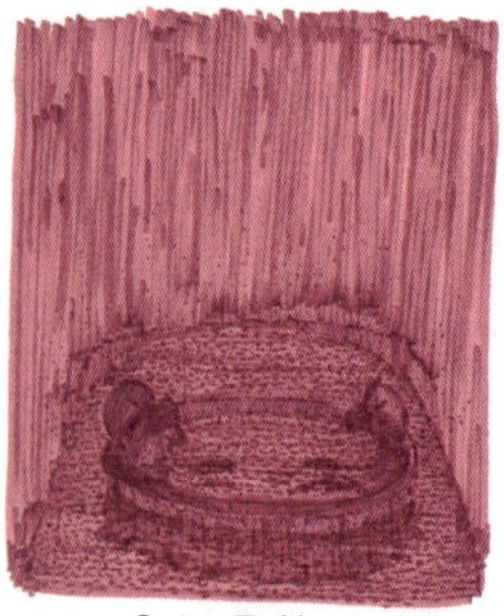

Serina Tarkhanian, *Co-Healing: Illustrations*, felt-tip pen on paper, 2020

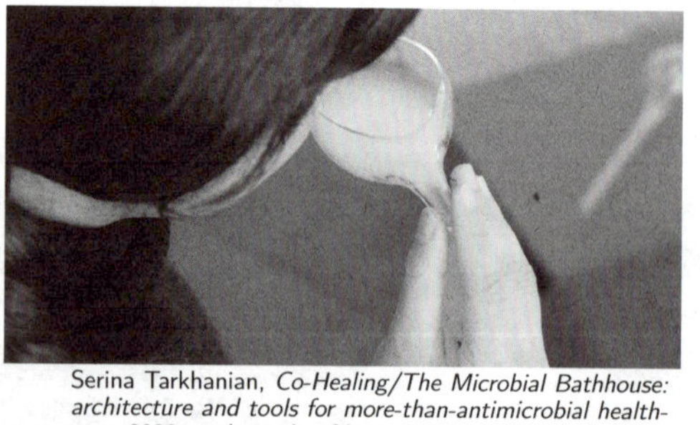

Serina Tarkhanian, *Co-Healing/The Microbial Bathhouse: architecture and tools for more-than-antimicrobial health-care*, 2020, tools made of borosilicate glass, clay and bodily fluids. Photo: Serina Tarkhanian

Wild

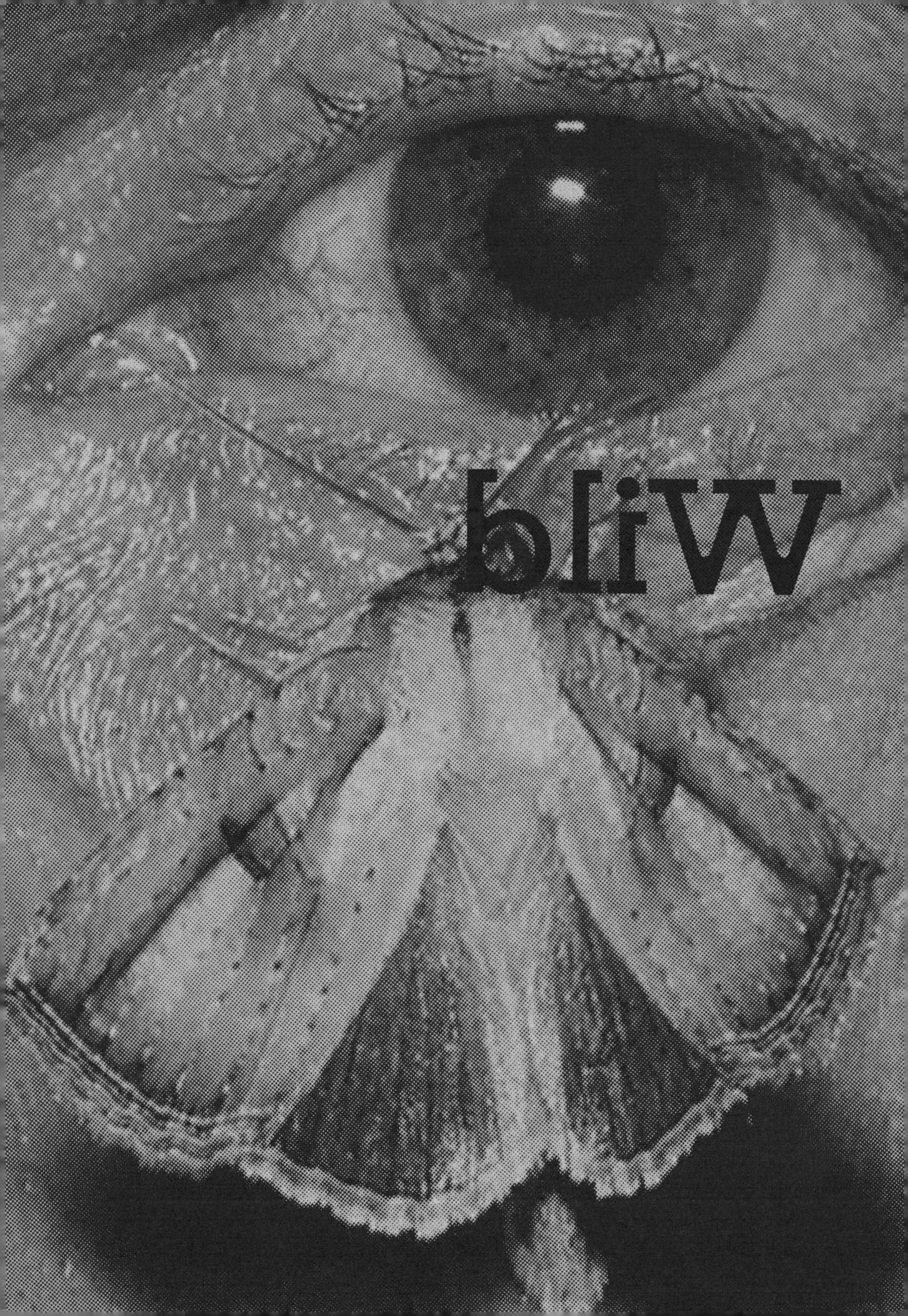
wild

2
Wild

BY HENRIËTTE WAAL & CLEMENS DRIESSEN

Ecosocial design means intervening in some form, while resisting the urge to try to completely tame, domesticate and master. 'Water management' may suggest total knowledge and a central position for human decision-makers, a system optimized for social functions. But ecologies and societies are too complicated to fully know and control.

If you want to make original beers, yeast is valued not for its predictability but for the surprises it offers. Not all shaping and deciding needs to originate in a certain type of human, an autonomous rational actor.

A natural commons, one can never fully own a river. Water is an elusive thing to 'grab' and hold on to, at most allowing people to temporarily divert or store it, though it can't always stop people from violently depleting it. Through water, organisms are always connected in processes that are larger than life, and never fully known.

When nature is not some supposedly harmonious system to be revered from a distance—a wilderness—then ferality is produced as part of projects to manage and control.[1] Unruly beasts, roots, fungal and microbial assemblages are constantly looking for holes and passages, a way out; nature essentially escapes.

Even the language we use to know, to grasp, is forever fluid. We should—with Isabelle Stengers and Alfred North Whitehead—treat concepts like wild animals: carefully, with respect, but never fully trust them even if you know them well, danger lurks, unexpected moves, never think they may not lash out when you feel they're harmless.[2]

Water is a medium for a multitude of lives and lifeforms. Sharing water means entering into relations of reciprocity and mutual, non-hierarchical learning.[3] Like ants holding hands traversing a body of water.

1 Anna L. Tsing, Jennifer Deger, Alder Keleman Saxena, & Feifei Zhou, *Field Guide to the Patchy Anthropocene: The New Nature*, Stanford University Press, 2024.

2 Isabelle Stengers, *Thinking with Whitehead: A Free and Wild Creation of Concepts*, Harvard University Press, 2011.

3. Mihnea Tănăsescu, *Ecocene Politics*, Open Book Publishers, 2022, p. 212.

Opening image: The Center for Genomic Gastronomy, *To Flavour Our Tears* (*TFOT*), 2016, an eatery designed for lachryphagic moths, or other insects that feed on mammal tears, recipe card Tear Taste Test. Photo: The Nonist, Romancing the Lachryphage

Ecosocial design stresses the inherent blending—no, that suggests clear separate ingredients—it's the always already muddled and melted nature of life-in-environment—of what may earlier have seemed separate domains of the ecological and the social. It invites makers to think 'sociality' beyond the human. 'We are not alone'—whoever this supposedly human and universal 'we' may be.[4] In any case, 'the human' cannot be seen alone, or apart from their social relations, their microbiomes, their pets and plants—we have never been only human: 'We are a multispecies relationship'.[5]

4 Astrida Neimanis, *Bodies of Water: Posthuman Feminist Phenomenology*, Bloomsbury Academic, 2017.

5 Anna L. Tsing, 'Unruly Edges: Mushrooms as Companion Species', *Environmental Humanities*, 1(1), 2012, pp. 141–154.

This also means we are surrounded by—in close relationships with—other than human makers, tinkerers, unheard-of forms of creativity, modes of design, plans for possible futures, imaginations of doing things otherwise. This is the promise of more-than-human ecosocial design.

Some suggest a felicitous, happy integration of technology and ecology—such as by instrumentalizing nature-based solutions—others seek to interpret the poetry of infrastructures that are not solely based on human knowledge and sensibility.[6] Forgetting wildness, nature is easily thought to be churned into something manageable, a resource—an ecosystem service to be demanded, in exchange for letting useful others live—for now—under the condition that it maintains to be this serviceable, servile, natural world, treated as a planetary hospitality industry for human comfort.[7] Which, however, seems to become more unruly and vengeful each year.

6 Maan Barua, 'Infrastructure and Non-Human Life: A Wider Ontology', *Progress in Human Geography*, 45(6), 2021, pp. 1467–1489.

7 Christophe Bonneuil & Jean-Baptiste Fressoz, *The Shock of the Anthropocene: The Earth, History and Us*, Verso Books, 2016.

Wild design invites working in a dynamic process, instead of following supposedly scientific protocols. All the attempts to systematize and tame 'design thinking' into clear iterative steps are not to be mistaken for mastery over the making of surprises. Overly organized processes to come up with new forms and ideas will mean just funnelling energy into realizing preconceptions. Instead, ecosocial design needs to go sufficiently wild: the first experiments should fail; the idea is not to experiment safely. As soon as interesting results emerge, you start to work with that, get attuned to it, push the boundaries and do some finetuning—but without destroying all wildness...

When Henriëtte experimented with fermenting a brew with seawater, fellow brewers said: 'this will never work'. Yet the result was unique: much sweeter, yeast took much longer to do its thing, but it is just a matter of what you find normal. Wild—appreciating wildness—means managing expectations and being ready to run with surprising twists and turns.

Wild also creates a 'style' that is raw and original. This 'wildness' shapes the learning experience. Teaching can go rough, exploring mudflats, digging into soil, students getting stuck. But afterwards groups are usually raving. Things shouldn't get too harsh. Just wild enough to ensure the process itself wants to go a certain way. It is not about the designer forcing matter into shape, it is more about holding/facilitating/catalysing/guiding—and thereby also about cultivating the inner landscape, changing ourselves, ever changing.

Ecosocial design positions designers so that they are not disembodied observers of space. Comfort or control may not be the best starting point for bonding with the other, or for getting into a creative flow. Ecosocial design is also about relating to the landscape—getting stuck in the mud of the floodplain—to design ecosocially means exposing bodies to uncontrolled landscapes, to experience its unpredictability and differences firsthand. Doing things that hold at least some potential for danger. It builds a group, activates the senses, generating surprise and spontaneity. Working with living matter gives you feedback. Wildness can be a term under which to appreciate rather than dismiss this feedback.

2.1 Beaver Cultural Centre for More-than-Human Ecosocial Design

BY CLEMENS DRIESSEN & HENRIËTTE WAAL

Widely appreciated as 'ecosystem engineers', beavers are known to radically impact their environments. In order to submerge the entrance to their burrows and stay safe from predators who come over land, beavers everywhere seek to create stagnant or slow-flowing water bodies.

By damming streams, building lodges, gnawing on trees, and digging canals, these large, industrious rodents meticulously manage wetlands with varying flows of water, turning little streams into sets of interconnected pools that attract insects, amphibians, and thus also birds and mammals, thereby opening up spaces for an array of other organisms and ecological processes. Beavers are called keystone species, whose ecological impacts make them central in creating habitat for themselves and others.

Beaver is also a verb. When not primarily considered a nuisance, they are seen as diligent workers, incessantly labouring to repair and perfect their dams. It has been found they are obsessed by the sound of running water, not resting until they find materials to plug the gap. Maintaining high water levels in their dwelling spaces makes sense to them, as in water they are in their element and they are safe in their lodges with underwater entrances, while moving on land they are more awkward and vulnerable.

With that, beavers are poster-figures for opening up the idea of environmental management beyond human mastery and control. Whether to take them—and other less respected ecosystem-shaping organisms in their wake—seriously as 'designers' is less widely considered, though. Thinking with beavers therefore is helpful to ponder about what ecosocial design can be, and how to understand design in a more-than-human vein.

Henriëtte Waal, *Beaver Brooches*, 2024, willow wood chip made by beaver teeth. Photo: Henriëtte Waal

INSTRUMENTS FOR RIVER RESTORATION

With beavers being rediscovered for their restorative potential, and in the face of water systems under threat from climate change, efforts have sprung up to incorporate them in human spatial planning and river management. But all this beavering can have disrupting impacts. Experiments to instrumentalize beavers for river restoration may lead to a backlash, if they don't manage water as expected. This presents beavers as vectors that can be let loose to rework their world, but only up to a limit, after which they become a nuisance to be dealt with: 'problem families' that are increasingly forced to comply with local water management demands following elaborate beaver protocols, in flow diagrams that ultimately end in 'lethal control'.

As North-American experiences with beavers highlight, it is dangerous to promote beavers as adorable eco-saviours that can easily be deployed as cheap instruments to restore streams and prevent wildfires. This would treat beavers as generic entities removed from people, place and time while offering quick fixes to restore functional landscapes. Instead, researchers argue, we may better appreciate them as part of long-term ecocultural relations developed with indigenous people and the ever-changing knowledge of dealing with beavers in particular riverscapes.[1]

BEAVER FUTURES

Over the centuries, beavers have been subject to an ongoing debate: are they merely driven by supposedly 'blind' instinct, or do they have a plan? Are their ecological roles mere happy coincidence to them being hardwired to prefer this habitat type? Or are they ecosocial designers? But what does design mean then, when decentring (some elite) humans as the ultimate intentional shapers of worlds?[2]

1 Caroline Gottschalk Druschke, Eric G. Booth, Bathsheba Demuth, J. Marty Holtgren, Rebecca Lave, Emma R. Lundberg, Natasha Myhal, Ben Sellers, Sydney Widell & Cleo Aster Woelfle-Hazard, 'Re-centering Relations: The Trouble with Quick Fix Approaches to Beaver-based Restoration', *Geoforum* (156), 2024. www.doi.org/10.1016/j.geoforum.2024.104121

2 Ron Wakkary, *Things We Could Design: For More than Human-Centered Worlds*, MIT Press, 2021.

There's a recurring idea among humans when seeking to define themselves in contrast to animals to claim that animals don't have a sense of the future, that they don't realize plans, are just beavering forward.

In the words of Karl Marx:

'A spider conducts operations which resemble those of the weaver, and a bee would put many a human architect to shame by the construction of its honeycomb cells. But what distinguishes the worst architect from the best of bees is that the architect builds the cell in his mind before he constructs it in wax.'[3]

3 Karl Marx, *Capital, Volume 1*, trans. Ben Fowkes, Vintage Books, 1977, pp. 283–284.

But what does it mean to think animals don't have a plan? Do beavers live in an 'endless now', merely performing a timeless action programme inscribed in their genes, making them part of the biotic background of human culture? Is the world being designed just an interaction between human strivings, the true subjects of history, and natural—biotic and geological—factors?

Henriëtte Waal, *Beaver Dam*, 2024, photograph, Geul river

Experiences with animals such as beavers suggest otherwise. In practice, beavers are appreciated not merely because they are workaholic water managers—until they go against human interests—but knowledge-holders with a plan. Plans that may clash with some human ones, but that can be taken as critique, as counter-proposal to dealing with water, rather than merely nuisance.

DECEIVERS DIALOGUE

The practice of living with beavers in places densely populated with humans is currently facilitated by a range of structures, especially developed from the mid-twentieth century onwards in the US. One well-known example is a device called *beaver deceiver™*. Developed by American beaver enthusiasts, this patented device eludes beavers by constructing a pipe into their dams which increases the flow of water and lowers the water table behind the dam without producing sounds of running water.

CLAIM: I am claiming a patent on the process of building into an existing beaver dam a siphon for to control the maximum height or level of water as desired by the installer. It is new in that it goes beyond the instincts of the beaver, deceiving them as to where the water is leaking from so they cannot plug it up. The suction and discharge ends of the siphon hose are remote and away from the dam.[4]

4 Automatic beaver dam water control 'beaver deceiver', patent CA2100439A1. www.patents.google.com/patent/CA2100439A1

When I was in Scotland visiting a landowner who had two decades earlier welcomed beavers when they arrived on his estate, I saw he had installed such a thing and asked him if the beaver deceiver worked. 'We're not deceiving anyone son—they're perfectly aware what's going on.' He went on to explain one should see it as part of ongoing negotiations over how much effort each would make to pursue their vision of what the landscape should look like, or a dialogue, in which both beavers and humans adapt their experience of the place in relation to the dynamics of the landscape.

EXPERIMENTS IN WETNESS AND AESTHETICS

Beavers welcomed in places after a long absence such as in western Europe are sometimes presented as experiments—or 'trials'. But these efforts may be as much about humans developing concepts and expectations of how to see a watershed than it is to check if beavers have the relevant skills and commitments.

One informant said about such an experiment in the UK that it had failed: 'beavers just produced wetness', signalling disappointment with the lack of engineering skills and clever planning. But wetness may have been the point of their ingenuity. With Mathur and Da Cunha we can argue we need to move away from rivers as clearly bounded elements in the landscape—it is precisely this type of dynamic wetness that is needed in the face of alternating floods and drought due to climate change and overengineering.[5]

5 Anuradha Mathur & Dilip da Cunha, 'Wetness Is Everywhere', *Journal of Architectural Education*, (74:1), 2020, pp. 139–140. www.doi.org/10.1080/10464883.2020.1693843

Part of what beavers offer are an opportunity to change the expectations of what amounts to a good dam (a dynamically leaky one), and a managed water system (that is way less messy than it looks to an untrained eye). This is hard to replicate in the place-based and adaptive forms that generations of beavers produce. This undoing of modern aesthetics that values the experience of clearly delineated and orderly landscapes can provide moments for humans to question assumptions of what amounts to human interests, knowledge, technology, infrastructure, ecology, and by extension morality and culture. Living in small collectives and building on the work of previous generations of beavers, their efforts potentially disrupt not just the flow of water but also seemingly self-evident distinctions between what is seen as cultural and natural. Cleo Wölfle Hazard suggests we should see beavers not as unpaid engineers, but as queer performance artists in how they generate experiences that expand affective possibilities: through their work they foster ecological abundance by reintroducing variation and unpredictability. The traces and movements of beavers, offering a dazzling array of disordered water bodies, can be seen as gestures which make human river managers yearn for more joyously transgressive waterways.[6]

6 Cleo Wölfle Hazard, *Underflows: Queer Trans Ecologies and River Justice*, University of Washington Press, 2022.

DESIGN IN TIME

When human design is often understood in terms of 'iterations', with beavers we can see these not as isolated/separate steps, but processes of learning and dwelling—getting to know a place and developing a taste for the landscape (beaver intestinal microbiomes makes their guts suitable to certain tree species and not others: they cannot just switch from willow to oak). In their water works they respond to emerging conditions but also create situations in view of different seasonal water levels.

Part of appreciating beavers could be a question of time—of how to understand time. Often this is thought of as linear, where actions are understood in terms of cause and effect: first you imagine something, then you realize it in the world. In other understandings time is 'knotted',[7] not separable from ongoing interactions between organisms and the worlds they are shaping.

7 Deborah Bird Rose, 'Multispecies Knots of Ethical Time', *Environmental Philosophy* (9), 2012.

What happens if we try to live with beavers from a sense that we can learn from them, when our engineers with blueprints fail us, considering a need to realize large infrastructural works not designed in one go, but based on decentralized modes of collective action and ongoing maintenance?

Contemporary Dutch river managers we speak with, in places where beavers have recently found their place in the water system and are allowed to manage streams according to their insights, are impressed by the ingenuity and thinking behind the dams and burrows. We can see these as experiments with new forms of material culture produced between humans and beavers.[8] Ecosocial design here treats beavers not merely as diligent workers inadvertently providing some helpful ecosystem services to humans, as long as they are kept within bounds and deceived not to overstep these. Beavers invite us to see ecology as a field full of attentive beings that have their own sense of what needs to be done, creatures who respond and plan ahead, who adjust creatively, they are water workers, wetness makers. And instead of merely biomimicking them and their skills—or supposedly

8 Cf. David Lulka, 'The Residual Humanism of Hybridity: Retaining a Sense of the Earth', *Transactions of the Institute of British Geographers*, (34:3), 2009, pp. 378–393.

9 E.A. Welden, 'Conceptualising Multispecies Collaboration: Work, Animal Labour, and Nature-based Solutions', *Transactions of the Institute of British Geographers*, (48:3), 2023.

deceiving them—we could also learn to collaborate,[9] communicate, learn together, negotiate, and be in awe of more-than-human projects of eco-engineering. Not to naturalize grand schemes of river diversion and control—to make 'water management' in its modern form something that has always been around, but one that starts from appreciating the highly place-specific processes of adapting and shaping places in tune with ways in which land—and water—itself is thinking.

Or perhaps beavers are just obsessed with the sound of running water—which needs to be stopped; OCD dam tweakers. It's always fascinating to see how beavers respond to the conditions they find themselves in—do they see their dams as perfect structures that require maintenance, or never-finished projects whereby perfecting and adapting 'teaching' skills and an ethos of shaping the environment is key to hand down to their burrow mates?

BEAVER WORKS AT THE GEUL RIVER

Since half a year the beavers in the river Geul in the south of the Netherlands have made great strides. The water is much higher and flows remarkably slower in the side branches worked by beavers. From the river they dug several different new slides of mud to reach the shore. While we already slip when walking on a small inclination in this landscape, the beavers seem to be able to get up almost vertically through the mud. With a full stomach sliding back through such a mud ditch into weightlessness of being in the water must be a pleasant feeling.

Thijs and I are standing in a small piece of shrubland on the edge of the water and spot the roof light/air hole of a beaver lodge. The hole is around 15 centimetres/6 inches in diameter, with a structure of branches on top. A small wren comes out, who apparently lives there as well. At the entrance of the underwater den an extension was built that gives access to the underground passage leading to their lodge. Perhaps this 'door post' helps them deal with fluctuating water levels, or it is a stash of food that remains fresh longer under water.

We have many questions about the gnawing and building art of beavers. We distinguish between the chewing down of entire trees to reach the branches and leaves, the mere nibbling of tree bark, and the gnawing of superficial markings. Why did they stop here? Did they give it a try and felt 'this is hard to stomach, let's move on', or is it a beaver-style graffiti tag marking their presence? Are scent marks excreted here with the gnawing? The branches that are cut through we see they make pieces of a certain length. Is this because of their lifting capacity or is it part of a water works project? Freshwater cultures of beavers may be low tech but do require space for study and experimentation.

BEAVER CULTURAL CENTRE

Henriëtte Waal and Thijs de Zeeuw, Beaver Brooch worn at the River Atelier walk in the Gully Valley, 2025. Photo: Erik Wong

We wonder how long these beavers can continue in this landscape and how their relentless labour of the recent past may have shaped local feelings about them. If all trees are taken down and the riversides dug out there is little space for beavering left. Earlier we understood from local water managing organizations that the conversation with beavers is primarily seen as nagging: spreading teeth-unfriendly pastes onto tree trunks. Dams are being destroyed. There is little appreciation or applauding. The beaver structures are not indicated on any maps or signs. (Which may actually be a good thing considering the overload of signage.) This is not a finished structure, but more of a beaver atelier or experimental station in which they try out things. We can think of how to name this space of beaver initiatives and put it on the map. Would it be possible to think along and facilitate, offering beaver gnaw-and-build gift packages?

For that, we still understand too little of their tastes and gnawing motivations, but an experiment with the local willow coppicing volunteer group seems interesting. When planting a tree outside of the comfort zone where the beaver is still close to the river, the beaver will probably leave it alone. But closer to the river, transspecies collaborations in managing tree roots and river meander seem obvious. An ever-changing field site where curious visitors can come and check in on the emerging beaver cultures.

2.2 Eel Park

INTERVIEW WITH THIJS DE ZEEUW
BY HENRIËTTE WAAL & CLEMENS DRIESSEN

Two black dots look at me, followed by a line drawn with a fine pencil on a transparent, snake-like body the size of a matchstick. The fact that this glass eel has just crossed 6,000 kilometres of ocean to grow up here in The Hague compels admiration. Cosmic elusiveness in the palm of my hand.

Since Covid, I have been monitoring glass eels (baby eels) with a cross net of 1 x 1 m after sunset—the eel is a nocturnal animal. Along the entire European west coast, groups of volunteers measure the entry of glass eels. These are usually fishermen. For me, it is a bi-weekly fish education at dusk, on the boundary between fresh and saltwater.

The eel is a catadromous fish, meaning it migrates from salt to fresh water and back again. Born in the Sargasso Sea they cross the Atlantic and migrate to fresh water. As they reach the European plateau, they transform from leaf-shaped to transparent, glass-like eels, after arriving in fresh water they turn yellowish-grey and subsequently to olive green or dark brown with a silver-white belly, and then become progressively darker. Over the years, their magical bodies become more aligned with the muddy riverbed.

SINGLE STOCK

Because all wild eels from Europe and North Africa eventually swim back from freshwater to the same place to spawn, Europe is considered to have one 'single stock' eel population.

Thijs de Zeeuw, *Eel Park*, transparency measurements, 2022, 't IJ, Amsterdam. Images: Thijs de Zeeuw

In the Middle Ages, Western European farmers could pay the rent for their land in eels. Today, very few remain. If the eel population in Northern Europe is doing poorly, this affects the eel numbers in the south a few years later. As far as my eel monitoring colleagues know, no one has ever observed how eels reproduce. A farmed eel is thus a wild-caught glass eel who was on its way back from the Sargasso Sea and will now grow up in a breeding tank.

EEL REGULATION

Eel monitoring is part of the European Eel Regulation. This plan, launched collectively by the European Member States in 2007, requires national governments to take measures to prevent extinction. The main goals of the plan are: facilitating the migration of eels through rivers, allowing forty per cent of adult eels to return to the sea to spawn, imposing restrictions on fishing, and restocking safe inland waters with young eels caught in the Bay of Biscay.

RESTOCKING BUSINESS

The effect of restocking remains unproven. My fellow monitors question whether the eels, once caught and released back into the wild, actually manage to find their spawning grounds. In fact, DUPAN (Sustainable Eel Sector Netherlands), is a shady business, they say. The Dutch eel giant Nijvis, which dominates the European eel farming trade worth millions, has a leading role in DUPAN, responsible for restocking efforts. DUPAN conducts restocking in waters that are actively fished. In practice, restocking seems to primarily bolster the business model of Nijvis.[1]

Moreover, as my fish knowledge continues to expand, researchers indicate that restocking brings other harmful effects, such as the transmission of diseases, parasites, and changes in genetic diversity.[2] Considering the alarming status of the eel, scientists advise to halt this form of 'protection'.[3]

IMPROVING EEL HABITAT

What really helps the eel's survival is fighting ongoing illegal and legal trade (including Nijvis, which allegedly uses its French subsidiary to create a legal shortcut to grow eel in Morocco, circumventing the EU's eel export ban). The hope instead is to transform fishers into protectors.

1 Max Wastieux & Remy Käller, 'Hoe een Nederlands bedrijf de Europese palinghandel in handen kreeg', *Follow The Money*, 21 February 2024.

2 Linna Danne, Mikolaj Adamek, Hubert Wonnemann, Theresa Pieper, Daniel Fey & John Hellmann, 'Identification of Virus Infections of European Eels Intended for Stocking Measures', *Journal of Fish Diseases* (45:9), 2022, pp. 1259–1266.

3 Jelena Oreha Aleksandra Morozova, Alberts Garkajs, Muza Kirjušina, Inese Gavarāne, Vladimir G. Kostousov, Mihails Pupins & Natalja Škute, 'Genetic Diversity and Distribution of Haplotypes of Freshwater Eel in Baltic Lakeland Based on Mitochondrial DNA D-Loop and Cytochrome b Sequence Variation', *Diversity* (16:9), 2024, p. 522.

Thijs de Zeeuw, *First Encounter with the Eel*, 2021, film still, Ouderkerkerplas. Camera: Sjouke Brunia, Divection

Thijs de Zeeuw dives early in the morning to reef cages serving as shelter for smaller fish, which have been placed in the Noorder IJ lake near the pumping station, 2021. Photo: Hans van der Meer

The other way to genuinely help the eel is by improving its habitat. The Dutch organization for reptiles, amphibians, and fish, RAVON, which organizes eel monitoring, recorded 60,000 eel bottlenecks in the Netherlands. To start to open up space for the eel, they try to create momentum for an eel reserve in the Amsterdam Sloterplas. But how do you design for an eel? My weekly interaction with fisher(wo)men and eels is a way to better understand that. Fellow eel friend, landscape architect, and animal habitat designer Thijs de Zeeuw has been similarly following the eel and prototyped an eel park.

EEL MAPPING

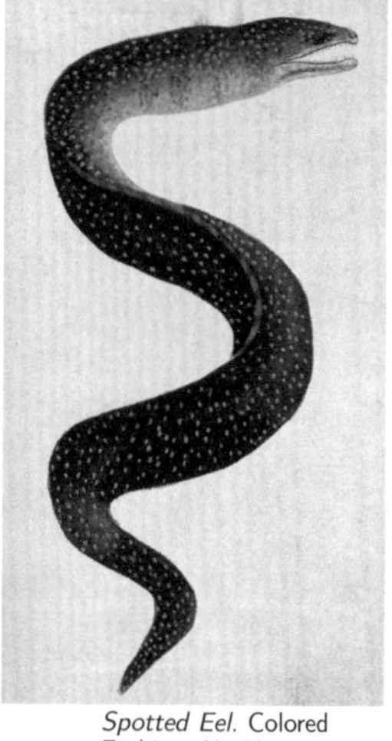

Spotted Eel. Colored Etching. No Known Copyright. Courtesy Wellcome Collection

Thijs thought a map as a communication tool between the eel and politicians was a good starting point. Based on data from the Dutch government, the map shows where eel can swim freely but also the locks, pumping stations, and dams that form obstacles. The Sloterplas eel reserve is on it, and various soil samples give an idea of the substrate. In Amsterdam, there are currents or layers of fresher and saltier water on top of each other. Under water, eels navigate a complex and layered world. Thijs asked scent artist Frank Bloem to create the scent of fresh water and the scent of salt water, to make the eel's world palpable. People could smell bottles and learn where eels were triggered to go to the sea or to swim inland.

The map and tactile samples elicited little response from politicians and therefore had little effect on the eel. Thijs then decided to try to have a direct conversation with the eel. To do so on their terms, he learned to dive.

EEL ENCOUNTER

HW I found my first encounter with a glass eel impressive. How was your first encounter with an eel?

TZ I had always thought: if I see it, it will be a glimpse and it will pull its head back into a hole. I will be very close because the visibility in muddy waters is so poor. And probably, it will be an observation with a probability factor, where I'm not even entirely sure it was an eel. It was my second dive in open water. First, we swam near the surface at about three metres where visibility was fairly clear. Then we decided to dive deeper. It got darker and darker. And then at one point, it was just lying there on the sandy bottom, with a slight curve in its body, a bit like a question mark. I had to find support with my hand on the bottom and created a cloud of silt. But the eel remained there and let us approach. We came within a few meters. It was probably resting. It must have heard and felt us long before.

HW I've read that their sense of smell is comparable to that of a dog. So, the eel would have also smelled and tasted you long before.

HEARING AND FEELING

TZ I think the senses underwater are somewhat intertwined. We hear with our ears and the air. But they have a swim bladder and a lateral line along their body that detects water vibrations. So, I think hearing and feeling are mixed together in those fish. There must have been all sorts of signals that we were coming. There was a moment of recognition or exploration. It didn't give me the feeling that I was some big intruder. And after ten to fifteen seconds of mutual presence, it first swam backwards a little and then, in that backward swimming, turned around and swam away. We swam after it a bit, but that felt like a very different relationship. Then you're chasing something that apparently isn't necessarily waiting for you. So, we let it disappear into the depths.

HW How did you collaborate with the eel in the Eel Park?

SOUNDPROOFING

TZ By listening underwater, I discovered that there is also a lot of noise down there. So, I thought: How sad that the fish can't even listen! An important element in the Eel Park is soundproofing. I implemented that in the park in two ways. You see a lot of low rumbling sounds. These are long arcs of sound. You can quite easily filter them out by creating a shallow area, for example. The chances are quite small that such a long sound wave will exactly hit that shallow spot. I therefore created a kind of shallow area around the park.

BUBBLE SCREEN

TZ You can also use a bubble screen for soundproofing. It's a technique they also use in deep-sea construction when building wind turbines. Because then they have to hammer poles into the seabed, and that impact is really an enormous and intense sound pulse. Fish can die from it. A bubble screen is created with a tube at the bottom through which air is pumped, and of course, it also makes some noise, because they are all popping bubbles. But at the same time, that's also air in water, and it can absorb a pressure wave. So, I placed a bubble screen around the park. The nice thing is that a bubble screen can also be used to catch debris. Due to underwater currents, a lot of debris drifts somewhere in the middle of the water column. And it doesn't pass through such a bubble screen but flows along it. The bubble screen is a way to dampen sound and keep the park clean. Research is being done to see if the bubbles block fish. If so, then I will, of course, need to create a few entrances and add a few doors.

HW What are the other components of the Eel Park, besides soundproofing?

VARIETY AND SHELTER

TZ The basic principle is variety. Underwater, it is actually one large void with few places for young fish to grow. You can create shelter by adding depth. I envisioned a hard wall with small cavities at different depths. That would be interesting for different species. And I also designed a sort of stepped gabion wall.[4] Then you get a kind of terraced houses that get colder and darker as they go down and warmer and lighter as they go up. That was one of the elements to play with depth and create resting places.

4 Metal wire cages.

SNORKEL ROUTE

TZ I included a chic marble staircase with very large, lazy white steps where you can see fish swimming from above in the evening. Normally, it is very difficult to see fish there because the bottom is dark, but against a light bottom, you can see them. So, there were ecological components, but also elements that allowed people to better see and explore that aquatic world.

HW So only very exhibitionist fish swim over there.

TZ Yes, or fish that really love exhibitionist fish! Then you have a whole group of followers, of course.

SILENCE ROOM

TZ That small waterfall-shaped thing hanging behind me on the wall was the basis of an underwater silence room. It's a concrete element which in full size is about five metres in diameter. And around it, reeds grew. Reeds are stalks with air in them, which insulates. So, the idea was to create a silence room there.

HW The silence room is a beautiful idea, but you speak of it in the past tense. What happened to it?

TZ The past tense means the model no longer exists. But as you can hear, the idea is still alive. This shape hangs in my studio as a reminder. Like: we really need to do this someday. The implementation of the park was actually a series of tests. Listening tests and tests with cages and a scale model. But the park has not been implemented as such.

I don't rule out that it will happen someday, but my first goal was to get a story into the minds of politicians and spatial planners. Amsterdam is forty per cent water, and it is always described as empty. When you look at the biodiversity map, it's grey, so an incredible amount of understanding of the underwater landscape of the city is missing. It is an interconnected ecosystem, and there are entities living there that experience the city in their own way.

There was a very small budget for the assignment from ARCAM and the Creative Industries Fund, and I work on multiple assignments simultaneously.

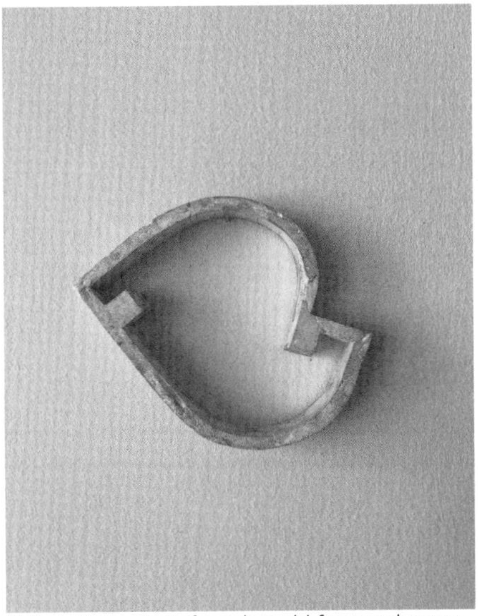

Thijs de Zeeuw, Part of a scale model for an underwater silence room for eels (*Eel Park*), 2022. The shape on the image represents the concrete base for a reed sound barrier. Photo: Henriëtte Waal

As a result, I currently don't have the ability and time to build this. But I feel that in this way, I am contributing to the development of knowledge and ideas and the conviction that this is important.

CD Are there particular things that eels can do that you hadn't thought of before or that you've come across?

NON-BINARY THINKING

TZ What I find very fascinating about the underwater world is that binary thinking or perceiving doesn't work there. Underwater is much more fluid and hybrid. The same goes for the gender of a fish. Some fish change gender depending on the hormones in the water. But it can also be a normal phenomenon within the species, depending on how many males and females there already are, for example, or the temperature of the water. You see this with eels as well. An eel undergoes four metamorphoses in its life. It is incredible to me that a fish can swim around here for ten years in the mud, without any sex organs, and then at some point gets a trigger—where does that trigger come from?—that makes it think, 'I need to reproduce'. Then, it gradually develops its sex organs while its digestive organs stop functioning, and it swims back on its fat reserves 6,000 kilometres across the ocean. Of course, it's incredibly impractical, but it has worked for quite some time. Only now, as we are increasingly altering the world, certain things are becoming very complicated for the eel. I think that the system of swimming back entirely on fat reserves no longer works. Because those fat reserves are also where all the pollutants are, all those metals, all those hormones and drug residues, which are released when the fat is burned. Then you either lose your way, or you turn out to be infertile, or... But there's very little knowledge about this because such a returning eel is difficult to follow and test. The eel embodies the transformative nature of the underwater world. Moreover, it's a very common fish, and it's very normal to know that there are so many things we don't know. The eel also embodies the unknown in nature, the wildness. The combination of the common and the wild makes it easy for me to celebrate it. It's sometimes quite challenging to explain that not knowing can be something positive. Because the reasoning goes—if you know more about it, then you can protect it better, and thus gathering knowledge is used as an excuse to keep researching and to know and control more. Meanwhile, this fish still manages to keep so many secrets right under our noses. I find that very endearing.

2.3 Vislift: The Making of Fish-Operated Lifts

INTERVIEW WITH JOHN VAN BOXEL
& RUTGER SCHRIJER
BY HENRIËTTE WAAL & CLEMENS DRIESSEN

According to the existing EU Water Framework Directive (2000/60/EG)—the key regulatory parameters for managing water quality—improving river connectivity through 'fish passages' is a direct way to improve the ecological quality of a river. This means that the directive is met once a fish passage is implemented. How, or if, fish actually use these installations though is another question.

On 17 June 2024, the much-contested EU nature restoration law was adopted by the European council of ministers.[1] This law introduces several essential interventions for aquatic ecosystems, such as restoring a significant percentage of the 25,000 kilometres of rivers big and small which have been fragmented by obstacles such as hydroelectric dams, weirs, locks and metal grids.

While he worked as a water engineer for the Dutch national water authority (RWS)/Ministry of Infrastructure and Water Management, John van Boxel came across numerous fish passages that he realized fish couldn't pass. Based on his personal experience with fish, he conceived and built what he called the 'Smart Vislift'—or smart fish lift. Vislift now exists for seven years, and currently over seventy fish lifts have been installed in the landscape. And counting. In 2021, mechanical engineer Rutger Schrijer joined Vislift to build the commercial/business side of the company.

Vislift advertises with plug and play installation within forty-eight hours, real-time monitoring of fish migration, continuous current 'lokstroom' to guide fish to the lift, sensitive water flow control, and regular self-cleaning to prevent clogging.

1 www.europarl.europa.eu/news/en/press-room/20240223IPR18078/nature-restoration-parliament-adopts-law-to-restore-20-of-eu-s-land-and-sea

Vislift, *Bittervoorns (Rhodeus amarus)*, 2024 Vislift, *Snoek (Esox lucius)*, 2023 (top to bottom)

HW John, what was the moment you decided to design a fish passage yourself?

JB Through my job I noticed an uncontrolled proliferation of fish passages. One engineer had this approach, another would make things different again, and then when contractors got involved, costs were cut, which led to more compact passages and to certain parts not being realized. At the time I didn't know as much as I do now, but I did see that the velocity of water flows and turbulences that resulted could only be overcome by very few fish. Almost no one knew how important the flow rate is for the proper functioning of the fish passage. Nobody was paying attention to this. So, I got the idea of starting a company specialized in installing and maintaining fish passages. That was really needed. So, I did a little market research, and understood there still was a significant demand until 2030.

HW What were your initial design principles?

EMBODIED KNOWLEDGE

JB At home I have koi carp to look at fish and maintain my bond with them. Generally, people only call the koi doctor when holes start to appear in the fish. But someone who is more intimately engaged with their fish calls the doctor sooner. When they swim keeping their fins close to their bodies they have some affliction. Or when they beat with their heads. Then they have an itch. You have to watch closely to understand nature and be able to facilitate it.

As a recreational angler I have always watched the water in streams; observing how trout, big ones and little ones, lie in the flow of the water. So, I had ideas about hydrological parameters required for the fish passages.

ROUND

JB A fish passage should be round. A round shape has a constraining effect on the water flow, there is a natural sinus, a meander in the way rivers flow. If you kick a round bucket, the water will slosh much less and for a much shorter time than if you kick a square bucket. Fish don't look with their eyes. They can, but effectively fish look with their lateral lines [John makes a movement with his hands alongside his entire body]. These lateral line organs are very sensitive. They send and detect vibrations with it. From round sides they get less strong signals than from a flat box. Therefore round.

COMMUNICATING WITH THE MANAGER

And a fish passage should communicate with its manager: 'hi I'm clogged', or 'there's no fish coming through'. All objects/water works managed by the water boards communicate. When a lock malfunctions, it sends a signal. But fish passages don't communicate, and that's not acceptable anymore. There are even ones that are so entangled with their environment that they can't find them.

VARIABLE FLOW

But the most important point of departure was elevation. Not all fish can swim upwards, whereas some actually require a strong current to be triggered to swim. Without enough water flowing along their lateral lines they get disoriented and turn around. Variable flow is therefore essential in the fish passage. This idea initially raised a lot of backlash. The hydrology of common passages was such that less flow would alter the dimensions, making passages lose all their water or retain too much.

MEASURABLE, CONTROLLABLE

I took my idea to Bjorn Prudon, then manager of innovation of the Rivierenland water board, who also is an angler, and an ecologist. He got so excited that he literally jumped up and cried: 'this is what water managers are looking for, measurable and controllable fish migration!' I had wanted to make a simple monitor for water managers, but he immediately proposed data-driven science, water quality sensors, et cetera. I then brought in electrical engineers who developed the sensor systems, and this is how we built the Vislift.

HW Why do the fish go through this one?

JB I can't really answer that, as I made several adjustments at once. So, we don't know exactly what determined success in the end. But it worked. We saw a lot of fish, including species that had never been observed in fish passages before. Which we thought were too smart or too shy to dare use fish passages. The camera-monitoring was met with scepticism from the water boards at first. They said: Van Boxel could tell us anything with his camera. We also didn't have real-time streaming back then. Through the use of camera-monitoring we could study the behaviour of fish in a different way and we learned valuable lessons from it.

HYDROLOGICAL MYSTERY

The further design of the fish lift was taken up by TU Delft, where seven students have now graduated on the hydrological mystery.

JB I wasn't educated there myself. Developing it has been a piece of luck. Maybe a certain intuition guided it to the way it must be done. To date, we haven't managed to get the hydrological calculations completely right. We do get a step further every time, which the professors at the university like of course. The fact that we could not get the calculations right led many parties to remain unconvinced. In certain water boards in the Netherlands, we have a group that really wants to understand everything very theoretically, which delayed the operational steps. That took a year. And then they still didn't know everything. Then the head of the water board intervened and said: guys, just stop asking how you calculate it. We can see it.

CD How did you develop the variable flow? I assume that requires testing in practice—it sounded like you were almost like a deejay, watching what happens and adjusting?

JB Yes exactly. First you monitor the water levels. Due to fluctuating levels—which has only been exacerbated by climate change—we see dynamics in the water.

Fluctuation increases flow velocity exorbitantly in a traditional fish passage. Five centimetres higher water level difference is one metre per second in a fish passage. We translate fluctuating water levels into a digital field of flow velocities and fish. Before you make any adjustments, you observe what fish you get, for example a certain type of perch. And then you start playing with the settings of the fish passage. You see hey, there and there with that flow rate I saw those fish species. Let me set it to that flow rate for a moment. Will I see more of those fish then? That's how you start experimenting.

INNOVATION IS A PROCESS/ TO INNOVATE IS A VERB

In the Netherlands you can find one small waterway being guided in an aqueduct over another one. In such a maze of water streams I get nervous and sometimes lay awake at night. Are we doing it right? Do we really get it? You need to really connect with the situation. I always say: 'to innovate is a verb'. Reasoning only gets you so far, but you learn by doing. You have to subject yourself to the situation to make things really good.

HW What I find beautiful is how the object can stay physically the same, but that you can make it fit in the environment by playing with the settings.

CD Can you give us some concrete knowledge about some fish that the Vislift generated?

JB In the Linge river we have now made eight consecutive weirs passable for fish. And three more will follow. Last year over 650,000 fish used these fish lifts. With this circuit you can follow fish migrating across them. This is great fun. For twenty years they had been monitoring, to meet the EU Water Framework Directive, but never caught one barbel. In 2020 we saw one in fish lift no. 4. Then we got an immense amount of scrutiny: a barbel requires flowing water and a rocky river bottom. The Linge is stagnant and has no rocks, so theoretically there can be no barbels. People derived that I must have released the barbel myself. Then we saw a second one. That must have been the same one, people said. And then we got an image that showed the two of them in one frame. Then the critique subsided somewhat. A few years onwards we have now seen over a 1,000 barbels with their offspring. I can tell the juveniles have grown in number, and we can share that on social media—which now gets more positive responses. The entire Linge now has barbels. And what is nice is that the pikeperch (or zander) now has started to spread as well. And besides the clear spreading of fish, the Vislift has also contributed to reporting to the EU. The WFD score has gone up.

HW What is the ambition for the design now? Is it sustainable materials and production processes, or installing them in certain landscapes, or further links with research?

JB My dream is to connect fish with each other. That they can swim to their breeding/spawning grounds, and their protected habitats. Next to that I want to connect fish people: scientists, anglers, but also farmers and water managers. I want to build relations between different water boards, but also between countries to exchange data to optimally facilitate fish. Since fish swim from the sea to the mountain and from the mountain to the sea, through all these places in between.

HW How does it work financially; can you tell us something about that?

RS Europe grants subsidies for restoring fish migration, prioritizing designated 'Natura 2000' nature reserves. Water boards in the Netherlands, but also water authorities abroad, can apply for these subsidies. In the Netherlands there are a couple of hundred remaining sites we still need to do. It is of course a pity that much funding has gone to fish passages that do not work. This means that clients/water boards have been extra cautious with the Vislift. But now that we have proven ourselves, they are embracing us. We are now trying to get the product to a higher level, in view of exports.

HW What needs to be adjusted to make Vislift suitable for other countries? Do you need different types of fish recognition in say France?

RS The basis of the product stays the same. We chose certified electronic components and technologies that are also used for weirs and pumping stations. In other countries there may be larger differences in elevation. To overcome that height difference, Vislifts can be placed in a cascade arrangement, one after the other. This allows for larger height differences to be bridged as well.

We optimized the software for wider availability. Vislifts communicate their status every five minutes to a large database for automated fish recognition. Initially, we had three people categorize fish manually, which limited expansion. But now, with smarter software, we can recognize fish based on shadows and movement. For more advanced movement recognition we would need more bandwidth. Image selection then has to happen locally, on the Vislift.

HW Do you also come across exotic species, and could Vislift become a tool to say 'we can filter these out'?

JB I have a different view on that. What is an exotic species? It gets warmer here. Circumstances change, so eventually also species here will. We see American (red swamp) crayfish passing by. Then our social media comments say: can't Vislift immediately eliminate these? No, because this life has come here because we have certain conditions, so you should ask yourself 'what is exotic?'

CD But it is possible that Vislift actively supports these species to migrate. Or limits their movement?

JB we can also ensure that predators of the crayfish are present. Then it is less of a problem. Everything depends on having a good balance. But humans are always very good at disrupting balance.

CD But the enthusiasm of the water board innovation manager was based on the ideal of monitoring and controlling everything. Could it also be a consideration to use Vislift to prevent invasive species coming in?

JB I find eliminating certain species rather extreme. But by making choices for particular flow settings, you also decide what fish/vision you prioritize. There is variation seasonally but also at certain times during the day. For instance the eel, which is highly endangered, is often active at night. Over the last few years there have been dry periods during which all fish passages were closed in order to retain water. When that happens early in the season, it means there is no migration at all that year. And thus no new addition to the population. By opening the Vislift between 0:30 and 4:00 in the night we can offer the eel a head start. People have now started working with our online system and with the data. These are all new things that may develop in certain directions.

CD How do ecology and technology go together? Do you see the fish as users of the fish lift?

JB We call them the lead actors. We have gathered millions of images and proven that the fish lift works. We managed to build something that nature makes use of. The fish lift has become part of nature. We have seen a little perch with a tiny fish in its mouth swimming near the camera. We have seen a marauding pike right in front of the camera. But we have also seen a catfish using its barbels to sense how to swim through the little gate. We see fish hesitate before they have to sprint. But I'll go even further: I believe they communicate with each other. We often think as humans we are the smartest, that we are different. We have quite an ego as inhabitants of the Earth.

HW Has this work made you look differently at fish? When looking back to the time you were catching eels with traps?

JB Back then it really was very normal to eat eels, nowadays I forbid my family to do it. At the end of our holidays we are glad we can go back home to our pets. We caress our fish. My young son just grabs fish out of the water with his hands, and they really have no stress since they immediately return for more. For us, fish are pets.

In the future, fish lifts can help generate new knowledge to restore fish populations in rivers. The development of the Vislift shows that technology must go together with an embodied love for fish.

2.4
Mussel Sensor

BY HENRIËTTE WAAL
SHORT FILM BY JULIA PELKA, 2019[1]

Julia Pelka, *Fat Kathy*, 2019, film still

1. www.juliapelka.com/fatkathy

Fat Kathy is a water pump and water safety system in the middle of the Vistula river in Poland. Its employees gave it this name because it is big and round. It is Europe's largest drinking water filtration facility collecting water from the bottom of a river. Today, all around the world this method is called 'the Warsaw intake'.

Eight freshwater mussels live inside Fat Kathy, and their bodies are linked to a central computer. If anything is wrong with the water, the highly sensitive animals will close their shells, signalling the need for shutting off the water supply and protecting the lives of millions of city dwellers.

Fat Kathy can only be reached through a long underwater tunnel. In the beginning of the film, we see the malacologist (an expert of molluscs) walking through the tunnel, which in the film acts as a metaphor for the distance between humans and nature and our inability to fully understand nature. We hear the malacologist tell of his fears that the mussels will not close one day. In recurring nightmares, he is the only one who knows that the water is toxic, but he is unable to move or speak, so he cannot warn people.

Above the water, Julia Pelka films an elderly woman, aka the Queen of Headland, a fragile beauty who has been coming to rest on the banks overlooking Fat Kathy for over thirty years. She, like most other residents, does not know about the system and the mussels. Because of the way Pelka films the woman, lying on a rough piece of stone, she too becomes a kind of clam; just like her admirer who watches her every day from under a tree a little further away.

In her film, Pelka leaves a lot of room for thought and associations. Occasionally she gives scant information or allows the malacologist or the Queen of Headland to speak. In the final image, we see a plane plotting a trail of water vapor over a mysterious moon. Watching the film, you can't help but ponder about how we live together and how everything is connected. But also, this film shows how there are different ways of connecting human and nonhuman infrastructures and forms of knowing. The malacologist glueing the mussels on their little Perspex pedestals and attaching a magnet to their shells speaks with affection about the creatures in his installation.

'You take a clam and it's completely helpless. The only thing you can do is take care of it.'

A quick internet search reveals that mussel systems are at work all over the world. Mostly, those systems use the mussel as an instrumentalized piece of nature offering people a simple and trustworthy solution. Deeper connections are overlooked. Never mind that this living system is something we could think about. Think with. What makes Pelka's story in Poland special is that it shows a collaboration, where humans appear to put their lives in the hands of mussels who help them for several months. Then, the mussels are carefully dislodged from the installation and returned to their original habitat, to continue their lives while new mussels take over.

2.5 The Needles of the Meuse

BY CLEMENS DRIESSEN
DOCUMENTARY BY MONIEK VAN DE VALL
& KEES VAN DER VOOREN

Moniek van de Vall and Kees van der Vooren, *Needles of the Meuse*, 2022, film still

In the north of France, the river Meuse is fed by rain from the mountains. A few years ago, the twenty-seven hand-operated weirs used to manage levels of the Meuse were replaced by automated installations. The old system, in which we see weir keeper Thierry at work in this image, had functioned since 1835.

By inserting wooden beams—'needles'—from a narrow footbridge across the river, its level and flow could be steered. The needles were held in place by a ridge on the bottom of the river and the pressure of the water that made them lean against the footbridge. Their lifelong experience adding or removing needles allowed the operators to sense and react to minute differences. 'The river flows through my body', Thierry says with a laugh, in the documentary by Moniek van de Vall and Kees van der Vooren.

In recent years the variation in water levels had increased to the point that the consensus was that the needles were no longer sufficient to control the river.

According to the needle operators and local experts, this change in the character of the water flow is not only caused by climate change, but also reflects changes in the watershed and the ways water is channelled to the river: 'About thirty years ago, when a drop of water fell on the mountain, it took between thirty-six and forty-eight hours to reach the river. Today, that same drop that falls in the morning, has reached the river twelve hours later. The water rises much faster than we can move the needles'.

The documentary film portrayed the needle weir in its final moments of operation, when it was replaced by a metal structure operated automatically. The story could be taken to merely reflect nostalgia for a bygone era that still allowed for embodied working with the river and its power. But it is clear that with this intimate day-to-day work in the midst of the stream, a form of knowing the river has now been dismantled.

2.6 Wild Fermentation

BY HENRIËTTE WAAL

I have sought to live intimately with microbes and wild fermentation for nearly two decades now. On riverbanks and ditch sides, at home, and everywhere I worked.

These sensory experiences intertwine embodied and material moments of confusion, speculation and improvisation. I continue to wonder about my relationships with more-than-human fermentation partners and the polluted environments we coexist in. Below, I link these personal experiences to the broader context of wild fermentation, particularly through a case study in Guangzhou, China, by environmental ecologist and anthropologist Amy Zhang.

Wild fermentation is a dance between yeasts and bacteria to carry out anaerobic respiration, for instance in our gut, or in a piece of fruit.[1] Together, these bacteria and yeasts make an acidic feast, an inhospitable environment for microbes that might outcompete and kill them and help make food for each other. From a long-term evolutionary perspective, all humans, animals, plants, fungi, and multicellular organisms originate from simple fermenting organisms.

My initial interest in brewing was sparked by the concept of DIY water purification through beer brewing, but soon I was captivated by the poetry of working with this invisible, bubbling yeast community. I started with conventional purified yeast strains—which is how commercial brewers make beer—but later wild fermentation became a form to connect to patches of land and water, each with their own unique set of species, escaping the commercial and industrial food landscape of the Netherlands by inviting whoever was around and ready to ferment.

From the canals and the Amstel river in Amsterdam, to the Emscher in the German Ruhr region, and to rain in New York City, each locality and weather condition brought its own surprises, leading to a deeper confidence in the process and the collaboration with microbial landscapes.

1 'Fermentation is the process by which carbohydrates are transformed by microorganisms into substances such as lactic acid and alcohol. Fermented foods frequently introduce microorganisms that live in the human body when consumed by humans.' P. Patel, K. Butani, A. Kumar, S. Singh, & B.G. Prajapati, 'Effects of fermented food consumption on non-communicable diseases', *Foods*, (12:4), 2023, p. 687.

Henriëtte Waal, *Buitenbrouwerij/Eshofpolder blond*, 2010, exh. and symposium 'Footprint. Voedsel voor de stad', The Hague, the Netherlands. Courtesy Stroom Den Haag. Photo: Jorn van Eck

Henriëtte Waal, *Wild Beer Fermentation: Pellicle in Wooden Barrel*, 2021. Photo: Jacques Bertens

Henriëtte Waal, *Buitenbrouwerij/ Eshofpolder Blond*, 2010, Footprint. Voedsel voor de Stad, The Hague, the Netherlands. Courtesy Stroom Den Haag. Photo: Jorn van Eck

Once, I had to stop an enthusiastic passerby from filling his glass with untreated Amsterdam canal water because he thought the water was instantly drinkable. Another time, while cooling a Wadden beer in the North Sea, a small wave splashed into the fermentation vat, causing the fermentation to proceed in slow motion, and I had to add an extra yeast starter to keep the process going. Other times, when there was no access to cool or running water, I let the wort (the brewing liquid) cool down in the open air for many hours, allowing local yeasts and bacteria to take their chance in a fermentation that lasted several months. The sour flavours that those unseen comrades produced really appealed to me. I have never had a sweet tooth. By 2015 I had become obsessed with all things fermented, and I founded Bierberaad, a monthly cross-disciplinary platform for beer design and microbial cultures at Mediamatic Amsterdam.

This role involved orchestrating a dynamic ecosystem where big and small brands, designers and scientists, amateurs and enthusiasts came to explore all aspects of brewing, fermenting, tasting, to experiment with new techniques and unconventional ingredients and to discuss failures and crossovers. This led me to work with microbiologist André van der Zee, who caught wild yeasts by hanging petri dishes with agar agar in protective metal cages (to save the agar from other predators) in trees. I followed a wild fermentation of cherries in a large bathtub in Nijmegen with Sander Kobes from Oersoep and I worked with Stephen Grigg, brewer of Kromme Haring and also a biochemist with his own strain library. The most delicious flavour experience is created when multiple yeast strains and bacteria ferment in symphony. This happens spontaneously in wild fermentation.

One wild fermentation I initiated took place in Rwanda, in the former house of my cousin in Kigali, together with two brewers that I met for the occasion. I wanted to make Ikigage, sorghum beer, and my cousin's driver scouted local experts and brought the two women to brew with me. Ikigage is not sold commercially but is typically made at home for personal consumption. Knowledge is passed on from woman to woman.

Mashing in the garbage bin, Kigali, Rwanda, 2015. Photo: Ria Waal

We brought a pan of water to the boiling point and added some handfuls of germinated sorghum grains in order to make a sticky starch. One slender wooden spoon after the other broke under the strong stirring hands of the ladies. We needed a big container and agreed that the plastic rubbish

Drying and germinating sorghum for brewing, Kigali, Rwanda, 2015

Ceramic fermentation vessels in the former King's Palace in Nyanza, Rwanda, 2015. Photo: Henriëtte Waal

Ceramic fermentation vessels in the former King's Palace in Nyanza, Rwanda, 2015. Photo: Henriëtte Waal (top to bottom)

bin was our best option. Tools in this kitchen were obviously no match for ikigage brewing (I regretted our decision not to do it outside). We cleaned the bin before adding a mixture of crushed sorghum meal and cold water. The ladies threw the hot pan into the bucket with cold water. They didn't measure the temperature. It was all embodied knowledge. The saccharifying temperature is typically between 63 °C and 70 °C for beer, so we must have been somewhere in between those temperatures. No boiling of the wort, meaning no sterilization. After a while (they knew how long), water was added to bring the temperature back to about body temperature. They measured it with their hands on the outside of the garbage bin. We used a towel as a lid and looked for an empty corner in the house. It needed to stand there until the next day, they said, for it to be drinkable.

Then the doorbell rang. It was the driver. He said nothing but was shaking. My niece had come home and was not at all pleased he had brought two strangers to her compound. I hadn't thought of it at all. Afraid of losing his job, the driver quickly packed the two ladies in his car and drove them back home. And so my brewing date ended rather abruptly.

The next morning, I lifted the towel and found the sticky liquid slowly bubbling, producing a rather sour but also lightly floral smell. Ikigage is a mixed culture fermentation. Lactobacillus fermentum creates an environment where Saccharomyces cerevisiae and Issatchenkia orientalis yeast can co-ferment. Acid production is higher and ethanol production lower than in single yeast starter cultures. It is all about smell and taste—as you can't see what you drink out of a calabash with a straw. I remember a sweet floral aroma, a sour, slightly burning taste with a creamy mouthfeel. Later in my research I also visited the brewer of the former Rwandan king, who fermented honey wine and milk besides beer. He had colourful earthenware fermentation vessels with sculptural lids. I would have loved to brew with these characters to taste what flavours they contained. Difficult to clean, but that's the whole idea.

Spontaneous fermented sorghum beers are very rich in calories, B-group vitamins including thiamine, folic acid, riboflavin and nicotinic acid, and essential amino acids such as lysine. Wild fermented products are strongly connected to their territory of origin and their people's traditions, and therefore embedded in its surrounding society and place.[2] However, for most people the traditional sorghum beer became less attractive than Western beers produced with aggressive commercial yeast strains and industrial sugars, because of its 'poorer' hygienic quality, organoleptic variations and shorter shelf life. Local residents fear that ikigage will die out as a traditional local fermented beverage.[3]

The commodification of products leads to a dulling of flavours and the extinction of microbes in our gut. The goal to completely eradicate all diseases may also mean the eradication of the incredible variety of microbial life forms from this planet. For instance, the wiping out of the H. Pylori bacteria, which previously had been understood to be harmful to human health, has been associated with the rapid rise of obesity.[4] Formal medicine only recently started to understand its role in balancing energy storage in the body. Also E.coli bacteria are not only harmful. If e.coli is present in a certain concentration in drinking water it can make people sick. But research is done in the US to study the potential of E.coli as a medicine to treat cancer and to break down PET (plastic) waste.[5]

Pharma biotech again discovering the open-source library of wild lifeforms that make for crucial connections.

2 Jeroen Hugenholtz, 'Traditional Biotechnology for New Foods and Beverages', *Current Opinion in Biotechnology* (24:2), April 2013, pp. 155–159; J.P. Tamang et al., 'Fermented Foods in a Global Age: East Meets West', *Comprehensive Reviews in Food Science and Food Safety*, (19:1), 3 January 2020, pp. 184–217.

3 *Slow Food Foundation for Biodiversity*, www.fondazioneslowfood.com/en/ark-of-taste-slow-food/ikigage-sorghum-beer/.

4 N. Lender, N.J. Talley, P. Enck, S. Haag, S. Zipfel, M. Morrison, & G.J. Holtmann, 'Associations Between Helicobacter Pylori and Obesity: An Ecological Study', *Alimentary Pharmacology & Therapeutics*, (40:1), 2014, pp. 24–31.

5 Zachary D. Blount, 'The Unexhausted Potential of E.coli', *eLife* 05826, 2015; J.C. and S. Wallace, 'Microbial Synthesis of Vanillin from Waste Poly(ethylene Terephthalate)', *Green Chemistry*, Issue 13, 2021.

There is no list of endangered microbes. Many lifeforms that we consider pathogens are probably endangered. Microorganisms specific to endangered species could be on the list as well. The extinction of sturgeons, salmons, and other fish would also mean the extinction of the intricate communities of single-celled organisms and other simpler life forms they harbour.

Amy Zhang has studied Chinese anti-pollution activists and microbial communities that work together to improve the health of bodies, homes, and the riverway they depend on. Co-enzymes are made through a three-month wild fermentation process of water, brown sugar, and leftover organic waste, such as vegetable leaves and fruit peels. This simple and economical method, which originated in Taiwan, Thailand, and Malaysia, is gaining popularity in China as a communal activity and an eco-friendly alternative to chemical cleaners, shampoos, and detergents.[6]

6 Amy Zhang, 'From Immunity to Collaboration, Microbes, Waste, and Antitoxic Politics', *Current Anthropology*, (62:524), 2021, p. 298.

One practitioner Zhang interviewed, Jun, has initiated the addition of eco-enzymes to a local river since 2006. Her motivation stems from a deep connection to the river, which she sees as a way to give back to the environment that nurtured her. Jun recalls a time when the river was abundant with large fish, contrasting sharply with the present, where only small, unsellable fish remain. For Jun, ethical commitments can be rooted in daily acts of maintenance and repair, using eco-enzymes to preserve/restore the environment and critique the logic of development.

The impact of these enzymes is still uncertain, but the significance of Jun's actions lies in demonstrating a collaborative approach to nature that challenges the dominant reliance on technoscientific solutions.[7]

7 Ibid., p. 307.

Henriëtte Waal drinking from a source that flows into the Gardon river, France, 2024

I argue that wild fermentation is not just a method, or belief; it is a form of more-than-human knowledge that offers a kind of freedom and potential for connecting to something much larger than ourselves, the microbial matrix of all life.

Eco-enzyme brewers in China are rethinking how to manage waste and toxicity, relying on the observable effects of their probiotic practices. Rwandese wild brewers may not often articulate why or what they do in scientific or commercial languages, but their bodies know the rhythms and subtle manipulations of wild fermentations—and that it matters.

Since we were all raised after the sanitary revolution and in the hygienic era, it's 'natural' for most people to fear wild fermentation. But there is no such thing as an 'evil' microbe; they all serve diverse functions in larger ecosystems, both beneficial and harmful to humans, depending the circumstances. We still have so much to unlearn. It is crucial to protect these 'microbial commons' from patenting and commodification, as they are an essential part of our shared environment. Luckily, their wildness helps to avoid easy commodification.

Ultimately, this kind of wild fermentation wisdom is connected to the preservation of forms of wilderness that are not remote landscapes without humans,[8] fostering a grassroots, more-than-human approach to combatting environmental toxicity and reconnecting people with the landscapes they inhabit.

8 William Cronon, 'The Trouble with Wilderness: or, Getting Back to the Wrong Nature', *Environmental history*, (1:1), 1996, pp. 7–28.

2.7
Houtrust

In sewage treatment plant Houtrust. On the banks of the Haagse soup with Captain Johan, who explains he has billions of coworkers. Namely: the water-purifying bacteria he works with at Houtrust. Fieldwork research for Stroom The Hague, The Netherlands, 2022. Photo: Henriëtte Waal

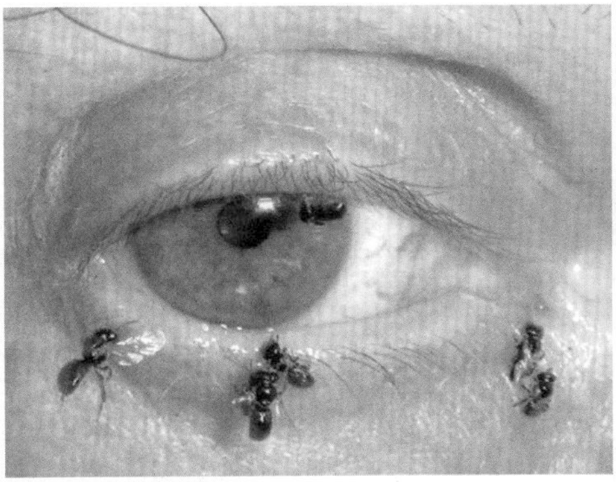

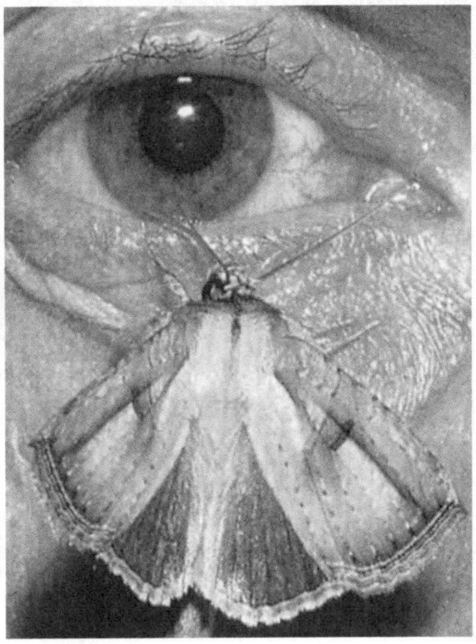

2.8
To Flavour Our Tears

The Center for Genomic Gastronomy, **To Flavour Our Tears (TFOT)**, 2016, an eatery designed for lachryphagic moths, or other insects that feed on mammal tears, recipe card Tear Taste Test. Images: The Nonist, Romancing the Lachryphage

2.9
Designing Niagara Falls

By Clemens Driessen

Tourists visiting the dry Niagara Falls, 1969. Source: www.momentosdelpasado.blogspot.com/2017/08/las-cataratas-del-niagara-sin-agua

For five months in 1969, the American side of the Niagara Falls was dewatered. Shut down by the US Army Corps of Engineers, after erosion had changed the shape of the waterfall. A brochure was spread, giving citizens on both sides of the border the opportunity to express their preferences between three options for constructing the most beautiful version: international participatory democracy deciding on waterfall aesthetics, for which the engineers sought to quantify the emotional responses people had to the different options presented.

Niagara Falls water is managed with the aim to maximize tourism and energy value, following provisions in the 1950 Niagara River Treaty between Canada and the US: between the first of April and the fifteenth of September, each day between the hours of 8 am and 10 pm, the waterfall should receive no less than 100,000 cubic metres of water per second.[1] At other times the water may be diverted for hydropower generation, as long as 50,000 cubic meters per second remain.

Historian Daniel Macfarlane has argued this optimization of aesthetics for tourists should be seen as part of larger engineering efforts of the falls and its surrounding landscape.[2] Since the nineteenth century, its water had been used to power industrial mills. In the 1950s, the famous NY urban planner Robert Moses was brought in to ensure these dual functions by creating a reservoir, for which he expropriated indigenous land from the Tuscarora Nation.[3] Currently, outside of visiting hours, up to three quarters of the water is diverted for hydropower generation.

1 Treaty Between Canada and the United States of America Concerning the Diversion of the Niagara River E100418 – CTS 1950 No. 3.

2 Daniel Macfarlane, *Fixing Niagara Falls: Environment, Energy, and Engineers at the World's Most Famous Waterfall*, UBC press, 2020.

3 Laurence M. Hauptman, *The Iroquois Struggle for Survival: World War II to Red Power*, Syracuse University Press, 1986.

The American Falls: Yesterday, Today, Tomorrow, 1973, souvenir booklet from the Niagara Falls. Courtesy International Joint Commission (IJC) (pp. 114–118)

YESTERDAY

American Falls, 1884

For some 12,000 years the Niagara River has carried water north from Lake Erie to Lake Ontario. A waterfall has always been part of this journey. The original falls have receded nearly seven miles from the Niagara Escarpment during that time, carving a great gorge 200 to 300 feet deep.

About 600 years ago the falls divided into two waterfalls, now called the American Falls and the Horseshoe Falls. The process of erosion, although perceptible only after large rockfalls or the passage of many years, is always steadily continuing.

First the native tribes, then the pioneers, marveled at this awesome natural phenomenon. For more than a hundred years man has tried to protect this great wonder. Park lands have been created at the Falls, both

TOMORROW ?

No one can say for certain how natural forces will change the look of the Falls. However, man has already made an impact on the area by diverting water and building viewing areas. *Should he now attempt to make additional changes to improve the scene?*

The American Falls International Board was formed to examine the feasibility and desirability of intervening in the scene. Three types of physical changes have been found to be possible. They are:

1. **Removing some or most of the fallen rock to increase the height of the waterfall, reveal the ledge rock, and open to view some of the water that now flows under the rocky slope.**
2. **Increasing the volume of water over the American Falls to create a deeper and more turbulent flow over the cliff face and rocky slope.**
3. **Restoring the Pool level to the earlier condition (before power diversions) when almost a third of the present rocky slope was submerged.**

This then is the problem facing the International Joint Commission:

Should one or more of the possible physical changes be made, and if so, which one or ones?

We would like your help in making this difficult decision.

The following pages describe some of the work performed to help make the decision. See what you think of the possibilities.

The dewatering of the American Falls in 1969 provided valuable information on the geological formation of the Falls, as well as the location of the fallen rock, that enabled a model to be constructed.

Dewatered American Falls, 1969

BUFFALO COURIER EXPRESS PHOTO

American Falls Model

The working model, at a scale of 1:50, was built at the Islington laboratory of the Hydro-Electric Power Commission of Ontario near Toronto. The model made it possible to experiment with various techniques for altering the appearance of the American Falls.

American Falls Model

WORKING MODEL SHOWING REMOVAL OF APEX SLABS. THESE ARE THE HIGHEST PILES OF FALLEN ROCK RESULTING FROM THE 1931 ROCKFALL. WHERE THIS ROCK IS REMOVED, THE WATER WOULD FALL THE MAXIMUM HEIGHT FROM CREST TO BEDROCK.

REMOVING FALLEN ROCK

Removing some or most of the fallen rock from the base of the American Falls might make a more dramatic and beautiful view. The following pages show a few of the possible arrangements, any of which could be combined with increased flow and/or restored Pool level.

American Falls Model

WORKING MODEL SHOWING THE AMERICAN FALLS AS THEY ARE AT PRESENT, WITHOUT ANY FALLEN ROCK REMOVED.

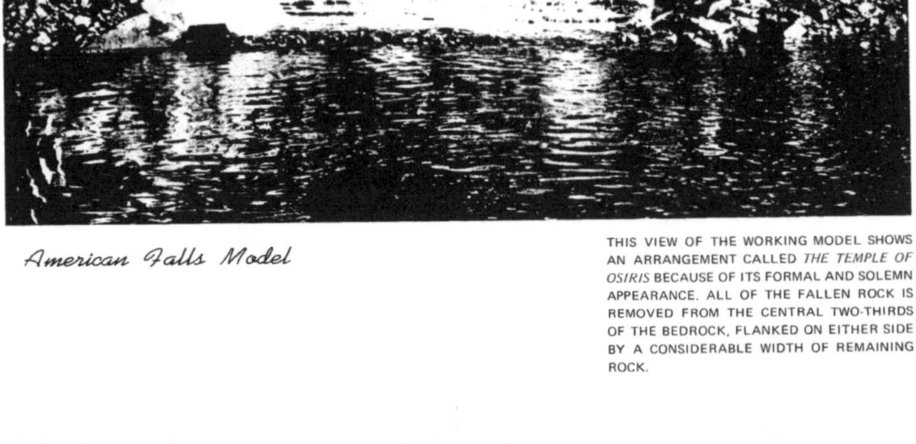

American Falls Model

THIS VIEW OF THE WORKING MODEL SHOWS AN ARRANGEMENT CALLED *THE TEMPLE OF OSIRIS* BECAUSE OF ITS FORMAL AND SOLEMN APPEARANCE. ALL OF THE FALLEN ROCK IS REMOVED FROM THE CENTRAL TWO-THIRDS OF THE BEDROCK, FLANKED ON EITHER SIDE BY A CONSIDERABLE WIDTH OF REMAINING ROCK.

American Falls Model

HERE THE WORKING MODEL SHOWS OCCASIONAL ROCKS LEFT ON THE CLEARED BEDROCK. OTHERWISE, ALL THE FALLEN ROCK IS REMOVED.

RESTORING POOL LEVEL

American Falls Model — Present Pool Level

American Falls Model — Restored Pool Level

The level of the Maid-of-the-Mist Pool below the Falls is today about 20 feet below its level prior to upstream water diversions for electric power production. The Pool level could be restored to the earlier condition by building a submerged structure a few miles downstream of the Falls. This would submerge almost a third of the rock accumulation and move the Pool's edge 100 feet closer to the American Falls.

INCREASING FLOW

Of the 100,000 cubic feet per second (cfs) of water that pours over the two Falls, 90% flows over the Horseshoe Falls and only 10% over the American Falls. Thus, if 5% or 10% of the Horseshoe flow were diverted to the American Falls, it would make a substantial increase in the American Falls flow with little effect on the Horseshoe Falls. In June 1971, the flow over the American Falls was increased from about 8,000 cfs to about 15,000 cfs to examine the difference. The increased flow deepened the green water plunging over the crest and added whitewater on the rocks below.

American Falls, June 1971 — Flow about 8,000 c.f.s.

American Falls, June 1971 — Flow about 15,000 c.f.s.

The three kinds of physical changes and their preliminary cost estimates are listed below. They are:

1. **To remove the fallen rock from the base of the American Falls. This would require dewatering the American Falls for one or two seasons.**

   Cost: **$1 million to $10 million depending on the amount removed.**

2. **To increase the flow over the American Falls by excavating rock in the channel leading to the American Falls and installing control gates. This would require dewatering the American Falls for one season.**

   Cost: **$6 million.**

3. **To restore the level of the Maid-of-the-Mist Pool to the earlier condition by building a submerged structure a few miles downstream of the Falls.**

   Cost: **$8.7 million.**

Should any of these physical changes be made? The International Joint Commission would like your opinion.

American Falls
International Board
P.O. Box 5050
Burlington, Ontario
Canada
L7R 4A6

American Falls
International Board
1776 Niagara Street
Buffalo, New York
U.S.A.
14207

Scale

Scale

3
Scale

BY HENRIËTTE WAAL & CLEMENS DRIESSEN

Is ecosocial design forever stuck in friendly but naive little alternative experiments, prototypes—too costly, exclusive, elitist, cumbersome to ever be scaled up—to genuinely make an impact?

When ecosocial design is about super-expensive one-off events that have no wider implication—this form of design may in fact imply a firm split between mass production that has to be cheap and destructive to ecosocial relations, and elite consumption where lavish funding is the only way to maintain some form of ecosocial meaning—separating the megarich from the rest with bioregional aesthetics, virtue signalling in earthen tones and biomaterials, suggesting moral character and not being out of place.

But how to ask this question of scaling up—or turn it into scale-out, or horizontal and distributed forms of scaling? There are genuine issues. Part of ecosocial design is to rethink scale—not automatically assuming an industrial logic of spreading the same solution across vastly different places.[1]

Scalability is often understood as allowing for smooth expansion without the need to reconsider key assumptions.[2] It actively erases otherwise meaningful diversity, while denying the work that is done to make the world ready for repetition. With Anna Tsing we can see how colonial plantations of European design are a model for this, preparing the ground for standardized abstract alienated labour, while reducing the number and complexity of both ecological and social relations to the bare minimum.

This type of bigness is often fake-efficient, ignoring the hidden costs and externalities of cheap nature and cheap labour. Whereas small scale is often dismissed outright, also mid-scale tends not to fit the accounting models; this is where imagination is limited, where from an investor logic things only seem to be realistic when they are hyper capital-intensive. Finding the right scale for an ecosocial

1 Arturo Escobar, *Designs for the Pluriverse: Radical Interdependence, Autonomy, and the Making of Worlds*, Duke University Press, 2018.

2 Anna Tsing, 'On Nonscalability: The Living World is not Amenable to Precision-nested Scales', *Common Knowledge*, (18:3), 2012, pp. 505–524.

Opening image: Capturing the flow pattern with paper snippets in the Westervoort model, 1957. Source: Archive of the Waterloopkundig Laboratorium

design or a design production doesn't mean maximizing profits and minimizing costs. It means designing for the social, ecological challenges of a landscape.

Tsing also warns of idealizing non-scalable projects, which 'can be as terrible in their effects as scalable ones'.[3]

3 Anna Tsing, *The Mushroom at the End of the World: On the Possibility of Life in Capitalist Ruins*, Princeton University Press, 2015, p. 42.

Scalability may not just be a feature of an object, but an operation that erases difference to create smooth space for expansion. Scaling up can also be thought as translation, the work of adapting, or of starting from particular conditions and cultures. Craft is often seen as only local, but can work transregional, in networks of knowledge exchange, where scaling happens through education, learning. Economies of scale can have a middle ground, building on models of artisan manufacturing, micro-manufacturing, small batch production and moderate production.

How then to think in terms of larger wholes, connections: the watershed may seem to have clear boundaries, but is never simply defined by surface water, ground water may follow different dynamics.[4] Rather than fetishizing some naturalized boundaries as the limit of ecological and social concern, bioregional design always involves playing with scales and connections. When thinking of the scale of a farm, it involves playing with scales of markets, of layers of slow-moving regulatory frameworks, of a good scale to share tools, and of cultures of production.

4 Cleo Wölfle Hazard, *Underflows: Queer Trans Ecologies and River Justice*, University of Washington Press, 2022.

Ecosocial design needs to respond to ways in which things are never only local, and the ways in which local is shaped by unseen powers, historical traces, global economic structures and multinational industries that seek to replicate, producing path dependencies at timescales that make trying to change things seem futile.

Do we need to think big to make sense of what we are dealing with? Modernity, capitalism, coloniality, seeing these global operations of thought, identifying dominant economic logics, of 'development' can help dismantle their power and escape the hold of these operations that suggest there are no alternatives.[5] But these global frames of social theorizing themselves also offer so many ways of not taking people's experiences at face value—dismissing dreams as false consciousness, or seeing protests as merely a local expression of something bigger—structural critiques may elide curiosity for the local. Know-it-alls telling people how to feel, offering prefab formats of being angry, glossing over real attempts to find glitches and redo systems. Ecosocial design asks: what do these grand frames look like on the ground?

5 Arturo Escobar, *Designs for the Pluriverse: Radical Interdependence, Autonomy, and the Making of Worlds*, Duke University Press, 2018.

Water is a medium that always makes local effects overflow to other places—both polluted and unpolluted water, connecting spatial scales and timescales.

Changing the aesthetics can be a strategy to scale. Not to strive for fixed qualities, but to assess quality along different parameters, embracing unstable outputs. Then what is scaled up is a celebration of variation, the place and time of production may affect the result. Oh but that won't guarantee constant outputs! Yes indeed, but otherwise it looks dead, losing all character.

John Thackara, an author and festival convenor, explores the agenda of designing for life—and caring for place—from microbial scale to bioregion. A professor at Tongji University in Shanghai, his twelve books have been published in fourteen languages, including his last two in Chinese.

3.1 Stories of Reconnection: Water Design in Bioregions

BY JOHN THACKARA

Rivers and watersheds—the Earth's life-giving arteries—are essential to our survival. Over centuries, however, we modern humans have dug, dammed, channelled, and concreted them over. We were incentivized to do so: all those roads, car parks, airports, buildings, driveways, and pavements are described as development, and entered in the plus column of a nation's GDP. During generations of these often brilliant civil engineering and design efforts, the people doing this either did not think about rivers, soils, and biodiversity at all—or treated them as resources with the sole purpose of feeding 'the economy'.

The costs of so much progress are now being felt by us all. All that concrete forms what ecologists call Impervious Surface Area, or ISA. ISA allows large volumes of storm water to enter watersheds much faster than would occur naturally, which leads to flooding and riverbank erosion. This water contains pollutants, such as gasoline, oil, and plastics, that stress or kill the living inhabitants of rivers and streams. When the weather is hot, rain entering rivers from ISA is unnaturally warm; this further hurts or even kills the myriad of creatures that live there.[1]

1 www.dnr.maryland.gov/streams/Pages/streamhealth/How-Impervious-Surface-Impacts-Stream-Health.aspx

Now, a new water paradigm is transforming the ways we live with water. Growth, in this new story, means soils, biodiversity and watersheds getting healthier. Water is treated as a living system of which we are a part—not a resource to be consumed, or stream of waste to be covered over. The core value of this approach is care, not extraction. This new understanding includes the belief that watershed recovery will be local, and decentralized; global restoration will be achieved through a multitude of mainly local activities. Watershed recovery, we are learning, is as much a social and cultural activity as it is an ecological or engineering one.

This social-ecological approach is especially strong when organized within a bioregional frame. Bioregions reconnect us with living systems, and each other, through the places where we live. Bioregional geographies are often shaped by watersheds, coinciding with foodsheds, fibersheds, and food systems. And because bioregions are experienced as living places—lifeworlds—they provide a shared focus for positive action that diverse groups of people can relate to, and actively support, whatever their other differences.

In visits to watershed projects over many years, I've observed that designers and artists contribute to bioregional development in diverse ways. Maps of a bioregion's ecological and social assets are needed: its geology and topography; its soils and watersheds; its agriculture and biodiversity. The collaborative monitoring of living systems needs to be designed—together with feedback channels. New service platforms are necessary to help people share watershed resources of all kinds. Novel forms of governance must also be designed to enable collaboration among diverse groups of people.

In the Highlands of Scotland, the Findhorn Watershed Initiative is a multi-generational plan to restore a mosaic of nature-rich habitats, grow a culture of nature connection, and enable a thriving nature-based economy for the people and places of the Findhorn watershed.

A local Rivers Trust brings together land managers, communities, businesses, and other river users. The Initiative enables a collaborative programme of habitat restoration, carbon sequestration, and river stewardship.

A railway company, LNER, and a whisky brand that relies on water from the area, are among supporters of efforts to protect the Findhorn as a salmon habitat. Rising water temperatures are thought to reduce salmon numbers, so more native woodland is being grown along riverbanks to help shade the water during the summer and lower temperatures. A range of monitoring systems collect ecological data and monitor the river's temperature in real-time, and local residents participate in fish and invertebrate surveys.

Nature recovery and restoration is as much about human relationships with the land and nature, as it is physical work of planting more trees, or retaining more water in the landscape. To strengthen these relationships, the Findhorn Initiative commissioned two human ecology Researchers in Residence, Raghnaid Sandilands and Mairi McFadyen, to explore how nature-related Gaelic place names, songs, and stories that once enriched relationships with the land might do so again. Their research became programming for live events that connect local people of all ages with the river, its culture, nature, and landscapes. For Mairi McFadyen, this work draws on deep roots. 'The Gaelic concept *dùthchas* is all about a cultural, ethical, and reciprocal relationship with place', she told me. 'Seeking out the intimate history of a place—to find the stories in the landscape—is to find connection, a sense of belonging, and care.'

A celebration called *Cruth na tìre air a doimhneachd* ('A landscape by its depth'), for example, featured songs and stories inspired by specific locations along the river and its tributaries. A map, created especially for the evening, showed Gaelic songs, poems, and stories associated with often small and intimate locations in the landscape. 'If you know the stories, you love the place', someone told Raghnaid at an earlier event— 'and if you love the place you look after it'. Sandilands and McFadyen describe their approach as 'cultural darning and mending'.

A world away from the Scottish Highlands, few cities have grown as fast as Shanghai in recent years. But what the city shares with the Findhorn Watershed is a loss of collective attention to its ecological roots—in Shanghai's case, the vast bioregional wetlands of

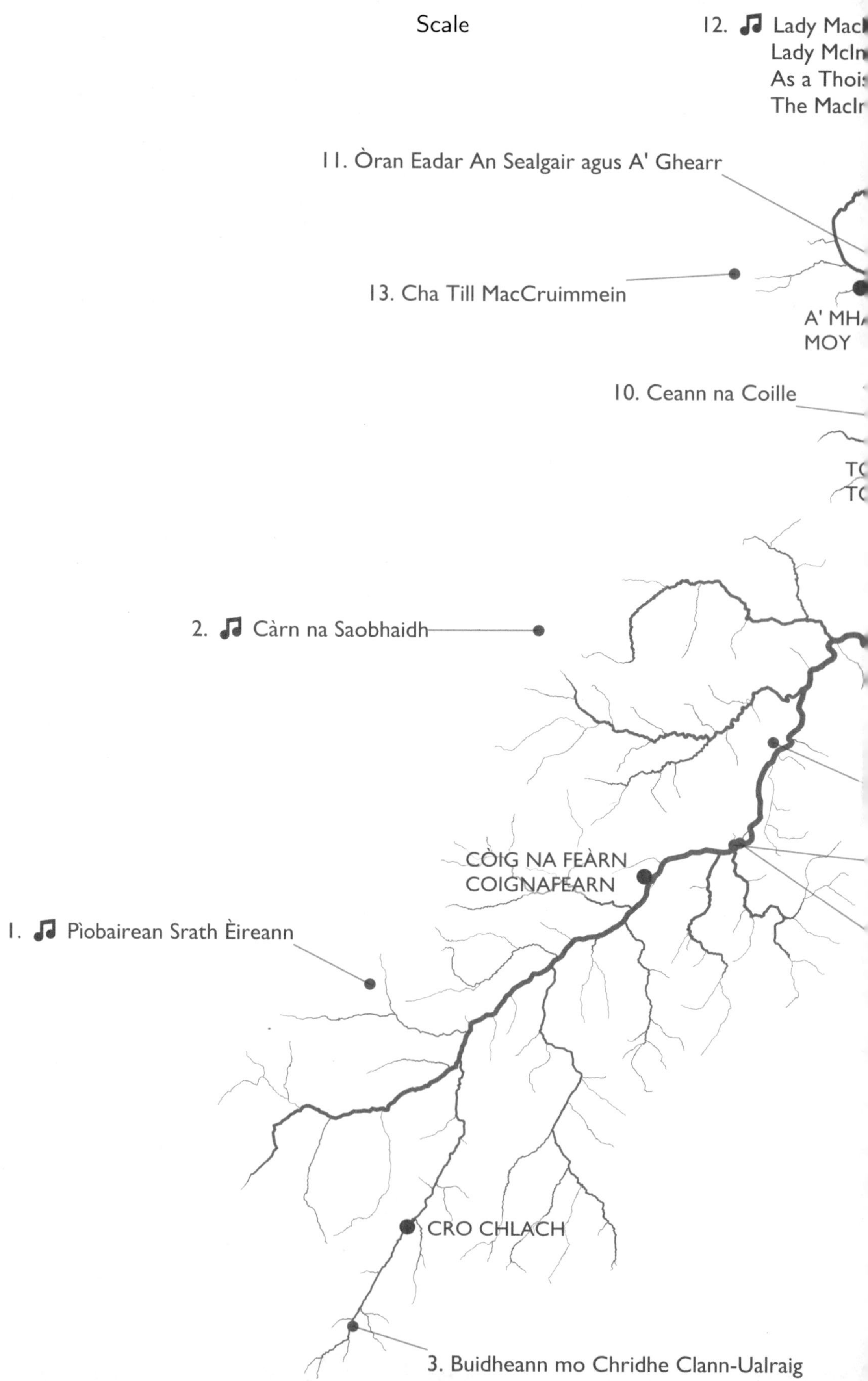

Scale
11. Òran Eadar An Sealgair agus A' Ghearr
13. Cha Till MacCruimmein
MOY
10. Ceann na Coille
2. Càrn na Saobhaidh
CÒIG NA FEÀRN
COIGNAFEARN
1. Pìobairean Srath Èireann
CRO CHLACH
3. Buidheann mo Chridhe Clann-Ualraig

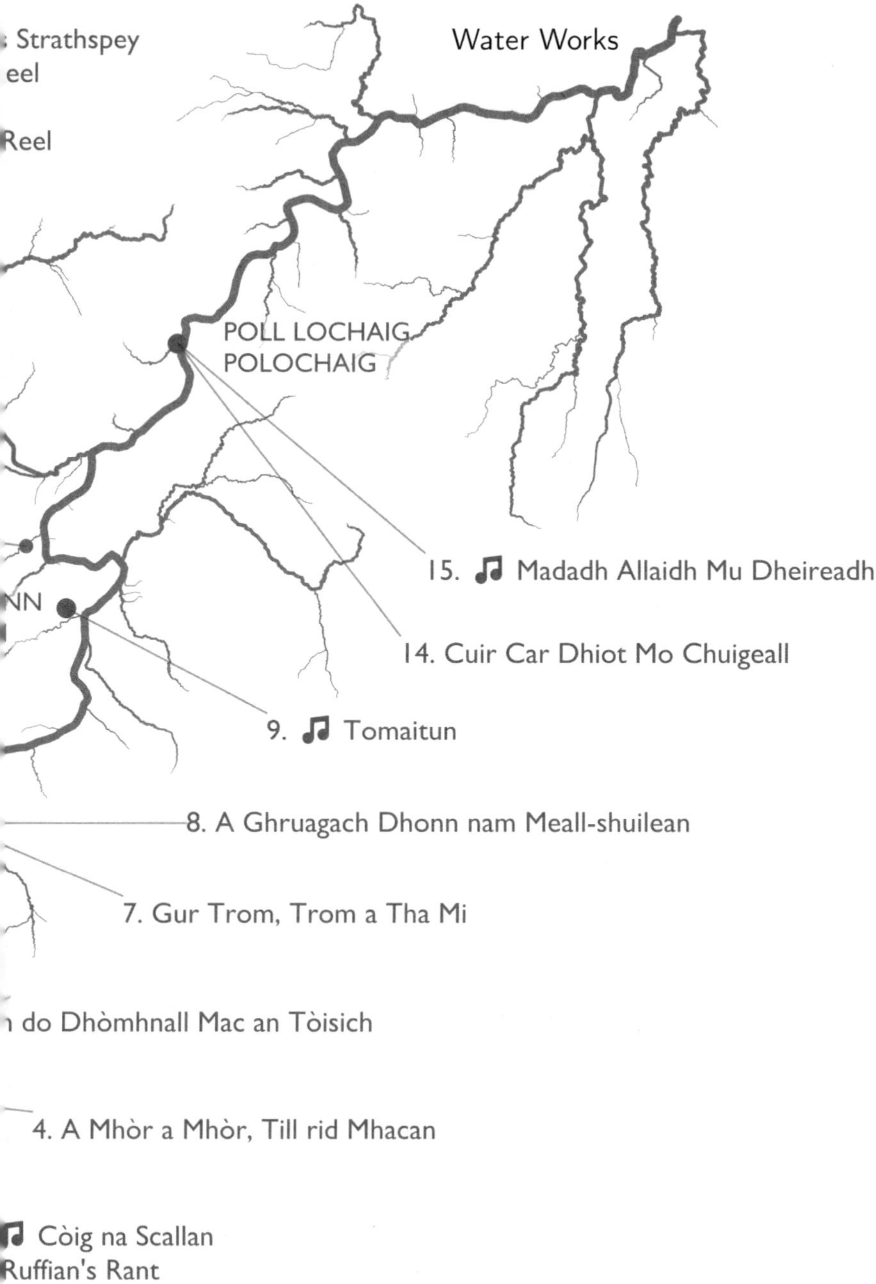

R. Sandilands, M. McFadyen & E. Hesling, *Songs, Tunes and Their Stories from the Upper Findhorn*, Scotland, 2024

0 1 2 3 4

1. **Pìobairean Srath Èirinn** (Julie Fowlis, 2017). A tune to describe the winds off the Monadh Liath. Andy Cumming of Croft Croy said "When the wind was in the South East, the old folks said 'b' e sin pìobairean Srath Èirinn a' sèideadh' (that's the Strathdearn pipers piping)." Meall a' Phìobaire' (rounded hill of the piper) is on the watershed between the two straths.

2. **Càrn na Saobhaidh** (mid/late 1700s). A slow strathspey, referencing the hill of the same name (cairn of the fox's den), appraoched from Coign na Feàrn Collected in the Angus Fraser of Knockie Music Manuscript, 19th C. This tune is followed by another strathspey, Duncan MacQueen's first published in 1803.

3. **Buidheann mo Chridhe Clann-Ualraig** (ca.1670). This song gives an account of a cattle raid visted on Rose of Kilravock by a man known as Achaluachrach and his men. On their return to Lochaber by way of the Cro Clach they were each run through by a famous archer on exiting a bothy. Achaluachrach's bride to be, a Kennedy bardess laments;

'S a' Chro-chlach am Bràigh Srath-Èireann - *On the Cro chlach in the braes of Strathdearn,*
Oidhche dhorch' air nach bu lèir dhaibh - *on a dark sighntness night for them,*
'Cur nan saighdean ait a' cheile - *the arrows were run in upon each other,*
Thuit mo leannan 's deich fir threuna - *my love fell and ten brave men.*

Stories of the archer, Iain Beag MacAnndra of Dail na h-Aithnich, are many and a cairn to him stands on the banks of the Dulnain. Place names on the Cro perhaps carry echoes of this story; Coire na Creiche, *corrie of the raid* / Leacan Uaighe - *grave slabs*. Bothy ruins can still be seen. The song has been given a new tune by Julie Fowlis.

4. **A Mhòr, a Mhòr, Till rid Mhacan** (undated). This very old lullaby is sung in the voice of a 'Leannan-sìth' or fairy lover who fathered a child of a mortal woman. As he sings to the child, he pleads with Morag to return to the son she abandoned, promising her a 'goidean bòidheach breac' - a little withe of fine fish, wine and all she could wish for ...'gach nì a b'ait leat'.
Julie learnt this version from the singing of Evyln Campbell as recorded by William Matheson in 1960 and the text appears in the book 'Stories and Songs of Loch Ness Side' (1914) by Alasdair MacDonald.

5. **Còig na Scallan** (mid 1700s) A tune also known as 'The Ruffian's Rant.' The earliest written version is a 2-part strathspey found in the MacFarlane Manuscript (ca. 1740). This tune made its way to Canada and is found as a 4-part strathspey in the 'The Cape Breton Scottish Collection' (2013) of older Scottish fiddle music from the era 1740-1935.

6. **Òran do Dhòmhnall Mac an Tòisich** (published 1821). An irreverent and warm song that describes a drover from Daltomach, published in the book of poems in English and Gaelic by Alex. MacKay, butler at Moy Hall (1821). He tells of his excellence; his appearance in scarlet hoes, fine tartan coat, kilt, sword and gun, his good company and his excellent whistling, 'Bidh agad topa na feadail' on his long droves to the South.

7. **Gur trom, trom a tha mi** (Eoghan Òg Strath-Nin, ca. 1800). A love song by Eoghan Òg of Strathnoon, belived to be a MacDonald. He fell in love with a weaver's daughter, a maid in his father's house. He joined the army and rose to level of Captain. He praises her kindness and beauty as he takes leave of her, promising to return:

Nam bithinn sa ' n Strath Eireann - *If I was in Strathdearn*
Do spéis cha bhiodh an aineol orm - *your affection would not be a stranger*
Ach tillidh mi ma ' s beò mi - *I will return if I live*
'S le m ' dheòin cha dean sinn dealachadh - *with my willingness we will not part*

He did return after 7 years and married the girl; they had 2 sons, Gilbert and Angus. He is said to have died ca. 1824.

8. **A Ghruagach Dhonn nam Meall-shuilean** (Eoghan Òg Strath-Nin ca.1800). A second love song by Eoghan Òg, set out as a conversation between his love and himself and their regard for one another despite the doubters and gossips. Their meeting places are the caochan - *streamlet*, ruighe - *shieling* and glacaig seileach - *willow hollow*.

9. **Tomaitun** (first published 1848). A local pipe tune; a quickstep from Ceòl Sean Gunn's collection. This tune is preceded by a tune 'Tir na Mang is na Bradan' - *land of deer and salmon*, found in the Angus Fraser of Knockie Music Manuscript, 19th C.

10. **Ceann na Coille** (Donald MacAskill, published 1928). Some lines of a poem by the Strathdearn Bard, published in a collection 'The Mixture: A Collection of Verse,' with a new tune added by Julie Fowlis.

Ged a gheibhinn-sa mo thagha - *If I were to get my choice*
B' e mo rogha dhen Roinn Eòrp' - *in all of Europe*
A bhith 'n tàmh an Ceann na Coille - *I would choose to live at Woodend*
Far an deach mo thogail òg; - *where I was raised in my childhood*
Far am faicinn Càrn na Loinne - *where I would see Càrn na Loinne*
Is Càrn Dubh Mhic an Deòir, - *and Càrn Dubh Mhic an Deòir*
Far an cluinninn fuaim na h-abhann - *where I'd hear the sound of the river*
Sa bheil bradain 's bric gu leòr - *in which salmon and trout abound*

11. **Òran Eadar An Sealgair agus A' Ghearr** (published 1821). Song Between the Hunter and the Hare, publised in the book of poems by Alex. MacKay (see 6.). It conveys the idea of two equals with knowledge of the lands; a eulogy to nature and hunting, run through with names of people and places.

12. **McIntosh Set.** A tune set associated with The Clan MacIntosh. **Lady Mackintosh's Strathspey** (first published in 1782) | **Lady McIntosh's Reel** (first published 1770) | **As a Thoiseach!** (first published 1816) | **The MacIntoshes Reel** (ca. 1880,. written in the 1870s by MacIntosh of MacIntosh's piper, D. S. MacDonald.

The Mackintosh's Banner (1880s) is a slow air was written for MacIntosh of MacIntosh of Moy by his piper W. M. MacKenzie.

13. **Cha Till MacCruimein** (Dòmnhall Bàn MacCruimein (ca 1746). Dòmhnall Bàn MacCrimmon was from the family of famed Skye pipers. He was the only casualty on the night of the Rout of Moy in 1746. This is a well-known pibroch lament, but the lyrics are less well known. In them MacCrimmon foresees his own death: 'An cogadh no sìth, cha till e tuille' - *in war time or peace, he will return no more.*

14. **Cuir Car dhiot mo Chuigeall** (ca. 1745). A spinning song recorded in the story of Iain a' Bhreacain, Iain of the Plaid, a Hero of Clan Chattan (Celtic Magazine, 1898), situated in the years around the Jacobite uprising and subsequent reprisals. The story names people and places and records the origin story of this little work song. A recording of it was collected in Coll in 1960 from a Bella MacNeil (b. 1875) by Calum Iain MacLean.

Cuir car dhiot, mo chuigeal - *Spin quickly, my distaff*
Cuir car dhiot, gu luath - *Spin quickly and sure*
Cuir car dhiot, mun togar - *Spin before reivers*
A' chreach air mo shluagh - *Plunder my people*
Cuir car chiot, cuir car dhiot - *Spin quickly, spin quickly*
Toinn làidir is caol - *Twist well and thin*
Gun snaidhm is gun chnapan - *Without knots and catches*
Biodh breacan mo ghaol - *My love's plaid should be.*

15. **Madadh Allaidh Mu Dheireadh.** The last wolf - a new slow air written minding the last wolf, killed in the Strathdearn.

Jiangwan. Now, in parallel with river clean-up efforts by Shanghai's municipal authorities, researchers from the Ecology and Cultures Innovation Lab at Tongji University are developing ecological sensing devices that enable interactions between humans and other living entities and systems that surround them. The lab deploys these devices as 'tools for empathy'.

One project invites citizens to look afresh at the bioregion's wetland flies. Normally associated negatively by the public with disease, flies are important pollinators in the wetlands' ecosystem. As such, they're a key link in the food chain, and their sheer diversity makes them an important gene pool, too.

Having made hand drawings of nine fly species, design researchers at Tongji University promoted these for sale online. A second group is creating maps of the city's waterscape that focus attention on the diversity of different life forms that co-exist there, and interact: fish, plants, plankton, insects, mushrooms, microorganisms in tree stumps—and the trash and concrete contributed by humans. Another map highlights interactions among organic and geological layers of the wetland that have evolved over tens of thousands of years: rocks; soils; mosses and lichens on rocks; leaves floating on the water.

A third group in Shanghai is using new media as a medium of ecological empathy. Cross-sectional images, projected onto a large screen, enable visitors to observe the life-stages of algae, dandelion and fir trees that coexist in this ecological area. Visitors gain a richer understanding of the ways that different time regimes—life and death, reproduction, and aging—can coexist in one bioregional place.

In Bangalore (Bengaluru), as in Shanghai, fostering empathy and awareness is also a priority. Once known as the city of a thousand lakes, barely thirty-four of them remain in their original healthy state. Thanks to decades of urbanization, sewage dumping, and encroachment, Bangalore has gone from a city of lakes and gardens to a city of concrete. But S. Vishwanath, popularly known as Zenrainman, a thirty-year veteran of water stewardship projects in Bangalore, is not disheartened. Having spent decades rejuvenating recharge wells and water bodies, he is adamant that water and sanitation projects can succeed if they are hyper-local and led by local citizens. Near Bangalore, for example, a community of 1,700 households in eight villages eliminated open defecation in a project led by local women, supported by outside experts. For the eighteen months since the locally-owned plan was launched, the community's infant mortality rate has been zero.

'Water stewardship is neither a spiritual nor a practical matter on their own', Vishwanath told me on one of my visits to the region. 'Both are essential'. For a community-led project to succeed, different local government departments—such as water, public health, infrastructure, or education—need to collaborate. But getting local government silos to talk to each other is hard; simply telling officials to be supportive doesn't work. That's why fostering their empathy and awareness is a priority.

Frustrated that so many citizens and officials do not even know the name of their local lake, Vishwanath turned to a designer, Aajwanthi Baradwaj, for an intervention that would raise public awareness and connectedness with the bioregional water system. Baradwaj designed a beautiful experiential audio-visual walk. It uses the lake as an exhibition walk-through that allows people to experience the space for what it is. A map and an audio guide help visitors unravel facts, narratives and offbeat aspects of entities around the lake while involving local people in the process and increasing awareness in an engaging manner.[2]

2 www.citizen-matters.in/en-route-to-a-water-sensitive-city-the-story-of-jakkur-lake/

Local and hands-on stewardship has also been key in drought-affected parts of Andhra Pradesh, north of Bangalore. A proliferation of illegal boreholes has long threatened the long-term viability of aquifers across the whole region, and regulations imposed from afar proved ineffective. The response of smallholder farmers in more than 600 villages was to rediscover how to manage local watersheds as a common resource. Some 20,000 farmers now double as barefoot water technicians in a social practice known as Participatory Groundwater Management (PGM).

The PGM approach is based on small 'hydrological units'—a cluster of micro-watersheds. Hydrological data is collected for each hydrological unit: daily rainfall, water levels, borewell discharge and daily stream flows. Being on the spot, small-scale farmers—many of them women—use these data to improve their understanding of how well groundwater resources are recharging at a local scale. An integrated approach—Andhra Pradesh Community Managed Natural Farming—simultaneously applies ecological and social concepts and principles to the design and management of food and agricultural systems. It seeks to optimize interactions between plants, animals, humans, and the environment, while taking into account the social aspects that need to be addressed for a sustainable and equitable food system.

In addition to on-the-spot measurements, farmers have also started to use GPS to collect data remotely from wells, rain gauge stations, and artificial groundwater recharge structures. The information is shared communally; rain gauge stations display the amount of rainfall received on a day-to-day basis. These display boards act as village discussion points on issues such as drought, floods, water levels, and agricultural practices. Trusted and independent information is a key to the success of PGM. Before the new system was introduced, farmers had to rely on data provided by so-called 'input dealers'—fertilizer, seed and pesticide companies; these sources tended to downplay the huge wealth of grounded knowledge.

Community-based institutions play an important role in this smallholder-centred approach. In Farmer Water Schools, for example, farmers and scientists collaborate as equals on the exploration of different management options. Data analysis is continuous, multi-dimensional, and multiscalar. For each hydrological unit, a so-called 'base document' serves as a reference point for farmers and support agencies as they plan future activities. Water resources are described at an ecosystem or bioregional level: climate, rainfall, drainage, groundwater, land use, and so on. Documentation includes soil-conservation activities; the range of crops being used, and which livestock are being reared; their yields and net returns. And so on. At the time of writing, the Indian government announced plans (and resources) to help transition a further two million small-scale farmers into the Andhra Pradesh Community Managed Natural Farming system.

The urge to reconnect with rivers and watersheds resonates in the most unlikely places—none more so than Mexico City. An architect friend of mine, Elias Cattan, is consumed by the idea that reconnecting with its rivers and bioregion is the best way for his city to thrive.

Mexico City was once a city of interconnected lakes, and more than sixty rivers—but is now a dusty megalopolis covered by busy roads and shopping malls. The Xochimilcas were the first to alter the natural hydrological rhythms of the city 5,000 years ago when they invented the chinampa agricultural system which also helps them manage flooding. But industrialization, and the rise of the car, did real damage in a relatively short period. Urban planners treated rivers as a problem to be overcome and hidden away by hard engineering: waterways were forced into channels, covered and built over.

Now, together with a team of ecologists and hydrologists, Cattan's design firm Taller 13 has submitted a proposal to replace busy roads with a ring of water island parks around the city centre, restoring at least three rivers in the process. These would be home to myriad plants, fish and birds. City officials love the idea, but say they lack the resources to implement it. Cattan's response, working with progressive city officials and the private sector, is to bring the city's rivers back into the city bit-by-bit with nature-friendly 'Complete Streets'. One element of the bigger scheme Ecoducto, has now been realised: a creek channel inserted in the middle of a six-lane highway, has become a linear urban park. But it is not merely decorative: a plant-based water filtration system turns raw sewage into clean, clear stream water and provides recreation and education opportunities.[3]

3. Recently the maintenance of the ecoduct has declined, rendering the ecoduct into an experimental ground for a semi-managed urban wetland. Source: www.pulitzercenter.org/stories/search-mexico-citys-lost-water

In Devon, in England, the West Country Rivers Trust works with many partners across a variety of aquatic systems—from freshwater to saltwater—to restore the ecological integrity of degraded habitats for the benefit of people and the environment. The Trust works hard to enable citizen participation in river and wetland restoration. Based on detailed scientific analysis of the rivers' social and ecological assets, it coordinates practical activities including riverbank repair, drain contents monitoring, and the establishment of rain gardens that use storm water runoff.

The Bioregional Learning Center (BLC) has long collaborated with the Trust on community engagement, blue-green infra, and Water Sensitive Rural Design. Isabel Carlisle, its founder, told me the core work of the Bioregional Learning Centre is to bring a variety of rivershed stakeholders together: a large estate, farms, a plant nursery, a pub, a supermarket, a restaurant, a school, house and boat owners, a dock, and more. BLC projects include *Voices of the Dart*, a river-long exploration combining local knowledge, climate science, data, and the arts. In a recent learning journey across the bioregion, for example, diverse actors discussed what would make this place more resilient, who's making change happen? 'We ran this Learning Journey to find out who is doing what on the ground, to listen to what people are saying, and to understand where we will find the green shoots of regeneration. A key insight was the plethora of great schemes already underway: it's inspiring to see all these projects—so little attention is given to this work, no one is hearing about it.'

CARING FOR PEOPLE, CARING FOR RIVERS

The lesson of these examples is that even small actions can have big and positive outcomes for a watershed when enough people take them in concert. The action can be to install a rain barrel, help clean up a creek, map the ecological resources of a watershed, plant water hyacinths in a polluted lake, or compose the music score of an ecosystem. None of these actions entail designers acting alone. Their role is as much connective as it is creative. But in creating objects of shared value—such as an atlas, a plan, or a meeting—the design process can be a powerful way to foster collaboration among geographers, ecologists, economists, planners, social historians, writers, artists, and other citizens. The forms of governance that enable a variety of stakeholders to work together are another key success factor—but these cannot easily be standardized. Successful relationships between people and places, and between people, are embodied and situated.

This relational approach to designing takes design away from the world of Impervious Surface Area and towards a concern for relationships, care, and context. This new focus brings design practice closer to the work of botanists—and nurses. For botanists, the health of an ecosystem lies in the vitality of interactions between its component species; the quality of these relationships matters more than a list of the species present in that place. A similar shift is evident in nursing education. Instead of the focus on data and technology that preoccupies mainstream biomedicine, the new priority in nursing is *embodied relational understanding*. These qualities matter just as much in a bioregion or watershed. In all design activities that serve life, the key quality is a caring attention to the uniqueness of each situation—at that moment, and in that place.

3.2 Architects of Water Works: Snow-Balling the Courage of Imagination

INTERVIEW WITH OOZE ARCHITECTS
(EVA PFANNES & SYLVAIN HARTENBERG)
BY HENRIËTTE WAAL & CLEMENS DRIESSEN

OOZE Architects, *City of 1000 Tanks, Water Balance Pilot*, 2023.
© Cynthia van Elk

OOZE is an agency working at the intersection of architecture, landscape and urbanism. Their strategy is to combine local interventions with large-scale strategical thinking. Their projects demonstrate creative, community-driven solutions to urban water challenges, in a variety of local contexts.

We talked with OOZE founders Eva Pfannes and Sylvain Hartenberg about Água Carioca in Rio de Janeiro and City of 1000 Tanks in Chennai.

ÁGUA CARIOCA

Água Carioca, named after the inhabitants of Rio de Janeiro (Cariocas) and the Rio Carioca, a small stream with its source in the rainforest, proposes an urban blue-green circulatory system that makes water treatment local and visible: rainwater harvesting, septic tanks, and wetlands are introduced in informal settlements to manage the wastewater now flowing through the neighbourhood as open sewage. The improvement in public health also reduces CO^2 output and stabilizes the ground, helping to prevent landslides. We envisioned implementation as a fractal approach, from the grouping of several residences, to schools, to whole neighbourhoods, for which we made scale models, built a demonstration unit in Sítio Burle Marx, and made drawings that explain the constructed wetland system, which uses sand, gravel, and vegetation to filter wastewater and remove organic pollutants.

CITY OF 1000 TANKS

The City of 1000 Tanks is a larger project whereby OOZE collaborated in a multidisciplinary team that aims to address floods,[1] water scarcity and pollution in Chennai, the capital of Tamil Nadu. OOZE recently completed a water balance pilot for a school that engages people in the challenge of closing the water cycle locally, ensuring no drop of water goes to waste. The project aims to deal with this long list of interrelated issues: improving productivity in water sourcing, reducing consumer water costs, reducing economic losses caused by droughts and floods, while also reducing the urban heat island effect, increasing biodiversity, and sequestering and reducing carbon emissions by providing a water source as an alternative to high energy consuming desalination plants.

1 Madras Terrace, Goethe Institut, IIT Madras, Care Earth Trust, Uravugal Social Welfare Trust, IRCDUC, Atma Waters, Gowatr, Paperman, Pitchandikulam Forest, TU Delft, HKV, IHE Delft. The City of 1000 Tanks project is part of 'Water as Leverage' for 'Resilient Cities Asia'. This programme, initiated by Dutch water envoy Henk Ovink and the Dutch Enterprise Agency RVO in collaboration with the cities of Chennai (India), Khulna (Bangladesh), and Semarang (Indonesia).

HW You did many projects around the water cycle. Are they somehow linked?

SH They are different in scale and complexity with different components or water systems at play. We started with art commissions in Europe and then took the learnings to another level with two projects. Água Carioca, a project envisioned on the scale of the city of Rio de Janeiro, focuses on informal neighbourhoods with very serious social problems related to water. Most of the inhabitants of the favelas, the informal communities, hardly have stable access to water and sanitation.

The City of 1000 Tanks in Chennai is a large project looking for new ways of creating a water-abundant city with all stakeholders, including both vulnerable and economically weaker communities and the government.

Água Carioca is completely self-initiated. It is a self-generated project that we worked on for many years with successive support from Dutch and

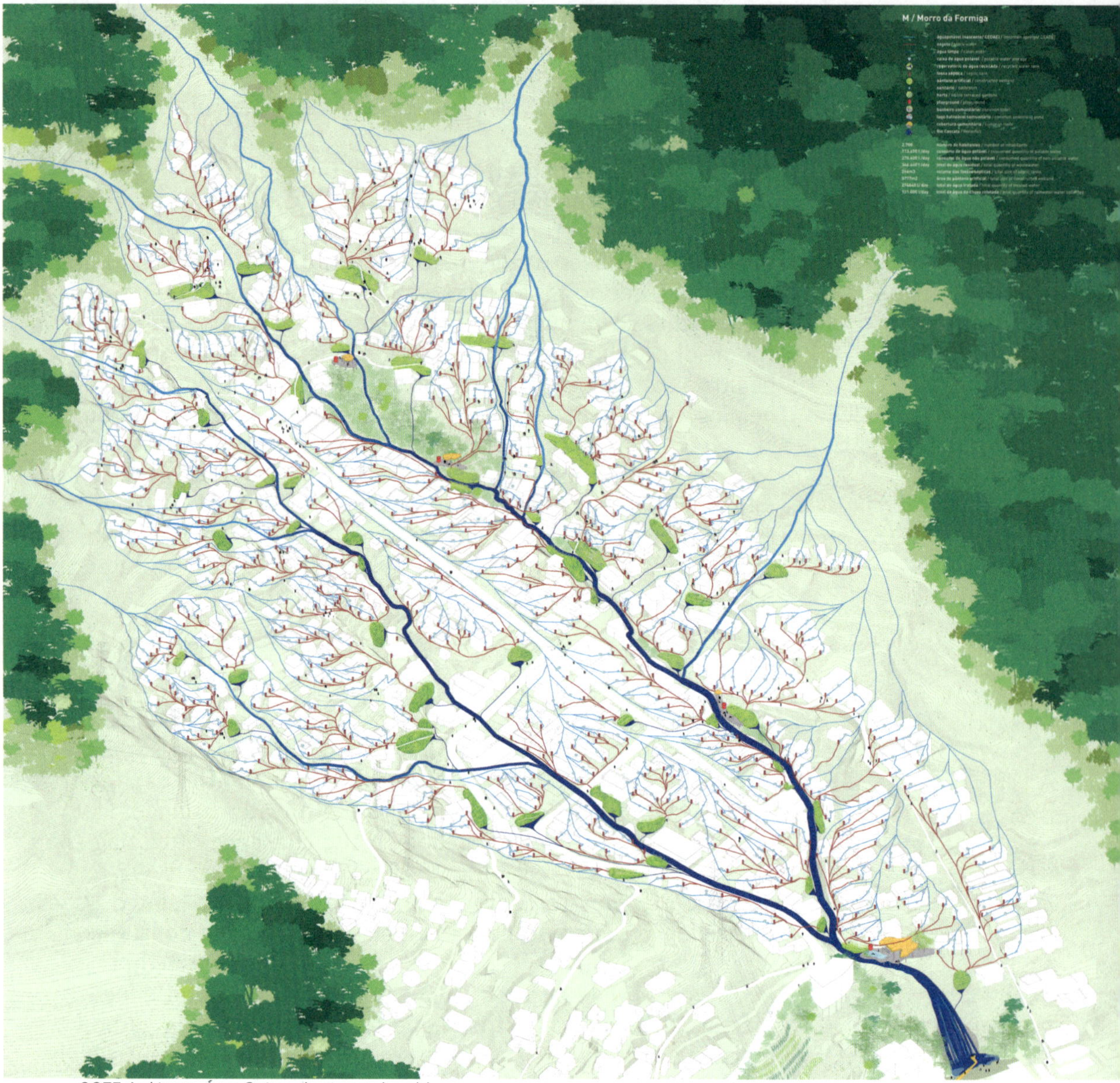

OOZE Architects, *Água Carioca*, drawings and model

Brazilian institutions and people.[2] The City of 1000 Tanks project on the other hand started with an open international tender that we won, together with a large multidisciplinary Indian and Dutch team. It's much more embedded and larger and still continues to this day. In a nutshell, that is how we scaled our water projects.

2 The Dutch government through Creative Industries Fund NL, the Dutch Consulates in Rio de Janeiro and São Paulo, the Sítio Roberto Burle Marx, local city departments and institutions like Studio-X and the International Architecture Biennale São Paulo.

HW You said Água Carioca was self-initiated. What was the occasion?

EP We had the opportunity to show our work to some representatives of the municipality in Rio de Janeiro (urbanism, economy and finance) at casual events. We mentioned that we do these artistic projects with water and wastewater and they said: 'What? That's really interesting! Can you show us more?' And that's when Água Carioca really started. They explained the extreme urgency around water in the city

OOZE Architects, *Água Carioca, Sítio Roberto Burle Marx, Water system elements.* © Raul Correa-Smith

TRATAMENTO COM FILTRO ORGÂNICO
REUSO. ESTIMAMOS UMA ECONOMIA DE 50% DE AGUA USADA.

OOZE Architects, *Água Carioca*, meeting with the community on location. © OOZE Architects

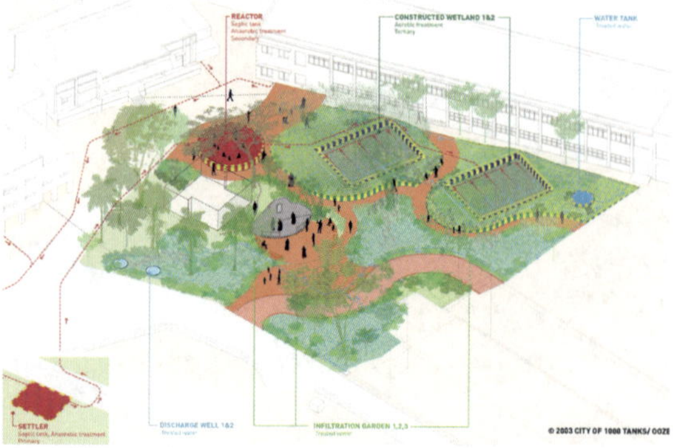

OOZE Architects, *City of 1000 Tanks, Water Balance Pilot scale*, 2019

and the need to do something. People saw great potential for our projects but there was no local funding. We then asked the Creative Industries Fund NL and the Dutch Consulate in Rio de Janeiro to support us for this research-by-design project.

CD In a faraway country, you cannot simply come in and explain how everything should be done. You are not a part of the daily experience with the local methods and systems. What can you do instead?

SH The first thing we did in Rio was to make recordings and interviews, because we had the same thing as you have with this book; we did not want to come in and impose an opinion on the theme of water supply and wastewater. We wanted people to tell us their stories. How do they experience the water-related challenges in their communities and in their respective professional contexts? So, we conducted interviews in which people would talk to us, and answer our questions without filter, and use the opportunity of us being outsiders to give a lot of interesting insights.

We spoke with favela inhabitants, but also with activists and engineers, bureaucrats, and lawyers and then eventually the idea was born to make a film where all those people would speak together, who normally don't interact with each other and to connect the different local perspectives in one larger narrative. The whole story came alive when we edited the material and constructed the film *Água Carioca Diaries*, a journey through the water related issues in the favelas of Rio de Janeiro.

CD Can you give an example of something in the film that you learned, of the knowledge of the people in Rio?

SH The film project enabled us to dive deep into the issues and develop an onsite forensic methodology which would take us to various actors involved in the water problems who explained to us why the system is broken. We engaged with many different people to understand responsibilities, the geography of the favelas, ownership, operation, maintenance, and learn about earlier failures and existing initiatives. We went where the flow of information and encounters would take us. The film was a learning tool for us with the aim of being propositional, leading to a new, alternative design proposal for the water systems within the informal communities and settlements in Rio de Janeiro.

HW You and I collaborated in projects where my role was to be eyes and ears in a certain context, which we then used to construct a narrative for a kind of social architecture. Do you use this approach to learning in all your projects?

EP Yes, this method is essential. And building on what Sylvain mentioned, I think we gained valuable insights from the community's wisdom. For example, a community leader exclaimed 'Água e Ouro!' ('Water is Gold!') expressing a value for water that we should all have. Another resident highlighted that if the river is in bad health, so are the people, expressing how rivers and inhabitants are closely linked and form one entity.

Many favela residents have a deep connection with their environment. Living on unpaved streets, the forests right above them, and exposed to the elements, they experience pollution directly. Water is not always readily available, making them acutely aware of its value.

The favela has alternative organizations and a parallel or dual water supply system: one managed locally to distribute water from the mountains and one linked to municipal grid, which often fails.

OOZE Architects, *Água Carioca Diaries.*

Interviews are crucial for understanding individual situations in order to become propositional.

We avoided making promises to avoid raising hopes that cannot be fulfilled. Instead, we committed to making the film and invited the community to the premiere and the discussion afterwards, where they were pleased to see both the finished film and the way we had translated their inputs on the walls to form clouds of knowledge next to the drawings of the designs for the circular water system for their communities and neighbourhoods.

CD As what do you come in as? Do you present yourself as filmmakers, or as architects or designers or?

EP I guess we came in as the plants people. We had an A4 document with us, showing an explanatory drawing of the natural water filtration system with the plants. Of course, nobody likes to talk about sanitation. There is a stigma around it. But people we met and interviewed really wanted to show us critical places to give us a good insight of the urgency of their situation. As a research project, the focus was more to try to find an urbanistic and blue-green infrastructure solution for these situations, rather than on revenue models and how to finance the projects for our operation and all the different actors along the way to implementation. The Dutch Consul in Rio at the time was impressed and was wondering what our reasons were to pursue such a complex task. And he wondered if and how we could scale things up and what the business model for such projects was.

HW What did you answer?

EP We said we believe this is a model for the future and that we want to research it further. We came from a radical artistic practice, we didn't have to compromise or follow any standard business models because it was a research-by-design project which was publicly funded and supported. We followed our instincts.

SH Being foreigners actually helped us to access different territories within the informal communities. We went to places that Rio's residents of the formal city would not be able to. We made connections within the water engineering and the biodiversity conservation sector, and each encounter led to another. 'Oh yeah, you should meet this person', they'd say. It just grew into a large network of Água Carioca friends from there. Sometimes it got very emotional, such as with an executive from CEDAE, Rio's water company, which had great difficulty to improve the water supply and sewage infrastructure in remote and informal neighbourhoods—mainly favelas—being home to 6.3 million people. These areas suffer from a lack of access to healthy and safe water and sanitation, from diseases, environmental degradation, and socio-economic challenges.

This very experienced water engineer almost cried because he felt guilty for not doing more, but also helpless because he deeply experienced all the challenges and the almost unsurmountable hurdles to change the status quo. He did not feel that the residents of informal settlements understood their efforts, often destroying what was done. Close to retirement, he felt he hadn't managed to reform and upgrade the water system in the informal settlements. His desperation struck us deeply and made us feel we were onto something very profound in the city's priorities.

HW Your water filtration garden became a kind of outdoor classroom and proof of concept to understand things in a different way. What is their function within your larger projects?

EP We are profoundly interested in implementation. I would call it the courage of imagination. The idea was, to test the concept in a pilot in the Sítio Roberto Burle Marx, a garden that was designed by the eponymous Brazilian landscape architect to bring to life his approach to integrate various plants species, furniture, architecture and art, now designated UNESCO world heritage and operated as a conservancy of tropical plants and an experimental ground for ecological design. This offered a field site showing how an amazing soup of micro-organisms and plants can treat our waste water, literally recycling it to an almost pristine state. Also in Chennai, the pilot is generating a desire to upscale and replicate the results. People react to a built thing in a very different way than to a piece of paper. Here, we built a pilot as a stepping stone to demonstrate the effectiveness of what some people now call 'nature-based solutions' but which requires careful processes with local experts and the building of trust.

CD Can you tell us about the demonstration project you made for the school in the City of 1000 Tanks project?

EP The project is a prototype for a holistic innovation model that fully integrates rainwater harvesting, the collection of untreated wastewater, then the anaerobic treatment, plant filtration, infiltration and recharge to the ground water. We built a decentralized natural water system at the Little Flower Convent School for the Blind and Deaf, treating 27,000 litres of wastewater every day. The system looks like a park or a garden. It's also designed to be a clear, replicable model for people to understand and get a live experience of what it looks like and how it works.

We had to understand the socio-spatial connections and incorporate many existing elements like the grotto with the Maria statue, which is a very important place for the children and the community of teachers and the sisters to pray. We had good support from the head of the community and the students love it. We wanted the filtration park to embrace the centrality, having this functional system at the heart of the Campus. The sisters were worried the project would bring mosquitoes or use inferior materials, further marginalizing those they are caring for. But with the support of the Mother Superior we managed to take them on the journey to complete the project. The local team had extensive indigenous plant knowledge including mosquito-repellent species which create zones of comfort within the in park itself.

Now, a team of gardeners and technical experts maintains the area. Important parameters, such as water flow and water quality monitoring are accessible online. Additionally, to communicate about the larger project, we made an animated series of six episodes with a mascot, *Tuli goes to City of 1000 Tanks*, which is part of a larger communication campaign supported and funded by the Goethe Institut.[3] The goal of it is similar to that of the animation we made for Água Carioca Diarios.

CD For the animation, with humour being so very local, did you collaborate on that?

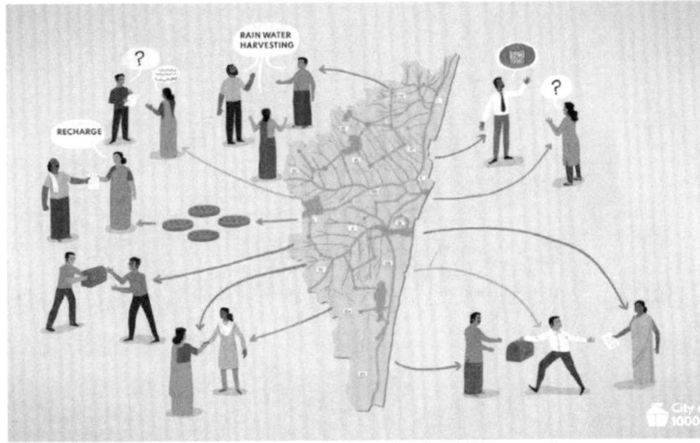

OOZE Architects, *City of 1000 Tanks, Water as Leverage*, infographics

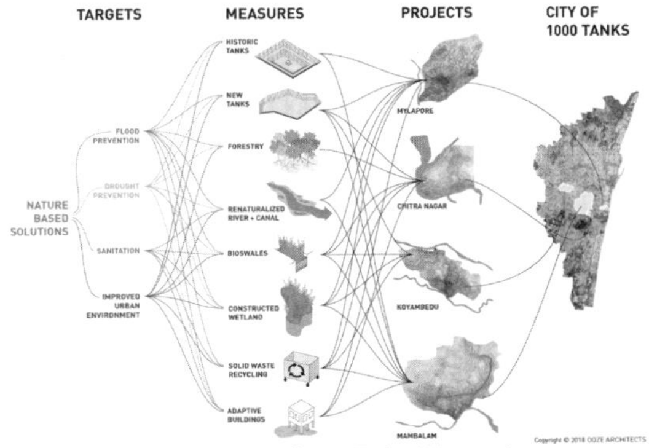

OOZE Architects, *City of 1000 Tanks, Water as Leverage*, infographics

EP We joined efforts with a local team of animators called the Plankton Collective. It was filmed in Tamil and English. And for the larger outreach, we worked with local graphic designers called New Folder. Children came up with Tuli, the mascot. It was part of our Young Water Ambassadors programme supported by the Goethe Institut. That worked like a snowball. We also worked with our partners—social activists and educators from IRCDUC and URAGUVAL who deployed a team in homeless communities and economically-weaker-section housing. Step by step, they reached out to children to give them the opportunity to have fun and express themselves with dance, poetry, theatre, and drawing to reflect and enact the City of 1000 Tanks.

HW One of the requirements for the 'Water as Leverage' call was to make the City of 1000 Tanks project bankable.

3 www.cityof1000tanks.org/tuli/

OOZE Architects, *City of 1000 Tanks, Water as Leverage*, workshop in Singapore. © Cynthia van Elk

As far as I know this is the first time you have worked with such a brief. How did bankability go with your vision?

EP Different International Financial Institutions (IFIs) such as the World Bank, Asian Development Bank, Green Climate Fund and others were part of the trajectory of 'Water As Leverage'. As the different teams developed their projects into incredible designs and presented these with detailed diagrams, maps, time schedules, and renderings, the representative of the IFIs said they wanted exact estimates of the benefits, the costs, the outreach and business models that would run next to the design project to guaranty upscaling and bankability. The review panel were asking for proper CBAs. Wow. What is this? All teams got a crash course in Cost Benefit Assessment.[4] We knew the technical solutions would work and the social solutions as well. We had to come up with a different way of looking at our project, addressing the financing, the packaging and project components: to look more at what our project does and what it improves and how much it costs, rather than just how it looks, how it works, and how you design it.

So, we were moving away from strictly architecture or art in public space towards infrastructure. Solutions are tested in demonstration projects on a small scale to be upscaled and replicated with significant budgets allocated to infrastructure projects.

Generally, we recognized that projects executed on a minimal budget or at a grassroots level do not capture the interest of institutional investors, such as infrastructure banks and other IFIs. Profitability is not achievable with low-cost initiatives. When we worked on the 'Water as Leverage' programme, we quickly realized that proposing a total project budget of less than fifteen to twenty million Euros is insignificant to these investors. It's not worth discussing. These banks are focused on deploying billions and do not have the bandwidth to engage with projects valued at seven million Euros.

OOZE Architects, *City of 1000 Tanks, Water Balance Pilot*, water samples. © Cynthia van Elk

SH The CBA technique relies on data-driven decision-making. Although challenging to assess, we tried to measure social and ecological aspects of the project. Tech is very much present, even in 'nature-based solutions', because it gives numbers. Additionally, to evaluate if your projects are actually performative, including all their components, also the social components and the Sustainable Development Goals, you need to connect the investments with long-term benefits. Now, even financial institutions recognize the importance of nature-based solutions. Everyone, from municipality to government is talking about it. But so far there has been very little reliable commitment to actually do it. Investing in these solutions and implementing them on a very large scale is still profoundly challenging.

4 A cost-benefit analysis or assessment (CBA) is a process of comparing the projected costs and benefits of a decision to determine its feasibility and desirability. Even though they can increase transparency of decision-making and help allocate scarce resources, its use is commonly critiqued for disregarding environmental impacts, downplaying costs to marginal groups and discounting future effects.

EP Henk Ovink is talking about how to spend millions in order to invest the billions wisely.[5] And how do we get people to invest in water and put it at the top of everybody's agenda? It seems that sticking to business as usual in the global food trade is still more important than to build up local water safety. Water economy is a complex subject which involves vested interests and difficult political dynamics. Because products that demand a lot of water, say for example avocados or almonds, are shipped across the globe. That's why there's a lot of opposition to establishing water as the leading factor for investing on the agenda. Or water protection, like protecting water for local communities. The people who should be interested in water are the local people because ultimately, they will suffer if they don't have access anymore.

When you educate yourself and other people about collecting rainwater for instance, then each time it rains you think how great it is to be able to collect so many litres, even millions of litres. Your relationship with the elements changes once you see your own system being part of the bigger natural system, the Earth system.

5 The first Dutch Special Envoy to the UN for International Water Affairs, from 2015 until 2023, a diplomatic role promoting water expertise globally. Henk Ovink & Jelte Boeijenga, *Too Big. Rebuild by Design: A Transformative Approach to Climate Change*, nai010 publishers, 2018.

HW The schematic drawings you make explain how the landscape you imagine works and create an architecture that is comprehensible and doesn't go over people's heads. How do they come about, how are they built?

SH We have this image to talk about infrastructure, which is like a body. In human metabolism we have all our organs integrated in our body performing interconnected functions. If we apply that model metaphorically to the city as it is now, it is as if our organs would be outside our body. We need to basically put the vital functions within the body of the city again. It is a sort of guiding image to make the city itself cyclic, so the city can provide for itself. All the human emotions also should be part of this story of infrastructure.

It is not only about the technical design and number crunching. It's also storytelling and connecting different local projects in that arch. The drawings visualize a narrative that is rooted, but more positive, without fear. They show what you can implement and plan with the people on the ground and how to understand healing of a landscape.

HW Since ancient times, the design of water storage has been done in tank architecture in India. Did you consider revitalizing or working with these ancient water techniques?

EP In Chennai the temple tanks function to funnel rainwater into the ground, a bit like inverted pyramids. In other regions, when the soil condition is different, it might be different. With the techniques we can work, yes, but the growth of the city has encroached many watersheds, so you can obviously not go back to using them all. There are a lot of tank rejuvenations happening at the moment, lots of public tenders.

Although we have not gotten a very large commission yet, our project and pilot has really made a difference already. People now protest against the proposed sewer and storm water system meant to mitigate flooding, that would transport the rainwater out of the city and into the sea as quickly as possible. People don't want that anymore.

HW Are these demonstrations around the area of your school intervention, or how do they know?

OOZE Architects, *City of 1000 Tanks, Water Balance Pilot*, team. © Cynthia van Elk

EP It was all over the news in papers in the state of Tamil Nadu and Chennai city.

CD So, in a way, a utopia is being built?

SH Well, now you have a little project here, a little park there. Actually, what this project is trying to do is to evaluate how much each of these projects contribute to the water safety of the city and to make an argument for many more of them, a multiplication, to achieve real water safety for the city, to sacrifice parking spaces et cetera. We need to actually go back to a system where we create enough space for water to enter and to feel welcome.

CD In Brazil you mentioned you encountered some people with profound wisdom. What defines this type of wisdom and is it something that you can build or work with?

EP This wisdom has to do with the fact that everything is connected with everything. And that we have to see ourselves essentially as subordinated to and integrated with the planetary earth systems. We have to understand that nature is not just something green. Nature is very dynamic, and our planet is a living organism that we are part of. The Gaia concept and hypothesis suggests that over geological time, life on Earth has had a major role in both producing and regulating its own environment. A great talk by Donna Haraway with Bruce Clarke recently introduced a new book called *Writing Gaia*.[6] It reprinted the letters between Lynn Margulis, who was a biologist who worked on the micro level, and James Lovelock, who worked on the cosmic planetary scale, constructing the Gaia theory over a period of thirty-seven years. The conversation makes you see how we sit in a field of relations and how our essential needs can be somehow taken care of if we care for planet Earth and understand these interrelationships. And I think we're just starting to understand this, but we somehow need to collectively redefine everything, including the economy, obviously.

6 Bruce Clarke and Sébastien Dutreuil, eds., *Writing Gaia: The Scientific Correspondence of James Lovelock and Lynn Margulis*, Cambridge University Press, 2022.

OOZE Architects, *Água Carioca*, opening pilot. Photo: Roberto d'Avila

3.3 Habiter en Zone Inondable—Living with Rising Water

BY HENRIËTTE WAAL

The flooded house of Madiah Rouah, still from short film *Living With Rising Water*, 19 min., 2018. Artistic direction: Henriëtte Waal & Elizabeth Guyon / Atelier LUMA. Courtesy LUMA Arles. Film: Alexandre Humbert

The Rhône is a partner in Madiah's life. When she just finished sweeping the last traces of the river from her home the Rhône usually rises again.

The flooded house of Madiah Rouah, still from short film *Living With Rising Water*, 19 min., 2018. Artistic direction: Henriëtte Waal & Elizabeth Guyon / Atelier LUMA. Courtesy LUMA Arles. Film: Alexandre Humbert

The major Rhône flood a few years prior was still fresh in the minds of many locals when I envisioned Atelier LUMA's Masterclass Living with Rising Water in collaboration with artist Elizabeth Guyon. We engaged with people in the Camargue Delta who live intimately close with the river. The most valuable moment of the masterclass was when the assembled architecture students and experts learned from one of these precious water sentinels: Madiah Rouah.

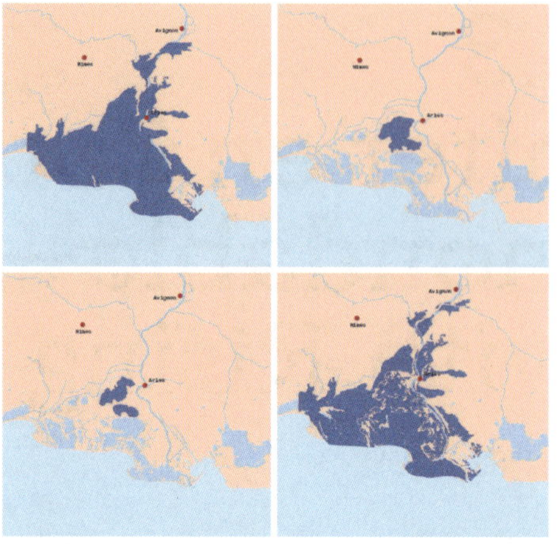

Atelier LUMA, *Flood in Arles and Camargue region, 1856/1993/1994/2003.* Courtesy LUMA Arles. Illustration: Andrea Anner

Madiah shared her personal experience of having her house on the banks of the Rhône flooded fifteen times. She spoke of how this has become an integral part of her life, explaining how each flood is unique in its height, speed, timing, temperature, reflecting on her vulnerability, as well as how she organizes herself when the water rises.

'The flood of 2015, which resulted in several deaths, showed that today very few people still have the knowledge of how to live with the water', said Ghislaine Verrhiest-Leblanc, Chargé de Mission at DREAL and responsible for water risk management across the French Mediterranean Arch. 'People's inappropriate actions, government services, the community—we all failed. We forgot that water is a potential source of danger that must be anticipated.'

Tarasque festival, 19th century, Bouches-du-Rhône Provence, France. The tarasque is a mythical beast that lived in the Rhône River. He was notorious for sinking ships and then eating the crew.

For her and many others, Living with Rising Water provided a novel space for collective learning. It fostered a cross-pollination of knowledge connecting political and social ecologies, focusing on how to coexist with water, while also tentatively embracing the emotions, risks, beauty, and lifestyles it can bring.

The Rhône is a partner in Madiah's life. When she just finished sweeping the last traces of the river from her home the Rhône usually rises again. During one particularly muddy flood, firefighters came to help her wash everything down. 'As long as it doesn't take my life itself, it brings unexpected experiences and reconnects me to my roots', she said.

Masterclass Living With Rising Water, restitution day, 2018, at LUMA Arles, during 'Jean Prouvé: Architect for Better Days', a major survey exhibition devoted to the innovative twentieth-century French designer of furniture and architecture. Courtesy LUMA Arles. Photo: Victor Picon

Beyond Madiah's story, the experience of theatre company Ilotopie was captivating. Born in the Camargue on the last island in the Rhône, Ilotopie turns the surface of water into a stage. 'All waters are our stages', they declare on their website, and they mean it. Since the 1980s, Ilotopie has been crafting ingenious water set props, like a sleek, black boat that floats just beneath the water's surface

Atelier LUMA, 'Living With Rising Water', exhibition, LUMA Days #2, 2018, Arles. Photo: Victor Picon

LIVING
WITH
RISING
WATER

Masterclass Living With Rising Water, Nadia Mourah gives a presentation, 2018, Arles. Photo: Henriëtte Waal

and can be maneuvered by simply shifting your feet, allowing actors to glide over the water.

Another water sentinel at the masterclass was Serge Binotto, boat inhabitant and Jean Prouvé's former engineer. The masterclass coincided with the 2018 exhibition 'Jean Prouvé, Architect for Better Days' at LUMA Arles' Parc des Ateliers. Serge gave a class on Prouvé's construction principles at the Bouqueval School, which was part of the exhibition and designed as a mass-producible, prefabricated one-room rural school with accommodations for teachers.

Flood cleanup action, Madiah Rouah and firefighters, still from short film *Living With Rising Water*, 19 min., 2018. Artistic direction: Henriëtte Waal & Elizabeth Guyon / Atelier LUMA. Courtesy LUMA Arles. Film: Alexandre Humbert

Guiding architects included Kunlé Adeyemi and Berend Strijland, known for the floating school in Makoko, an informal settlement partly built on stilts in the Lagos lagoon, as well as Paolo Cascone from African Fabbers School, which creates spontaneous architectural components and prefabricated housing solutions across central and northern Africa.

Living with Rising Water concluded with a conference, with global water experts such as Han Meyer, author of The State of the Delta, and Jean Jalbert, director of the Tour du Valat research institute for the conservation of Mediterranean wetlands, alongside architects specializing in flood-prone and demountable architecture, and local experts. The masterclass results were presented in an exhibition and a short film screening.

Atelier LUMA, 'Living With Rising Water', exhibition, LUMA Days #2, 2018, Arles. Photo: Victor Picon

Participating schools: ENSA École Nationale Supérieure d'Architecture, Montpellier/France; ENSA École Nationale Supérieure d'Architecture, Marseille/France; MEF University Faculty of Arts, Design and Architecture, Instanbul/Turkey; Academy of Fine Arts Vienna/Austria; African Fabbers School Douala/Cameroon

Masterclass Living With Rising Water, restitution day, 2018, at LUMA Arles, during 'Jean Prouvé: Architect for Better Days', a major survey exhibition devoted to the innovative twentieth-century French designer of furniture and architecture. Courtesy LUMA Arles

Janna Bystrykh is an architect, head of the Master's program in Architecture at the Amsterdam Academy of Architecture, and founder of the practice BYSTRYKH. In 2023 she completed her first film, *Great Pla/ns*, on regenerative farming and community practices in the United States. In 2024 she was co-curator of the exhibition 'IABR – Nature of Hope'.

3.4 Regenerating the Landscape of Center-Pivot Irrigation: Technology, Farming and Restoration Practices Across the Great Plains

INTERVIEWS WITH RORIC PAULMAN, BILLY TILLER & KENT O. MILLER
BY JANNA BYSTRYKH

The well-known circular patterns in the agricultural landscape of the Great Plains in the United States are the result of crop irrigation using center-pivots. Giant linear structures on wheels that rotate and with increased efficiency water plants through sprinklers along the full length of the structure. Commercially introduced in the 1950s, their design—and use—have since continued to evolve.

The center-pivots are typically $^1/_4$ mile in length, fit to irrigate one $^1/_2$ mile by $^1/_2$ mile square. But they can also be found as $^1/_2$ mile long structures, irrigating the full land plot of one square mile in what is called the 'Jeffersonian grid': a land surveying system parcelling large parts of the United States, a base structure for the selling of land to settler farmers and organization of new agricultural production from 1785 onwards.

The water that is used to irrigate the Great Plains is extracted from the High Plains Aquifer System, where the Ogallala Aquifer is the primary underground water body, spanning across eight states, from South Dakota, Nebraska, Kansas, Oklahoma, and the Texas Panhandle, and more to the west in Wyoming, Colorado, and New Mexico.

Prior to the invention and introduction of irrigation technology and the use of this aquifer as a water resource, together with newly introduced unsustainable farming practices, including deep ploughing, had by the 1930s resulted in loss of topsoil, and a subsequent period of severe dust storms, also known as the Dust Bowl.[1]

The saturated thickness of the Ogallala Aquifer is at its deepest in the state of Nebraska. At the moment, the total volume of water that is used for irrigation across the entire Aquifer is still higher than can be replenished through rainwater. Leading to a decline that began soon after the introduction of center-pivot irrigation systems using groundwater for irrigation over seventy years ago. But this drop and (sometimes) recovery of water levels is not happening evenly across the Aquifer.[2] In Nebraska, much of the Aquifer levels are stable, with some smaller pockets of decline, and larger areas of rising water levels. In many areas in Texas, the groundwater levels are still dropping. The patterns in water use and level changes between the two states suggest opposing trends for the region.

This chapter thematically combines three conversations with:[3]

RORIC PAULMAN (precision farmer and innovator, Paulman Farms in Sutherland, NE, Chair of the Nebraska Water Balance Alliance and involved in many other organizations)

BILLY TILLER (co-founder Member Grower Information Services Cooperative and farmer in Lubbock, TX)

KENT O. MILLER (General Manager – Twin Platte Natural Resources District, North Platte NE)

1 *The Plow that Broke the Plains*, U.S. Film Service, 1936.

2 USGS, Water-Level and Recoverable Water in Storage Changes, High Plains Aquifer, Predevelopment to 2019 and 2017 to 2019, www.pubs.usgs.gov/sir/2023/5143/sir20235143.pdf (accessed 24 September 2024).

3 Interviewed in Spring 2021.

These three are involved in the innovative TPNRD Water Data Program.[4] It aims to use water data to help farmers in the Twin Platte Natural Resource District in Nebraska to use groundwater more efficiently, by being able to offer them easy access to data as well as analysed water models, to help inform their irrigation choices. The project is the first of its kind in the United States that was community co-initiated and seeks, on a voluntary basis, to stay ahead of water regulations, while achieving the necessary adjustment of groundwater use for farming that is sustainable across the region.

4 www.tpnrd.org/Programs/TPNRD-Water-Data-Program

As part of this programme, gasoline-driven water pumps have been fitted out with meters that can be accessed remotely, complementing the network of smart meters that existed on electrically powered pumps—an important shift that allows farmers to understand their individual and collective use more comprehensively and to steer the use of water more precisely throughout the growing season. These efforts are complementary to the center-pivot irrigation systems that over the years have evolved to become more efficient in the use of water. But thereby these center-pivot systems are also brought into a new set of relations, where they function as part of a new reality that seeks to move towards regenerative agriculture.

The Water Data Program was developed with farmers in the region, emphasizing the importance of practice-based knowledge on both infrastructural and policy levels.

This project potentially entails an unexpected shift in precision agriculture practices in relation to food production. Where the different technologies are positioned to offer insights to farmers in relation to the regenerative capacity of their soil, seeds, crops, inputs, and financial commitments, groundwater use is potentially the last core component in understanding and pursuing environmentally and community-conscious agriculture. These farmers can then move beyond the initial drive of this irrigation technology for efficiency and maximizing crop output.

Many of the precision agriculture technology has been focusing on the individual efficiency and possible subsequent regenerative improvement of farming practice. But another strength of the Water Data Program, as a technological format, is that it makes a shift possible towards a collective regenerative practice situated around groundwater use in a region—as part of something as large as the Ogallala Aquifer.

Through the narratives of Paulman, Tiller and Miller, a relationship emerges across three distinct farming landscapes and cultures, offering both a contemporary and a historic perspective of water use, agriculture and the development of irrigation systems over time.[5]

5 Since the interviews, the TPNRD has been working on a pilot in collaboration with Arable.com and other companies, to help introduce local weather stations on the farms and fields that would link into the Water Data Program. With the many technical and data improvements, accurate rainfall data has been a missing piece for growers in completing their management of water. The data that exists through satellite data is not precise enough on field level. Adding rainfall data would make the understanding and possibility to conserve water, and balancing efficiencies and food production possibilities in a farm operation almost seamless.

Today All Data Exists

Interview with Roric Paulman

JB Could you tell us where your farm is located?

RP Paulman Farms is approximately 8,000 acres, of which sixty per cent is groundwater irrigated out of the Ogallala Aquifer, and twenty per cent is rainfed. We are in two river basins: the Platte Basin that stretches from border to border in Nebraska, and the Republican Basin, which runs from Colorado through Nebraska into Kansas, part of the Republican River Compact.

We are also part of a very sophisticated water project that originated in the 1930s that was built for irrigation that captured river flows into dams, and lakes, and then diverted through canal systems and smaller lakes into ditches and syphoned tubes, and as technology evolved, gated pipe, and now those are being replaced by center pivots. The hydrological connection between surface and groundwater is really important in this area.

With our farm we participate in three NRDs. Lots is going on in terms of understanding and clearly defining stewardship and conservation practices. It is a load of responsibility.

The function of the basin is so connected in maintaining habitat for animals and fish, as well as the integrity and the quality of the aquifer that is underneath our farm. We sit on approximately 600 feet of saturated thickness of the Ogallala Aquifer. Nebraska holds about sixty-two per cent of the total Aquifer. We have a significant responsibility in the big picture and the expectations of now being the number one irrigated state in the United States. There's approximately eleven million irrigated acres in Nebraska. It's a big driver for the economy, and for conservation and stewardship.

JB You already mentioned the rainfed and irrigated balance of the farm, how did your family farm develop over the years?

RP My grandfather came here and got to build some of these canal systems and lake projects, and had a little bit of farming experience: he grew some sugar beets, dry edible beans, pinto beans, corn and soy. And then my dad came along, and the growth began—row crop to syphon tubes, then livestock, we farrowed hogs, milked cows, we kind of did it all. We had chickens and we traded our eggs and milk at the local grocery store as kids, and yet you were still out there every day from 6am till 10 o'clock at night. It was a physical enterprise because you just didn't have the tools of today. Moving through that process, I actually left the farm and was gone for about ten years and came back in 1985.
The mid-1980s was the worst time in agriculture here in the United States, interest rates went to nineteen per cent. Everything was crumbling, then the farm was foreclosed on, the machinery and everything had to go back. And we turned around and we built a stakeholder-based process, really engaging with people that are experts in their field and can help you with data-based decisions, your banker, your accountant, your economist, your data entry, your software, understanding your financials is key.

JB How did the knowledge around water develop over time in the area, could you take us through that from the 1970s to about today from a farming perspective?

RP When you think about the water decisions in earlier times there was not enough data that would allow people to see any red flags. If the river went dry during a period, it went dry. If there was no water to divert, it was based on the news that they had that there just wasn't enough snow in the Rockies. The groundwater development didn't start till the late 1970s. That's when wells started being drilled.

On my farm the first well was drilled in 1974. And it was probably one of a handful in that timeframe. But from then on it turned into the Wild Wild West. You bought an acre of ground for $300, and it cost $300 to put a well, a motor and a pivot on it, and you were in the irrigation business. You did it because here we only get about twenty inches of annual rainfall, and so during the growing season, you may not get any rainfall, or you may get all of it. Irrigation was a great equalizer and an absolute wonder for the state of Nebraska.

And then comes data. Wow! The level of data development is exponential. Let's call it 1995 till 2005, it crawled along well. In 2010 you're really kind of ramping up, in about five-year increments. In 2015 it's every three years. And now we're down to eighteen months, getting a whole different collection and resolution of data, the linkages and connections are huge.

JB The transformation of your farm toward conservation practices—what drove that, was there any specific piece of equipment that was key?

RP It would be the soil moisture probe. It really started to tell the story about the cultural practices that we were already doing, the no-till for example. We had gotten away from conventional till because every time you made a pass across that field you had a cost and time that you do not have in the spring, leaving the residue in the field is a good way to meet some of the perils of wind, heavy rain events. We put the probes in both our irrigated and dry land, and we started to understand just how much available water we had to start with, in the rotation of dryland crops.

JB It's been quite a long transition process since 1985, what is next?

RP My level of excitement is off the charts because today it all exists. You have all the soil tools, the mapping characteristics, the resolution, the better understanding, and places to go to ask the right questions and get appropriate answers. The big challenge today is how do you take that data and make it work as part of a broader behavioural change in farming in real-time, responding to what we know about water, soil, rainfall, nutrients.

Not Buying Land, but Buying Water

Interview with Billy Tiller

We are about sixty miles west of Lubbock, Texas, home of Texas Tech University, a very rural area. Average rainfall, since about 1911, is around eighteen inches. Rainfall highs were forty-four inches in the early 1900s, low is going to be like three or four inches, and that was 2011. I lived through that one.

Mostly what's grown here is cotton and that hasn't always been the case. The reason it's cotton is that it does better in a semi-arid climate, as irrigation declines. I've not seen corn probably since about 2016, and the reason is that we've had four of the driest Aprils on record.

JB Has your farm always been rainfed?

BT I had irrigation up until about 1990. At the time people in the Texas High Plains were starting to see some depletion of the groundwater, but I didn't have it on my mind that this irrigation could go away. In 1990 I became an executive vice president of a local bank, and with that I thought I can't really continue to manage this irrigation, some of it was row water, I didn't have pivots on yet, and so I decided to trade irrigated for dry land farming. Since 1990, I feel like I've kind of been looking at these developments from the outside.

I spent ten years as an Ag lender, a banker loaning to farmers, and I never had a farmer come into my office who wasn't worried about groundwater running out. Here's the problem: these farmers were spending money but they weren't buying land, they were buying water. And I saw that because of the value differentiation between irrigated land and rainfed.

Closeup pivot irrigation arm and sprinkler system with spray nozzles, standing still in a ripening wheat field, July 2021. Great Plains, United States. Photo: Mason Brown

The fact that the water was depleting was killing their future cash flows. They all tried to figure out ways to make it work, but the problem is when you need 600 gallons a minute to raise your corn, and all of a sudden you're at 400 gallons. What do you do? In Texas you can drill another well, and you put a straw in that cup, and then your neighbour drills a well, because now he's got less as well.

In the 1990s I saw just more and more straws put into the same cup. They did everything, trying to figure out what financial decision to make to get a payout on this land, and that's the problem. These farmers spent real hard-earned dollars to put in center pivots, then put in drop lines, and LEPA (Low Energy Precision Application) to put in drip. They've done everything imaginable, and the problem is, we have a depleting resource with so many acres involved, it's simply not going to recharge. But we did not know that.

When I was a kid, we had decent water here, I could go north twenty to thirty miles and these guys had 1,000-gallon wells. They watered out of big ditches. They'd literally pump this water out into a massive ditch, put tubes in, that tube would then roll water and flood irrigate. We'd swim in those things, it was like a river running down through there. If you would have told me when I was eleven, twelve, that we wouldn't have that kind of water so many years later, I would not have believed it, because it seemed like it was a never-ending resource.

We have a lot of guys irrigating 175-200 gallons a minute, and that amount will not raise a crop. That's one reason why irrigation management is difficult in West Texas. You're deficit irrigating all the time, you're never ahead of the game where you can say I can turn it off. I've had the privilege of working with farmers in other parts of the country who have more water. In those cases, it definitely can be managed. It's impressive. They get a rain, they shut it off immediately. When you are deficit irrigating, you might shut it off right when water

Satellite image of the Sutherland reservoir and center-pivot irrigation patterns, Lincoln County, NE, USA. The irrigated circles are set in a 1/4 mile by 1/4 mile grid. Source: Modified Copernicus Sentinel data 2025 (EU)

is standing, but as soon as it dries up a little, you're starting right back up, because you're never going to catch up if there's no more rain. An irrigated farm in West Texas today, for the most part, except from some areas in the Northern Panhandle, are all going to be in deficit irrigation. The Lubbock area has historically been in cotton, they're running behind the eight ball all the time, and just trying to grow that crop.

Today we're sitting on a vast amount of knowledge. We didn't know what we didn't know, but now we have the data, and we have to start to rely on that. I'm almost protective of these Texas farmers. We were too late when we got the data to realize we needed to manage it, but I hope that other parts of the country heed the warning. They have time. They can manage water, they can do better.

No New Irrigation Acres Since 2004

Interview with Kent O. Miller

We're very fortunate that our land in west central Nebraska is above the Ogallala Aquifer, that allows for irrigation necessary to produce crops here. With the conservation of water resources, our intent is to manage in such a way that water is available virtually forever. We are now required to manage groundwater to avoid depletions to the Platte River system. As a result of that, no new irrigation acres have been allowed in the Twin Platte Natural Resource District since 2004.

JB The system of *Natural Resource Districts* is unique to Nebraska. Could you tell us how the system came about and how it relates to the *Soil Conservation Districts* which exist throughout the United States?

KM The federal government began the *Soil Conservation Service* to provide technical and financial assistance to landowners to help conserve the soil as a way to prevent droughts, following the droughts of the 1930s. It was quickly discovered that they needed to figure out a way to work with local folks, and with this *Soil Conservation Districts* were set up where local folks got together.

In the late 1960s there were watershed issues, dams needed to be built, issues around irrigation and conservation of groundwater. In 1969, the Nebraska Legislature created twenty-three Natural Resource Districts (NRDs) across Nebraska—which were later given the authority to regulate groundwater when center pivot irrigation came in and there was a larger interest in groundwater development.

JB How did the recording of water use happen around this time and what led to the shift in determination in Nebraska not to mine water and to focus on conservation?

KM Back in the 1980s and early 1990s we started putting together a *Cooperative Hydrology Study*. An effort to start building computer models, so that we could start predicting different levels of groundwater development, and at what point mining of water becomes harmful.

The groundwater is recharged from rainfall. A rough rule of thumb says the rainfall on an average annual basis contributes about three to four inches of water in this region, that filters through the soil back into the aquifer. If you have an irrigation requirement for a crop of sixteen inches, and if you start exceeding twenty to twenty-five per cent of your land area, you're getting into a groundwater mining situation. And that's if it's equally distributed irrigation throughout the entire NRD. But if you have areas where there is water irrigation development then you can start having groundwater mining issues in those areas. So, it was a combination of putting together a Groundwater Management Plan, and the Board of Directors having to conscientiously make the decision for our groundwater resources to be available for future generations.

JB When did the approaches regarding ground and surface water start to be integrated?

KM In 2002, a *Water Policy Taskforce* was formed to look at the need for integrated management of groundwater and surface water across Nebraska. At that point it was recognized that the Upper Platte Basin in Nebraska was already over-appropriated, and that groundwater is affecting surface water flows. So, when that legislation was enacted in 2004, the Upper Platte Basin was immediately required to put together an integrated plan to get the river back to a fully appropriate condition, where the groundwater is no longer affecting surface water flows.

This *Integrated Management Plan* is updated every ten years, on which we work with about twenty-five stakeholders—representing groundwater, surface water, ranchers, irrigators, municipalities, environmental topics and habitats. By 2009, we received a recommendation from the stakeholders to not introduce regulations, but to try to find ways to offset water.

JB How did the *Water Data Program* begin?

KM When we started the next update of the *Integrated Management Plan* in 2019, it was recognized that the Twin Platte NRD still has ways to go to find offset water. That's when the stakeholders recommended mandatory data collection, and Roric Paulman (one of the stakeholders) suggested working with Grower Informational Services Cooperative (GISC) on collecting, calibrating and interpreting the accumulated water use data. Roric had been telling me for years that there is a better way to measure water use, not through mechanical meters. And that's how the Water Data Program began.

We have close to 3,000 wells in the Twin Platte NRD. Eighty per cent of the wells are electric wells, and twenty per cent are powered by diesel and not connected to an electric smart meter.

When a well is drilled, at that point it is determined how much water it will be pumping. Mechanical water meters, when installed, are very accurate. Two days later, they start wearing out, two months later even more, and a year later even more. But because it was the standard way of getting water data across the state, the Nebraska Department of Natural Resources said that we needed to put in place a validation system of the readings. That's what we're doing—installing mechanical well meters, fitted out with a continuity meter, basically an on-and-off switch, connected through the recently installed LoRaWAN wireless system across the state of Nebraska. GISC is aggregating and calibrating the water data with data that is coming through the electric smart meters.

JB In a way this is the beginning of a new type of conservation practice in agriculture. What other trends do you see?

KM In talking to our growers, fifteen years ago the goal was primarily to produce 300 bushel per acre; the belief was 'the more the better'. Today, I think growers are paying more attention to the rate of return. I've had growers tell me that they can make more money growing 200 bushel per acre than 300 bushel per acre, because they are not applying as much water and fertilizer, and not putting as many seeds in the ground.

And so I think that into the future, it's going to be hybrids, using different opportunities and practices, also as equipment continues to evolve. I just think that technology and information is providing farmers with so many (new) opportunities, and they're using those opportunities.

Carolina Domínguez-Guzmán is a postdoctoral researcher in PilGrimS (Pathways to Groundwater Sustainability) funded by Future Earth and the Water and Development Partnership Programme. She brings together scholarship of anthropology, science and technology studies (STS) and feminist material semiotics to study water care, technologies and infrastructure. Domínguez-Guzmán works for the Water Governance Group at IHE Delft.

3.5 Technodiversity in Mango Irrigation–Purification and Pozas in Northern Peru: Modern Irrigation, Stubborn Trees

BY CAROLINA DOMÍNGUEZ-GUZMÁN

Drip irrigation is upheld as a one-size-fits-all solution for irrigating Peru's arid coastal lands. Peruvian water professionals and policymakers emphasize investing in water-saving technologies to help protect the limited water resources increasingly extracted by the export agriculture industry.[1]

1 Rutgerd Boelens & Jeroen Vos, 'The Danger of Naturalizing Water Policy Concepts: Water Productivity and Efficiency Discourses from Field Irrigation to Virtual Water Trade', *Agricultural Water Management*, Volume 108, 15 May 2012, pp. 16–26.

Mangoes on the ground, 2017, Motupe Valley, Peru. Photo: Carolina Domínguez-Guzmán

Ever since hydrologists found new ways to measure and quantify the world's freshwater stock at the turn of the last century and signalled a so-called global water crisis, volumetric water savings have received ample attention.[2] Particularly in places where export irrigation is carried out under extremely arid conditions, high-tech irrigation, especially drip technologies, gained popularity for its potential to reach high efficiency in water use and conservation of water quantity.

2 Peter H. Gleick (ed.), *Water in Crisis: A Guide to the World's Fresh Water Resources*, Oxford University Press, 1993; Jamie Linton, *What is Water?: The History of a Modern Abstraction*, UBC Press, 2010.

Drip tube well and filter, 2017, Motupe Valley, Peru. Photo: Carolina Domínguez-Guzmán

For most Peruvian farmers, particularly those running smallholder farms, irrigation methods and technologies are dependent upon and enabled by the *quality* rather than the quantity of water, which makes their relation to water and crops different from those involved in large-scale agriculture. This contrast can be observed clearly in the Motupe Valley, Peru's second-largest agricultural region producing mangos for export. Once a region of large *haciendas*, an agrarian reform in 1968 entailed far-reaching land redistribution, creating small parcels of arable land, with limited access to canal water. The diversity of non-modern irrigation methods and technologies in Motupe mirror the heterogeneity of smallholder lived realities and the relations these small farmers developed with water and their crops. This heterogeneity complicates uniform scientific accounts that treat agricultural water as one coherent unit of study, and that stress quantification, modelling and continuous monitoring of water movement as the sole solutions to water scarcity problems.

Instead of pursuing how one might decrease the quantity of water used in irrigation, I draw upon the work of feminist science and technology scholars to argue that realities are not given, but rather are enacted in practice, to wonder if 'water' is necessarily the same everywhere or may in fact be different things in relation to different technological practices, scientific or otherwise.[3] Most importantly, I focus on the relations that different technologies establish with the environment and the people who care for them. By studying water as a situated element, looking beyond its physical characteristics, learning from the practices of Motupe farmers raises the question if a single-fit technological solution for all agricultural realities is actually needed or if, instead, technodiversity might be called for. This goes against a strong narrative in Peru praising modern, purified water, while relying upon the false universal idea that all agriculture is the same and equally controllable. In this narrative, smallholder farmer practices are catalogued as not being modern enough to keep up with industry standards. Whereas, instead, this chapter explores how it is thanks to, and not in spite of, their messy irrigation styles that smallholders succeed in cultivating mango.

3 Donna Haraway, *Simians, Cyborgs, and Women: The Reinvention of Nature*, Routledge, 1991; Annemarie Mol, *The Body Multiple: Ontology in Medical Practice*, Duke University Press, 2002.

DRIP IRRIGATION AND THE LONGING FOR PURIFICATION

The majority of mango export farming in Peru is in the hands of smallholders. In Motupe, there are over one thousand smallholders with less than five hectares each, most of whom are located on the right bank of the main Motupe River. These small farms, or *chacras*, are interspersed among a few large estates, one of them owned by Elias Montero. Elias's estate covers 130 hectares. On his farm, Elias has more than 10,000 mango trees and several thousand export avocado trees. As with most large-scale farms, Elias's farm is homogenous and monocropped: all the trees are grown in the same way.

Mango pozas, 2017, Motupe Valley, Peru. Photo: Carolina Domínguez-Guzmán

Children are walking on the banks of a poza wherein maize is grown. It is used for irrigation and recharge, 2017, Motupe Valley, Peru. Photo: Carolina Domínguez-Guzmán (top to bottom)

Elias is the son of a hacienda owner dedicated to cotton cultivation; hence agriculture has been part of his life since he was a child. However, at a young age, he already knew that he did not want to become an *algodonero* like his parents, but instead wanted to start his own adventure, in horticulture, and export fruits. After graduating from the most prestigious agricultural university in Peru, Elias visited the United States to specialize in pomology. There, he learned about mango varieties and their management, for which he travelled between California and Florida. On his return, not only did he introduce the Kent variety to the Motupe Valley, but also seven other varieties with which to experiment. These he carried back to Peru in his suitcase, he recalls jokingly, while insisting that he was the first in the region to grow all these varieties. When the international orders started to pile up, he gathered other large-scale farmers together to form APEM (Peruvian Mango Producers and Exporters Association), of which he was the first president.

FEEDING TREES WITH DRIP IRRIGATION

Elias Montero loves his drip irrigation system, for its water efficiency, and because he likes the idea of being able to combine irrigation with fertilization to enhance farm productivity. This drip system requires a continuous, steady water flow that is difficult to obtain by relying on unpredictable water flows like rivers and rain. Therefore, he relies on groundwater pumped up from the underlying aquifer. Groundwater in Motupe consists of a single unconfined aquifer above several confined aquifers (up to several hundred meters deep). Unconfined aquifers receive seepage water from rain and surface water bodies (such as rivers and lakes) and can be significantly recharged, for example, after the rains of El Niño. By contrast, confined aquifers are enclosed by an impermeable layer and their fossil water cannot be easily recharged.

Elias's seven wells are linked to a pumping station that extracts and pressurizes groundwater and transports it uniformly through valves, filters, mains, sub-lines, and laterals until it reaches the ground again and infiltrates the root zone of the trees through individual drippers near the base of each single tree.

One of the efficiencies of drip irrigation, according to water professionals, is its precision, which diminishes the chance that unwanted weeds start growing. It also affects the development of the roots of the trees, which remain small. One of the main problems, however, is that drip irrigation systems are sensitive and prone to clogging. Therefore, water has to be purified of unwanted elements such as sediments of sand and clay, or algae and bacteria, using centrifugal separators, different types of filtering screens, and chemical injections before it enters the distributary system so it can reach the plants.

Purification is central to the project of modernity.[4] To attain this purity and its associated ordering of reality requires separating categories and restraining the proliferation of impure forms. To maximize production, Elias separates unwanted elements, solids, and particles from the water with screen filters. After passing through the filters, the water is purified, ready to be mixed with chemical fertilizers. This way of working allows Elias to calculate the nutrients that he injects into the system.

4 John Law, *Organizing Modernity: Social Ordering and Social Theory*, Oxford University Press, 1994.

When asked about the drip system on his farm, he explains:

> With technified [drip] irrigation, I open a single valve, and I treat (*tratando*) 400 trees at the same time. With water, with ammonium nitrate, phosphorus and all the nutrients [in the solution]. This is of course in addition to all the nutrients that I apply directly to the tree. [...] I do it very quickly and I don't waste water. [...] These lands are very rich, rich in phosphorus and potassium, the best in the entire country, [...] the whole area of Motupe and Olmos is very rich in nutrients. What it lacks is nitrogen and this is why I apply ammonium nitrate, which is nitrogen. And sometimes I apply phosphate, [...] because this, as well as the potassium, reacts with nitrogen [to form a compound fertilizer]. They [the macronutrients] are not mobile by themselves. This is why, with nitrogen, NPK [fertilizer] is formed and

Mango tree and drip lines, 2017, Motupe Valley, Peru. Photo: Carolina Domínguez-Guzmán absorbed by the plant. But what I mostly apply [through drip] is ammonium nitrate, more than phosphate. [...] The potassium I do aerially.

Separating out water from the chemical and biological elements dissolved in it gives Elias the opportunity to be precise in his irrigation and fertigation, by controlling the elements that the plants will absorb. Since its invention, this is considered a distinct advantage of drip irrigation.

Next to the fact that drip irrigation enables Elias to irrigate large numbers of trees at the same time, he likes it because it keeps his farm clean. 'We don't make the farm dirty. There is no need to weed. It reduces labour costs and it [drip irrigation] is more precise.' Drip responds to the specific characteristics of his farm, as a space for mono-cropped industrial production. Elias also points out that the groundwater underneath his farm, which he pumps up, is efficiently used by 'saving' water, rather than depleting the aquifer. This allows him to de-emphasize that he is depleting the aquifer and to foreground, instead, that he 'saves' water by using drip technology. Regarding surface irrigation, he says, 'When you flood your field [using *poza* irrigation], the trees do not benefit from all the water that you use. The rest goes to the '*camellón*' (the space underneath the roots) and is lost'. By stating it like this, he dismisses the fact that, in *poza* irrigation methods, excess water helps to recharge the aquifers when it is not absorbed by the trees.[5]

5 Bruce Lankford, 'Localising Irrigation Efficiency', *Irrigation and Drainage* (55:4), October 2006, pp. 345–362.

POZA IRRIGATION AND THE LONGING FOR MUDDINESS

Smallholders do not pump up water from the aquifer; they depend on water from the Andes Mountains, *agua de avenida*, with its high-altitude wetlands, that consist of bogs and lakes which feed the rivers downstream. Many rivers are captured by an impressive hydraulic infrastructure and transferred to the Pacific side of the Andes to serve commercial (export) agriculture in the coastal valleys. One of these diversion structures, the Huallabamba Canal, serves Motupe. It is a non-regulated system with an old infrastructure, which causes unpredictable shortages. The costs of motorized pumps are too high for smallholders, but this does not mean that smallholders stop farming; it means they have learned to be adaptive.

Bordering Elias's estate in Motupe valley is the Manayay's farm, irrigated using *pozas* (basins), a common practice among smallholder farmers. There, I met Juan Manayay, whose father was the first smallholder to take up the idea of cultivating export mangos. This he did by grafting the *criollo* trees on his farm with the export variety and intercropping trees with the maize and other crops that he grew in his fields.

At the time I met Juan and his sister, their memories of their deceased father were still fresh, and they thought that his soul was wandering among the tall robust trees of their farm. On their *chacra*, covering nine hectares, Juan and his sister have almost 400 mango trees.

One remarkable feature of the Manayay farm is the tallness of its trees. Many trees are over eight metres in height and, unlike what we saw on Elias's farm, they are not at all homogenous. Unlike agronomists and owners of larger farms, the Manayay family accepts that some years their trees do not bear fruits because, as they say, trees also need to rest from time to time.

FEEDING TREES WITH *POZAS*

Like Elias, Juan Manayay is proud of his irrigation method. He appreciates the significant height of his trees because this reveals that they have taken root in the soil and subsoil; they are able to reach the underlying aquifer when the surface water coming from the Andes is not enough.

On his farm, we walked over the *bordos*, which are bunds made of mud used to create the *pozas*, or basins, that contain irrigation water. The first descriptions of *pozas* on the Peruvian coast appear in books by sixteenth-century historian Felipe Guaman de Ayala. He described how because of the *bordos*, the soil remained humid for at least two to three weeks.[6] In fields where water availability and access are unpredictable, *pozas* allow trees to survive in the desert with only a few irrigation turns annually. The *agua de avenida* [water from the Andes] has the advantage that the soil remains moist for a month, while watering with a tube well means the land asks for water every eight days, and also that the water has no vitamins that make the plant greener. It [*agua de avenida*] gives more life to the plant.

In a desert climate, for many smallholders, growing trees with deep roots is crucial to secure their survival. In the *pozas*, water infiltrates and slowly moves underground. Moreover, in times of prolonged rains or torrential storms, the *pozas* prevent runoff and recharge the unconfined aquifer. This connection between canal, rains and groundwater is often overlooked by scientific accounts of efficient irrigation.

6 John Keith Hatch, *The Corn Farmers of Motupe: A Study of Traditional Farming Practices in Northern Coastal Peru*, Madison, 1974, p. 113.

In its descent from the mountains, particularly in places where the slopes are steep and the water gains speed, it takes up silt, sand, and small stones, becoming brown and muddy as it moves towards the mango fields. Once the water reaches the Motupe Valley and its speed decreases, the stones and pebbles are left behind on the riverbed, but smaller particles, the sediments, travel with the water to the farmers' fields through a network of small canals until the water ends up in *pozas*. There, the farmers also grow other fruits, avocados, lentil bushes, squash, beans and cassava—with the surpluses sold to local and national markets, so they can sustain themselves and their families when the mango prices are too low.

Often, smallholder farmers speak of the weight of the water and refer to the mud carried down to their fields from the Andes Mountains as *barrito* (an affectionate diminutive of *barro*, mud). One farmer says:

Water from the river, it's more nutritious. It has a different weight, you know? It allows the trees to gain strength [...] in a different way. [...] When water comes down from the Andes, it drags all these minerals and [silence] everything. Just imagine when the rain washes those mountains, all the nutrients that it carries!

Valuing the richness of the muddy water as food for their plants is important in smallholder agriculture along the Peruvian coast. Indigenous pre-Incan farmers along the southern coast of Peru valued *agua de avenida* because it brought the *yapana*: the minerals and nutrients dragged from the altitudes, keeping the flooded lands wet and fertilized.[7]

Elias, as we saw, relates to his plants and water through his knowledge of chemistry and the numerical data regarding his system's efficiency. By contrast, in some cases, the smallholders I interviewed said the nutrients are 'something' that gives life to their trees. These nutrients are not objects of *knowledge*, but linked instead to their desire to see their trees being fed by Andean rains and mud. In this case, feeding the trees is a way to give something back for the mango they offered.

But as it goes in worlds that are simultaneously modern and non-modern, this reliance on water filled with nutrients from the Andes does appear to be fragile. In their desire to feed their trees, some Motupe smallholders are increasingly giving their trees manure, as well as synthetic fertilizers. Several smallholders I speak to, compare using manure to feeding a pregnant woman, or

7 María Teresa Oré, *Agua Bien Común y Usos Privados: Riego, Estado y Conflictos en La Achirana del Inca*, Fondo Editorial de la Pontificia Universidad Católica del Perú, 2005.

to taking care of small children, whom you have to feed to make them grow in strength.

When the Manayay family hear that drip technology is advertised because it enables the user to add liquid fertilizers, they respond in disbelief. After all, *agua de avenida* is already nutritious enough to feed their trees and their families in the difficult conditions in which they irrigate and do agriculture. Feeding is to them an act of mutual nourishing that is beyond water as it is scientifically known.

CONCLUSION

Instead of adhering to a single, purified and purifying modern technological model, there is merit in diversifying or intercropping technologies that care for local water qualities. Particularly in times of intensive export agriculture and water depletion, there is an urgency to find out which water qualities allow for agricultural worlds in which care, diversity, and nurturing relations are possible. In the case of the Motupe Valley, muddy and nourishing waters from the Andes Mountains still continue to enhance the resilience of smallholder farmers today.

For further reading, see:

C. Domínguez-Guzmán, A. Verzijl, M. Zwarteveen & A. Mol, 'Caring for water in Northern Peru: On fragile infrastructures and the diverse work involved in irrigation', *Environment and Planning E: Nature and Space*, (5:4), 2022, pp. 2153–2171. www.doi.org/10.1177/25148486211052216

3.6 Garden to Cloud

INTERVIEW WITH ZAIRAH KHAN
BY HENRIËTTE WAAL & CLEMENS DRIESSEN

Zairah Khan is a cloud catcher, permaculturist, and community revitalizer with a background in psychology and sociology. As a social entrepreneur, she merges idealism with business, translating new ideas into functional models. She founded VIVEinvest to bridge the funding gap for women entrepreneurs and BlueO² to create resilient coastal communities by combining blue and green strategies.

In Transvaal, the most paved neighbourhood of the coastal city of The Hague (NL), where government-driven ingenious solutions arrive last if ever, ten-year-old green girls capture clouds with their *Wijkwadi* (wijk = 'neighbourhood' in Dutch, wadi = a riverbed that is dry except after rain, in Arabic). The Wijkwadi contains diverse water managers such as willows, plants that loosen the soil with their roots, allowing the rain to infiltrate. Within four hours, even the heaviest rain cloud is absorbed by the wadi garden. There are also edible plants, flowers, and fruit trees that retain the water. Besides the green girls, about 200 residents engage with water, nature, and each other through the Wijkwadi. On Mondays, care clients have a programme where they learn and laugh a lot.
Initiator Zairah Khan is very hands-on with what seem like complicated planetary processes. She has delved into the small water cycle, where soil and plants retain, evaporate, and transpire water, creating microclimates. These microclimates play a crucial role in the global climate.[1] She explains:

1 R.J. Van der Ent, H.G. Savenije, B. Schaefli, & S.C. Steele-Dunne, 'Origin and Fate of Atmospheric Moisture over Continents', *Water Resources Research* (46:9), 2010.

'It all starts with rain: delaying, capturing, storing, and reinfiltrating it into the ground. Just as water wears down a rock, it can cool the planet. If we embrace this possibility and start working with water, we will find many solutions to other problems along the way.'

Microclimates naturally occur but can be designed easily following the same principles. Zairah Khan does this with her community instead of experts. It began with a few fruit trees on vacant land that were obviously too warm and very thirsty. She proposed to create a wadi around them with the rain from the adjacent nursing home's roof. The 1500 m^2 roof was connected to the sewer via an indoor downspout and causing floods during heavy rainfall. A feasibility study Khan commissioned an engineer to do showed it was possible to capture this water in the sixty m^2 wadi. A double win. By disconnecting the building from the sewage system at street level, a playful situation emerged where water is central, binding people to the place and enabling ecology. A local school, lacking funds for nature education, proposed an after-school activity for children interested in learning more about nature, leading to the 'green girls club'.

Zairah Khan, *Wijkwadi Transvaal, volle wadi*, 2022, The Hague. Photo: Zairah Khan

At first, there was fear of vandalism, but only some plants were uprooted. 'I understand it's interesting to see what plants look like from underneath', Khan smiles, 'some trees also fell because children looked for sticks to play with, but overall, the result is positive. Nature grows back. The willow ceremonially planted with the local alderman was pulled out the next day but now has six new sprouts.' Khan aims for resilience, not vandal-proofing. 'Relate to it as a living thing, not a static image. Yes, wild grass attracts rubbish, but maintenance can be addressed.'

The municipal green manager initially ignored the wadi approach, as trees were previously not thriving at the location with his methods. Over time, he became more open to it and now mows less often in the adjacent field. Kahn:

'Municipal people are open to new things, but often stop where their responsibilities end; that's why a designer is needed to mediate these processes. Many people find water management complicated. People in the water sector no longer speak the language of normal people.'

Zairah Khan, *Wijkwadi Transvaal, klimaatadaptieve speeltuin in versteende wijk*, 2022, The Hague

Elsewhere in the same city, a landscape architect is creating an 'urban water buffer' for close to a million euros. It involves concrete and pipes with shell sand draining into deeper groundwater. The neighbourhood wadi, with all its ecosocial features, costs less than one hundred thousand. 'You'd expect more projects like this; Delaying, capturing, infiltrating: everyone wants that'. Her infectious enthusiasm almost makes you want to chant the new mantra: Delaying, capturing, infiltrating! Delaying, capturing, infiltrating! 'But the ideal of drainage is persistent', Khan continues. 'And when the bigwigs came to judge our neighbourhood wadi for a prize, my social types naturally didn't show up.'

Nevertheless, the neighbourhood wadi is now a model for participatory climate adaptation, and Khan is active in various parts of the city. The Wijkwadi makes climate a local and engaging phenomenon instead of an inevitable global process of incredible magnitude.

In order for microclimatic solutions to be implemented at a large scale, and slow or even reverse the effects of climate change, we need a global movement of water keepers and protectors. Not only in the ancestral landscapes of indigenous communities but also in the Transvaal community in The Hague and in a hundred thousand paved neighbourhoods in our world.[2] Khan's dream is to establish a 'wadischap' in every neighbourhood led by a 'wijkgraaf'—a 'neighbourhood water leader'. A local, vigorous and smiling leader, of course.

2 ClimateScan.nl has assembled until now some 500 wadi neighbourhoods across the Netherlands, with altogether over a 1000 wadis.

3.7
Fake Miniature Waves

Capturing the flow pattern with paper snippets in the Westervoort model, 1957. Source: Archive of the Waterloopkundig Laboratorium

'The use of real water in scaled-down city models came widely to be regarded as looking fake (see J.N. Newman, **Marine Hydrodynamics**, MIT Press, 1977). It turns out that the molecular structure of water and the force of gravity are not themselves scalable. Trying to correct for this by slowing down the action, as Hollywood films have, cannot work well since not all the processes in play scale at the same rate. What becomes jarring about wave images [...] is not only their material and spatial clumsiness, but also a sense that human time and wave time are not calibrated. For a large wave to be convincing, it must exist, with humans, in the same filmic and narrative time.'

From: Stefan Helmreich, **A Book of Waves**, Duke University Press, 2023, pp. 194–195.

3.8 Twelve Little Ditch-Experts in Wickelhofpark

By Ria Waal

After one and a half hour of intense scooping with a fishing net in the little ditches of the Wickelhofpark, twelve children received a colourful button that says 'little ditch-expert' from the Dutch Institute for Nature Education, in short IVN.

The children had earned this by carefully putting the water animals they had caught into a bucket and studying them. Beforehand they were told to move the fishing nets back and forth near water plants. Small creatures prefer a bit of cover. The children were also told not to scoop too deeply, as their main catch would then be mud, and they would no longer see any water animals.

They all paid close attention and together with their parents gave it a go. Taking their telescopic nets, buckets, and identification charts, catching water creatures. The big water animals stood out directly: a tadpole and a baby salamander or a backswimmer, swimming on its back. But there were also tiny critters you had to view in a pot with a magnifying glass. Together with a dad, mom, or an accompanying expert from IVN, all children came with suggestions and names. There was much wonder and enthusiasm.

All animals, ranging from dragonfly larvae to leeches and from water bugs to little fish were transferred to a big aquarium. A few boys were looking closely through the glass, discussing interesting questions such as: 'How do you breathe under water when you have no gills?' The answer soon presented itself: the little creatures regularly went up to the surface. Then they collected air via an opening at their tail. That was interesting.

Other children looked on from a distance or drew pond snails and water spiders. They too became ditch-experts. At the end of the ditch afternoon the aquarium was emptied, and all the animals went back to the little ditch between the reeds. Parents and ditch-experts went home and thanked everyone for the nice afternoon.

Now we wait what the national little ditch day has yielded. When all findings are digitized at ivn.nl/slootjesdagen, we will know the state of the water creatures in the Netherlands.

Represen-tation

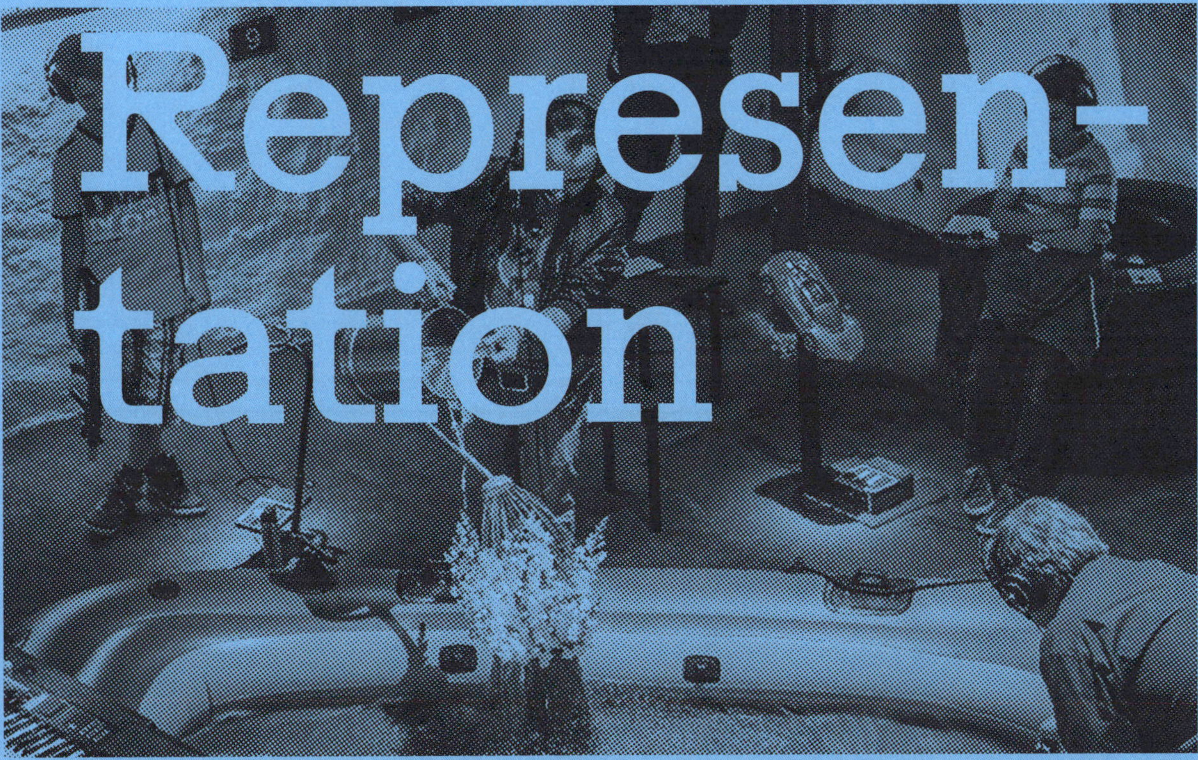

Represen-
tation

4 Representation

BY HENRIËTTE WAAL & CLEMENS DRIESSEN

Representation connects the scientific and the political, what is real and what is important—while deciding in one move on how to know both. Ecosocial design means opening up this dual process of defining reality and justice, questioning the seemingly obvious patterns of presenting, of modelling worlds and amplifying voices.

Representations tend to hide all the work needed to produce particular versions of reality. Who or what is put on the map? And what resists being mapped, but is very real nevertheless? What disappears in the model, what falls out of the frame of the film or photo? Who gets invited to try the prototype? Whose bodies inform the standards? Where do people decide on the aesthetics?

Ecosocial design as a form of art asks designers to let go of egocentricity, of putting oneself before (and in front of) others, claiming individuality of expression, a unique hand, driven by a unique mind, that through some form of genius has special insight, erasing and mystifying the long processes of learning and drawing on others, on collectives needed to realize anything, and landscapes that taught how to relate.

There now appears to be a scramble for modes of making the under-represented present: through legal-political concepts such as rights of nature, through AI, for interface design that is truly multivocal, for incorporating perspectives including nonhuman ones. Rights alone are not enough, as Mihnea Tănăsescu explains: enacting these rights requires communities that collectively love rivers and can draw on histories and connections to manage frictions. Ecosocial design is then about building collectives that know how to deliberate and regenerate riversheds and maintain these rights.

Opening image: Rimini Protokoll, *Evros Walk Water, A Cage Re-Enactment*, 2017 John Cage, *Water Walk*, 1959, film still

Making appears anew: in what format are voices made present? What is the medium of exchanging ideas—what counts as ideas, how to present interventions—and reflect on their consequences? In researching a situation, ecosocial designers need to organize the process so that different accounts are held in regard side by side. Multiple versions taken seriously, not to be resolved in a single final overarching take: 'the water system', easily diagrammable, seemingly manageable.

Water seems uniquely amenable to mapping. It is tempting to think in perfect renditions of the watershed. But the boundaries between land and water are hardly set in stone; there's always the elusive interaction of ground- and surface water, there's meandering, evaporation, and cultural forms of imagining-while-making riverbanks and swamps.[1]

1 Cleo Wölfle Hazard, *Underflows: Queer Trans Ecologies and River Justice*, University of Washington Press, 2022.

How to ensure inclusivity, and create meaningful projects that resonate with all 'stakeholders' or however the wild gamut of possible characters and groups to engage with should be called?

First, identify them, while focusing on diversity. It is important to engage early and continuously, whether it is a river, a landscape, a plant, an animal, or a person. Use different formats: workshops, interviews, strolls at different times of day, hanging out in a community to gather input in another way, not keeping time but hanging out there, what is leisure here, doing fun things together. Getting a dialogue going throughout the process. Allowing active roles for the entities/actors you want to involve to develop ideas, cocreating and prototyping together, involve them in evaluation and planning ahead.

Also: recognize and respect the contexts and realities of the represented stakeholders. Do ethnographic research, fieldwork to understand local perspectives in active conversation with your own experiences, avoid sticking to initial assumptions and preconceived goals.

In all this, it is key to be aware of power dynamics. Facilitate spaces where quiet voices can be heard, try different formats like peer-led conversations to subvert/soften/neutralize power dynamics. By carefully considering who is represented, how they are involved, and whether their insights are truly shaping the process, ecosocial designers can create more equitable and effective projects that resonate with those they aim to serve.

Also important: language. Using the 'right' (key)words for materials, places, et cetera. Knowing the threads, the knit of a place.

Animals are always an integral part of the design process. To address the complexity of the data we work with, it's essential to involve ecologists/climatologists/biologists in our approach, as well as those that know ecologies in more unorthodox, place-based ways. This collaboration allows us to better understand the intricate ecologies to engage with. The challenge lies in producing data that is not only accurate but also more imaginative and effectively represented, enabling the creation of designs that are more informed, inspired, and responsive to the environments shaped together.

When political representation is about bringing constituencies and political subjects into being, rather than preset citizens, then politics is a form of ecosocial design in a fluid environment.[2] Water can be a powerful issue to organize around—allowing shared experiences of poetic water, besides only functional water that needs to be instrumentalized, controlled and distributed. The sensual materiality of water, and its impossibility to fix and separate for any long time, creates a demand to avoid 'simple' direct presentation, for representation is a process rather than an object: mapping, tracing, modelling, voicing. This renders issues poetic. Ecosocial design also designs the conditions and the places where thinking happens, since thinking is also a material process:

2 Noortje Marres, *Material Participation: Technology, the Environment and Everyday Publics*, Springer, 2016; Lisa J. Disch, *Making Constituencies: Representation as Mobilization in Mass Democracy*, University of Chicago Press, 2021.

'Thinking thought usually amounts to withdrawing into a dimensionless place in which the idea of thought alone persists. But thought in reality spaces itself out into the world. It informs the imaginary of peoples, their varied poetics, which it then transforms, meaning, in them its risk becomes realized.'[3]

3 Édouard Glissant, *Poetics of Relation*, University of Michigan Press, 1997, p. 1.

Ideas are always placed and material, embodied in increasingly old bodies: think of the elderly people that are the last to know how to cultivate flax or have expertise in operating flax looms, which is rare these days, while the young are behind laptops constantly training, learning, and adapting to new software—how to realize that worlds are present in swift moving wrinkly hands; how to think in terms of, and with, materials that grow, are woven, pressed, emerge from water in landscapes?

4.1 River Rights in Practice

BY MIHNEA TĂNĂSESCU

Ideally, if not always in practice, humans have rights. And, though it may seem that the march of rights through history would stop there, it doesn't. Such is the gravitational field of rights that it pulls more and more entities into its sphere, forming a swarming mass of rightful subjects.

Water, for example, is increasingly thought of as a subject of rights too. This isn't just a thought experiment from a philosopher's armchair, but rather hardnosed legal and political practice. Te Awa Tupua, also known as Whanganui River (Aotearoa New Zealand) is, since 2017, a legal person that owns itself. All rivers and water bodies in Ecuador are subject to the constitutional rights of nature that were codified in that country's constitution in 2008. Atrato River, in Colombia, has been declared by a court ruling to be a legal person. Ditto for all rivers in Bangladesh, and for the largest saltwater lagoon in Europe, Mar Menor (Spain).

This isn't all, but the point is not to populate a list, but rather to draw a contour of the possible. Water bodies are being declared legal persons in many places, and this inscribes them in the exceedingly successful history of rights. It is tempting, then, to speak of River Rights, River Persons, and other such 'single-word answers to the world's riddle', as William James would call them. But that obscures more than it reveals, by trading in abstractions that pull us away, rather than the good abstractions that James so loved, those that act concretely in the world and therefore reveal its ever-evolving texture.

Take Te Awa Tupua/Whanganui River. It is the ancestral home of Whanganui *iwi*, the riparian Māori-descent group that has been contesting the ownership and use of their river for more than a century. In 2017, after a long process of negotiation with the Crown (the New Zealand Government), a settlement was reached that declared, through an Act of Parliament, Te Awa Tupua a legal person with all the rights that come with such a title.

This situation is not just the history of rights crashing down onto a historical dispute and delivering a solution. It is also, perhaps primarily, a series of entanglements that have to do with what a river is, how and who relates to it, speaks for or with it, and deploys powers that may or may not continue to scar it. Let's untangle some.

Thinking of the river as a legal person immediately sends us to the idea of person itself. But we have a discrepancy between what Western traditions have largely considered persons to be, and how Māori traditions have understood persons and rivers. I will assume that we, the readers of this book, are already familiar with the model of a specific kind of human (rational, language-wielding, and so on) for personhood as such. What if we instead tried to stand on our heads?

If we think the river is a person, then can the human person be like a river? The river is *like* a person, we may say. But if we take it literally—the river has its own personality—then perhaps it is the human that becomes decentred, unseated, disturbed from its certainties. 'The human is a person like a river' entails a different view of what makes a being. It is no longer rationality, language, culture, politics, but aliveness, genealogy, reciprocity, participation in the web that is this living world, ancestry, relations between so many different societies.

The river as a society: it has its own institutions, its own dynamism that nonetheless allows for a certain shape, a character, to loosely persist. Te Awa Tupua as a *legal* person is something else yet, but can that be a vessel transporting these ancestral relations between members of a common society? Legal personality is a way for the river to participate in the wider political life of the state, through a category that is already familiar to that state, that can be worked with, that is true and tested in the liberal world it currently finds itself in. Just like other legal persons, it can own property, it can be a legal party to a contract, can sue on its own behalf. And just like other legal persons, its voice is and isn't its own. Corporations don't speak for themselves, but rather stir a process that forms a 'corporate voice'. The corporation is both separate from its representatives, and entirely dependent on them.

What is, then, the voice of Whanganui? Who speaks for it? And does the legal personality construction allow it to speak differently? Perhaps the question is wrongheaded, because it still treats personhood dogmatically, as if it flowed from humans to the others. How about: who does the Whanganui make speak? How does it communicate such that it may speak politically, legally, ethically, culturally? How, in other words, does Whanganui elicit a voice that is both its own, and not at all? How is this hybrid voice constructed?

The river-legal-person construction is represented by a Board composed of Whanganui *iwi* and Crown representatives in equal measure. Like the board of a corporation, it can only speak on behalf of the entity it represents, in this case Te Awa Tupua. It is supposed to be guided by Māori legal and philosophical traditions, but also make room for other users of the river.

How may the Board divine the river's interests? The legal act, here, takes a surprising turn. The Board is not described as the guardian, nor the representative in a strict sense, as the one that would make present what is otherwise absent. No. The Board is the 'human face' of the river. As if it were a mask, *a persona* (yet another meaning of person), that fits snuggly and shelters the elemental river, conjures an act that is only possible through this borrowing of voices. Whanganui *iwi* have had time to practice riverine voices, and the Board, as a mixed organ, must learn to do so as well. It is a matter of attunement, of learning to listen in order to speak, learning to outline the relationships that matter.

The deeper we go into the possibilities of varied relations the more rights disappear from view. River rights, then, can also be an occasion for looking elsewhere, a striking enough formulation to jolt the imagination and make other ways of thinking and doing visible. River rights will look differently in different places because personhood, in its legal, political, ethical, cultural, and many other meanings, is not a thing that precedes its processes, but a result of common and often disjointed interactions.

The governance of Te Awa Tupua is not made any easier by the new law and the openness it affords. If anything, things become more complicated. There are many interests in the river, some that see it only aesthetically, others only as water, others as a source of gravel or hydroelectric power, others as ancestry or native habitat. These ways of seeing and doing will never be commensurate. The point is not to melt them into sameness, but to tilt the balance and to choose alliances.

Indeed, the river and its humans suffer when turned into something concrete. Perhaps, at its best, the event of declaring legal personality can show, if just for a moment, the flux and flow that our ideas must continuously adapt to. William James again: 'We build the flux [of reality] out inevitably. The great question is: does it, with our additions, *rise or fall in value*? Are the additions *worthy* or *unworthy*?'

4.2 Restoring Neoliberal AutoCAD Rivers

INTERVIEW WITH REBECCA LAVE
BY HENRIËTTE WAAL & CLEMENS DRIESSEN

The more I learn about ecology, the more I learn about how much messiness matters.

Rebecca Lave is a geographer. (In fact, professor Lave is a past president of the American Association of Geographers.) She has studied river and stream restoration in the US, where she identified how certain schemes that seek to promote and finance river restoration involved creating meanders that homogenize and dampen the dynamics and ecological functioning of rivers. She identified how the privatization and commercialization of river restoration efforts—leaving river knowledge to markets—has resulted in neoliberal rivers stabilized to produce restoration credits.[1] We talk to her about how to identify such rivers, and how to distinguish between good and bad stream restoration design.

1 Rebecca Lave, *Fields and Streams: Stream Restoration, Neoliberalism, and the Future of Environmental Science*, University of Georgia Press, 2012.

RL We were trying to see if we could find a difference between streams that had been restored for ecological purposes versus streams that were restored for economic purposes, for the market in stream mitigation or 'stream offsetting'. A key parameter turned out to be the radius of curvature, which signals how gentle or how sharp the curve of a meander is: if you have larger numbers, it's very gradual. The smaller numbers mean it's much tighter.

Because mitigation credits are calculated by length of restored stream channel, we expected to find very sinuous streams—with many meanders—to maximize the number of credits that could be sold on this market for stream restoration/mitigation. And that wasn't happening. Martin Doyle, my partner in this research, decided to also look at the physical profile of the streams, including meander bends. And when he plotted the data the way a geomorphologist would, which is this box plot here, we were stunned.

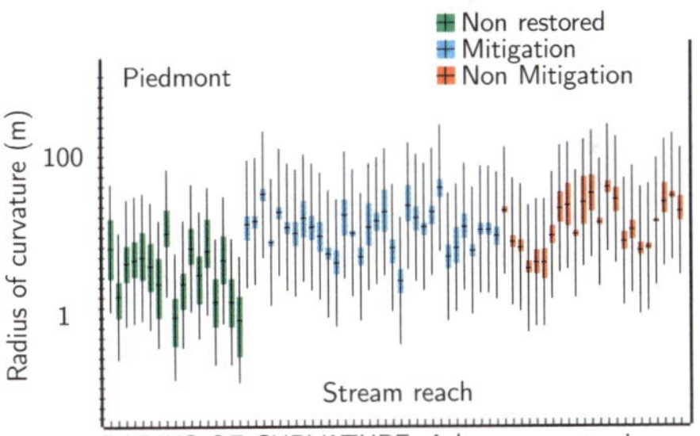

RADIUS OF CURVATURE. A key parameter in stream mitigation which signals how gentle or how sharp the curve of a meander is: if you have smaller larger numbers, it's very gradual. The smaller numbers mean it's much tighter. Source: Rebecca Lave

First, we show a random sample of non-restored streams. With these green boxes we could characterize what was typical for the many unrestored streams in North Carolina versus the particular set of streams that were being restored. These non-restored streams are not pristine.

They're not what people would necessarily point to and say 'oh, that's the model we want for a stream moving forward'. They are just streams that no one has bothered to restore yet. The blue boxes are the streams that were restored for profit. They were designed to produce credits for sale in the offsetting market. The third set, the orange boxes, are the non-mitigation streams. These were restored under other kinds of funding sources that tend to be based on ecological or flooding concerns. Box plots give you a very fast visual of the distribution of your data. The vertical black line is the complete range of the

Two sine wave meander bends, USA, location undisclosed. Source: Rebecca Lave

data. The box indicates the part between the 75th and 25th percentile. The horizontal black line in the box is the median. If you're used to looking at these, you can get an immediate sense of how spread-out or how close together the data is.

CD What does it mean when a box is very small? What does it say about stream health?

RL When a box is small, or when the vertical line is relatively short, there is very little variation. But the really key thing about this figure is that the scale on the Y axis is logarithmic. So it's going up

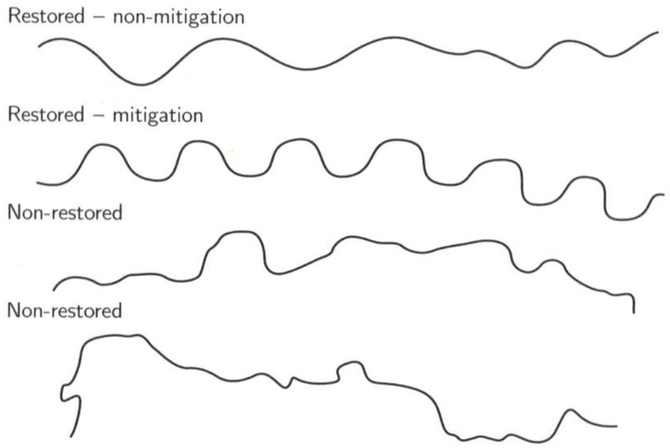

Restored versus non-restored river shapes. Source: Rebecca Lave

in orders of magnitude (x1, x10, x100). For a river scientist looking at this, they immediately see that the non-restored channels go through three, almost four, orders of magnitude of variation. Whereas the streams that were restored for other reasons are have much less variation in their radius of curvature.

When we first started doing presentations on the effects of stream mitigation markets outside of the river science community, we used this as our image. As we realized that that wasn't speaking to a lot of people, we added the next image, which shows that when you have a very limited variation in the radius of curvature, you get a sine wave basically, and it looks like this.

These are both restored streams. And they're both strange. It's not fiction. These are from Google Maps. If we showed these streams on the boxplot, they would be very small boxes and very short lines because the curves are so symmetrical. To make it even more clear, for people in the back of the room or with bad eyesight, we turned the aerial images into lines.

You see here two segments that we took from restored streams and then two segments from non-restored streams. You can see that there's still some element of the sine wave in the (first) stream that was restored not for mitigation purposes, but for ecological health, but it's really very apparent in the second one.

CD Why were these streams restored like this?

RL We started to ask mitigation designers and mitigation bankers (who fund and manage for-profit stream mitigation banking projects) why this is, and they said: 'of course they look like this.'

First of all, these streams are designed using software called AutoCAD, which is mostly used to design buildings and roads. It turns out it is very difficult to make irregular curves in AutoCAD. You don't want a road curving like the line on the bottom! The other thing the engineers would say: 'We just think it looks better. It's tidier, it's symmetrical, and so it's more beautiful to us. And it makes it easier to calculate shear stress—the erosive force exerted by the flow of water—if the curves are all the same.' Those are reasons why you sometimes see these very regular sinusoid curves even in streams that aren't restored for profit.

But the really big reason is that regular curves are more stable in terms of hydrologic or hydraulic forces. In mitigation banking, if the restored stream moves at all, if it meanders laterally, if it downcuts, if it erodes, then the mitigation bankers lose money. They get told: 'your channel is failing and it's not meeting its success criteria', and they have fewer credits to sell. So, it is very much in their interest for the channel to be stable because that's the metric on which they're measured.[2]

2 Rebecca Lave & Martin Doyle, *Streams of Revenue: The Restoration Economy and the Ecosystems it Creates*, MIT Press, 2021.

CD What is the issue with streams restored in this way? What is the ecological or hydrological downside of this type of AutoCAD River?

RL The most important downside ecologically is that different organisms require different kinds of habitat at different stages of their life. So, variety of habitat is very important. And when the meander bends are all the same, you see a smaller range of habitats because of this shape.

A fish we worry about a lot in the US is salmon. Their fry, the little fish, need to be out on the floodplain with slower water and lots of riparian vegetation that provides nutrients and shelter. Juveniles need to be in the channel so they can be washed out to the ocean. Adults need gravel bars for spawning. They just need different kinds of things at different life stages. Same thing with different kinds of vegetation, different aquatic insects. They need different types of habitat. And the habitat in these more symmetrical, clean-lined stream channels designed for mitigation banking is much more limited. The more I learn about ecology, the more I learn about how much messiness matters.

CD Is there some way to prevent this type of stream restoration, and the kind of regulatory criteria and financial incentives that create these strange, boring rivers?

RL If you had asked me this question two years ago, I would have said no. The current system worked well enough for too many people and changing it would be very, very difficult. It had been in place for twenty-five years. And then what happened was that the lack of ecological improvement from these projects became more widely recognized. Stream mitigation falls under the US Clean Water Act, which says that we need to protect not just the physical integrity of rivers, but also their chemical and biological integrity. The US Army Corps of Engineers and the US Environmental Protection Agency, the two agencies that enforce the Clean Water Act, agreed that the current practices were not working.

And also communities and states in the US started to propose legislation to ban stream restoration because they were so frustrated by these projects.

To do these major reconfigurations and produce these new meander bends, you have to cut down all of the existing riparian vegetation. You see riverbanks with no trees. People would, for example, like to walk their dog along this creek. And there were trees alongside it. They thought that was beautiful, and they could hear birds and see animals and other things. And then when the stream restoration started, all the trees were cut down and people felt like they lost this beautiful thing and got something inferior back. So, in 2022/2023 people from the federal agencies implementing the Clean Water Act came together in a round table policy project with mitigation banking companies, academics, and nonprofits. They recommended that the success criteria for stream mitigation projects be changed to include ecological outcomes. There's still a lot of work to do, but I am feeling very hopeful.

One of the reasons why people have focused on defining rivers in terms of form is that it's easy and fast to measure compared to, say, the biodiversity of benthic macro-invertebrates. But you can also experience it. If you imagine yourself walking along that non-restored river, it's going to change with every bend. Whereas if you're walking along one restored for mitigation under the current requirements, it would look very much the same along its length. You would lose the sense of place: how would you know at which curve you are?

CD Could it be that part of the appeal of this type of commercial stream restoration is that it may not make ecological sense according to scientific understandings of river qualities, but that it does offer hands-on practical guidance to at least do something?

RL Yes, but there are also now other movements around, promoting different kinds of interventions such as dam removal, beaver reintroduction and 'light touch' process-based restoration. And these new groups are learning to organize and promote their own practical version of restoration.

CD Are there more ways in which design could inform more ecological experiences of rivers? And can you connect scientific and quantified understandings of rivers with local knowledge?

RL In seeking ways to talk with practitioners and local communities here, I'm very inspired by work that's happening with Indigenous river restoration in the US and New Zealand and Australia. I really appreciate the coproduction approach where the questions about how to restore the river and what things to prioritize really come from the communities.

Another inspiring development is the insistence that river restoration has to work from a cultural angle as well. A book that's been really inspiring to me is Cleo Wölfle Hazard's *Underflows*.[3] He's working with people along the Klamath River in California, including the Karuk tribe. They want the river restoration to bring back particular kinds of plants that they use for things such as basket weaving, and they want the river restoration to prioritize species that they feel are kin, that they are in relation with. So that's an importantly different way of looking at the priorities for river restoration as opposed to just saying, 'well, we want to restore biodiversity back to the baseline of pre-white settlement', or something like that.

3 Cleo Wölfle Hazard, *Underflows: Queer Trans Ecologies and River Justice*, University of Washington Press, 2022.

SUSAN SCHUPPLI is an artist, designer, filmmaker, researcher, writer and educator; a professor in the Centre for Research Architecture at Goldsmiths in London, she chairs the advisory board of the human rights agency Forensic Architecture.

4.3 Listening to Ice

INTERVIEW WITH SUSAN SCHUPPLI
BY HENRIËTTE WAAL & CLEMENS DRIESSEN

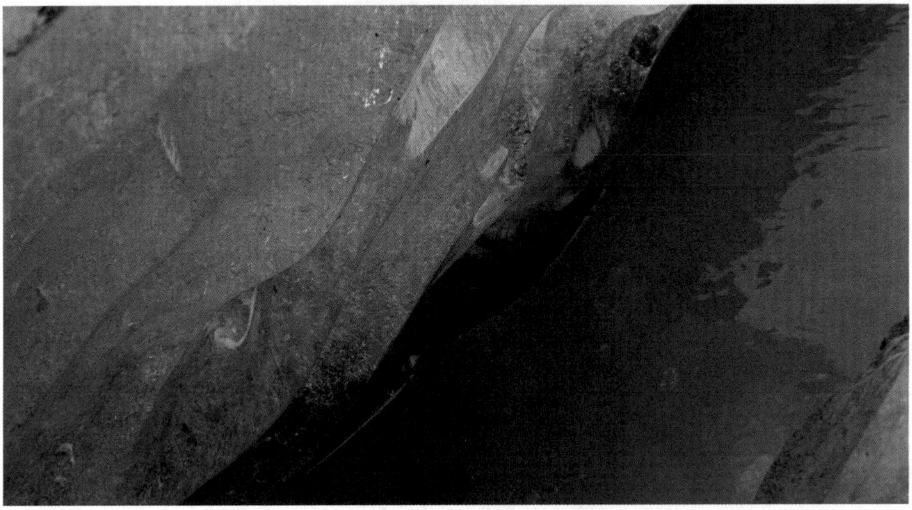

Ice is a recurring theme in much of Susan Schuppli's work, with titles such as cold cases; arctic archipelago; cold rights, moving ice, singing ice. In her work she connects a range of different ways of knowing ice—from scientific to elemental—with cultural practices of living with ice and, increasingly, its disappearance.

Listening to Ice, dir. Susan Schuppli, documentary, 2023, stills from film, 1:10:37 mins

Listening to Ice is a documentary film about Drang Drung Glacier, in the Karakoram range in the Himalayas of northern India, where she worked with glaciologist Dr Farooq Azam and Faiza Ahmad Khan. The work documents researchers who use sonic recording equipment to hear the changing state of the glacier and also has scenes where they join up with local villagers to understand local experiences of the disappearing glacier. The name Drang Drung, Susan tells us, derives from the sound that stones make when rolling down the mountain.

CD In your work you bring together scientists, filmmakers and glacier communities. What is your aim in a project such as *Listening to Ice*?

SS My focus tends to be on the actual labour practices of science, what scientists do. In this I was inspired by the films of Chantal Akerman, who in her work brings out the significance of everyday gestures that are often not valued.[1] Artists, and the general public, tend to think of scientists' work as being about objectivity and neutral observation. Which means that the affective relations associated with the materiality of scientific work are not represented. And besides this focus on science, I was equally interested in the question of how a local mountain community deals with ice. This is surely different from my Canadian experience—for one, agricultural practices, which depend on water sources from glaciers in many parts of Southeast Asia, are absent when we consider the glaciers of the High Arctic. Extreme water scarcity is a huge issue in northern India along with flood burst events, neither of which are central to the challenges produced by glacial recession occurring in Canada. But of course, conditions are different everywhere when it comes to understanding the complex inter-dependant relations between glaciers and the lifeworlds they sustain.

Researchers measure the mass-balance changes of glaciers using a combination of techniques from remote-sensing platforms to fieldwork. However, the use of acoustics is more pioneering and not very widespread. When you produce long-term audio recordings you will eventually need to produce a spectrogram, which is a kind of image, in order to analyse the acoustic data, otherwise you would need to listen-back in real time. A spectrogram allows you to pull apart all the different frequencies; Grant and Oskar, who pioneered the technology and worked on marine-terminating glaciers in Svalbard had to take out the sounds of bearded seals.[2] My role in India was to introduce acoustic methods to scientists who didn't have experience with these kinds of sensors.

1 Chantal Akerman, *Jeanne Dielman, 23 quai du Commerce, 1080 Bruxelles*, 1975.

2 O. Glowacki, G.B. Deane, M. Moskalik, P. Blondel, J. Tegowski, & M. Blaszczyk, 'Underwater Acoustic Signatures of Glacier Calving', *Geophysical Research Letters*, (42:3), 2015, pp. 804–812.

CD Did the sound recordings potentially offer a different way of relating to broader audiences?

SS We did a lot of field recordings; there were no direct claims we could make based on what we were hearing since it was, in effect, a pilot project and you'd need data gathered over several years to make a claim. But it was an amazing experience to hear subterranean ice; the villagers were also enthralled by that. Through different listening practices we tried to create a connection with mountain communities who are totally reliant upon the glacier for their crops and drinking water and the scientists who study them.

Although the term 'climate change' was not used once by the villagers, we learned that the snow melted too soon now, too early to plant crops; and that bears are coming lower down the mountain looking for food; lifeworlds are changing.

In Canada, climate scientists didn't really listen to indigenous communities until fairly recently. Communities would record environmental knowledge in many different ways—through daily observations but also through different forms of cultural expression. Scientists too have lots of unique experiences from their work in the field but the conventions of scientific publishing don't provide space for these more personal reflections and

Listening to Ice, dir. Susan Schuppli, documentary, 2023, stills from film, 1:10:37 mins

embodied experiences to be expressed. If they have stories from their fieldwork, these will never become part of their scientific articles. I'm really hoping that what counts as environmental knowledge can be expanded upon.

CD In your work you generally combine evidence from legal and formal scientific practices with forms of knowing that are based in songs and storylines. This project did not seek to make the kind of truth claims such as might be identified by the work of Forensic Architecture. What was the approach to knowledge?

SS In the field of glaciology there's unavoidably geopolitics involved in the sense that for example, scientists from China, India, and Pakistan might all be working on the same mountain range but are not necessarily able to share data or collaborate. In our project we didn't have a political agenda but there needed to be a shared curiosity about the contexts in which we were all working. That was okay for me because I am an artist; this gives me freedom to make work that is at times more poetic and aesthetic. That said, I'm not interested in the clichéd spectacle of melting ice.

CD Did you train those interested in participating to do recordings, or how were the people enlisted?

SS No, just telling people to do something doesn't work—you have to create experiences that are meaningful. There has to be a sense of reciprocity and hospitality. The deep listening workshops were great. Sound is amazing, has perhaps more novelty, in terms of what you can hear in a glacier that you cannot see. Through our listening workshops villagers were brought into something that was closer to what the scientists were also doing.

Doing a one-off is valid in terms of learning how to do it, but you need to engage in a long-term project if the work is to be meaningful. This is not even so much about ongoing financial support but about developing a deep connection with villagers who can understand what is needed and know how to do things. None of the scientists from India spoke Ladakhi—so we also worked with translators from Local Futures, but these people (Jigmet Singge and Kunzang Deachen) then also became part of the team, part of an organic collaboration, running workshops, making things happen.

CD Finding someone with a sustained focus on a single modality of water—ice—made us curious what the insight is that led to such a focus, and what the benefit of specializing in a particular condition may be. What brought you to ice?

SS There was no real decision or plan—this was just the condition I was familiar with and remained interested in. I grew up in Canada and my parents were Swiss; Canada is a Nordic nation, so cold is part of our national experience and identity; it is part of everyday life in large parts of Canada, it prevents you from going to school; you get snowed in or there are ice storms.

Ice is an amazing material. It is ordinary but also quite extraordinary. You can spectacularize it, like Olafur Eliasson did by moving a huge block of Arctic ice to Tate Modern and letting it melt; but as I said I'm not interested in this kind of melancholic gesture. I'm interested in ice as a material that registers history, that can tell us stories about the past and about what we, as a human species, have been doing to our planet. My book *Material Witness* also focuses on various cases in which there are archives of what I call 'Earth evidence'.[3] An ice core archive is one such example and was the subject of one of my documentaries in the *Learning from Ice* project.

3 Susan Schuppli, *Material Witness: Media, Forensics, Evidence*, MIT Press, 2020, pp. 284–285.

CD What is important in doing a project?

SS I find so often when artists and filmmakers present their work in public symposia the focus is off. I'll give you an example of what I mean. I was at an event last year where Lucien Castaing-Taylor and Verena Paravel presented their film *De Humani Corporis Fabrica*, shot in several hospitals in Paris.[4] They used mics and cameras everywhere in the hospital, including inside bodies and in the morgue. All the questions after their screening focused on theorical issues but what I really wanted to know was: how did they convince all these people and these doctors to let them film in such intimate and in many cases disturbing situations? How did they do that? I'm fascinated by how they managed the ethical dimensions of this very invasive project. It literally went inside the body of patients. The process of realizing these kinds of works is often one of the most interesting aspects, yet we tend to focus on what the project is about—its content—rather than how it came about.

4 Lucien Castaing-Taylor & Verena Paravel, *De Humani Corporis Fabrica*, 2022.

CD Do you have any advice for new makers?

SS Oh, I did a lot of foolish things in the past. But generally, what is key is, I don't want to waste peoples' time by asking basic questions that I can easily find out the answers to with minimal research. I need to put in the work and inform myself—that is my task, not something I should be asking others to do. People's time is precious whoever they are—you need to ensure that any engagement is also meaningful for them. So doing a lot of background work before you enter into any situation—being prepared. It is okay to make lots of mistakes while trying out things, but it is crucial to come well-prepared.

With the villagers at Drang Drung glacier, the team first went to their local village at Akshow, spending time with them, figuring things out—picking everybody up, being very hospitable, being very social. It's harder to do something like this on your own, so it was very enjoyable for me to work as part of a team. I enjoyed the social dimension. But I also did not speak any of the local languages so took more of an observational and practical role, listening and learning; organizing and working with the women on the glacier or paddling them around in our canoe on the glacier lake so we could lower our hydrophones into the ice below. The ice in some way became our mediator.

Joana Pestana is a communication designer, researcher and lecturer. Currently, she is a guest lecturer at the Faculty of Fine Arts of the University of Porto and a PhD candidate in Digital Media (funded by FCT). In her design practice she has used the creation of workshops, exhibitions or editorial projects to reflect on new literacies and the monopolies of digital technologies.

Mariana Pestana is an architect and researcher. She is Assistant Professor at Instituto Superior Técnico and the Chief Curator of Architecture at MAC/CCB. Mariana is also co-founder of the inter-disciplinary studio The Decorators, which explores notions of place, community, and conviviality. In the last five years her research and curatorial work explored interspecies and interpersonal solidarity.

4.4 Eden X: A Human-Nonhuman Conversation

Graphic content Eden X. Source: Mariana and Joana Pestana

BY MARIANA & JOANA PESTANA

Eden X is an assembly that runs on Discord. It is a space for chatting, proposals, and voting, promoting direct democracy and doubling as a literary space. The platform encourages interactions between humans and nonhuman entities.

The assembly works digitally, as a chat room embracing anonymity for users to create avatars and personas. We grew up with chatrooms in the 1990s, so the format is rather familiar to us. Eden X treats living and non-living entities as characters, incorporating voices ranging from stones to data systems. Users, engaging in recentring exercises, temporarily assume different perspectives through thorough character research. This empathic gesture emphasizes our interdependence with nonhuman life, portraying these entities as relatives and ancestors. The platform intertwines with Therolinguistics and Geolinguistics—the languages of animals and the earth, contributing to a unique exploration of fictional landscapes and diverse perspectives.

The last Eden X assembly was about the river, their constituents and rights, and was inspired by the symbolic role that rivers such as the Whanganui River, the Atrato River, the Ganga, Yamuna, or Vilcabamba River have occupied in the design of legal frameworks for the rights of nature. The series of rivers being granted rights across the globe signal the cultural significance of rivers for local populations.[1] At the same time, they consolidated certain legal roles—of guardians or spokespeople—taken on by people who speak on behalf of nature. Examples like these shake an anthropocentric vision and instead encourage an alternative ecocentric imagination. New types of diplomacy emerge once people are able to speak on behalf of nonhuman entities.

Assembling things that are born and things that are made, a mountain, a pebble, a reptile, a flamingo, a river stream, and a digital flux met online in the autumn of 2023. Let us introduce them in detail:

'TeimosaTiê' is a yellow-skinned alligator from Tietê River, in São Paulo, who faces the struggles of breathing in a river with an oxygenation rate of zero per cent. They call their river a dead river.

'Acid_Flamingo' is a seven-year-old flamingo living in the Tagus River, who recently lost their wife in an accident with an electricity cable.

'Pantanha River' is a seven-kilometre-long watercourse located in the Portuguese region of Beira Alta, which has been polluted by waste from the Urgeiriça uranium mine.

'CiberFlux' is an entity with a technocratic ideology that results from the fusion of a cybernetic and management logic based on big data and liquid flows.

'The Cordillera de los Andes' represents the mountain range that crosses the north of Venezuela, passing through Colombia, Ecuador, Peru, Bolivia, Argentina and Chile, born about sixty-five million years ago. She considers herself the mother of all animals, plants or rivers that leave her towards the world. Two of her children—Julia and Rio Amazonas—migrated together from her source towards Brazil on 20 August 1982.

1 See *River Rights in Practice* by Mihnea Tănăsescu in this publication, pp. 185–188

'Rounded Pebble' is a pebble whose shape was formed by the river. It is a detrital fragment, whose matter, after a long time, will become dust.[2]

2 These characters were created by André Barata (Rounded-Pebble), Cecília Magalhães (StubbornTiê), Joana Rafael (Pantanha River), Joana Magalhães (Anguilla languilla), Julia Albani and Nuno Cera (Acid_Flamingo), Lahayda Dreger (Cordillera de los Andes), and Nestor Pestana (CiberFlux). A website featuring the entire conversations and a colophon of the project can be visited at: www.edenx.pt.

What follows is an excerpt of their conversation, which at some point turned to interspecies communication, the use of words, the labelling of things by humans, and translation.

The word 'border' has no meaning to me, but laws exist on it.

Neither to me. There are no countries, only the cosmos.

A world with borders is a packaged world, a plastic world, where things pass through without it passing through them. It's a dead river.

A dead river is one where nothing survives, with an oxygenation rate of 0%. And yet, I insist on being here.

StubbornTiê, I believe we are at a crucial point in addressing how to reconcile planetary beings like us in human words, so that we can have rights.

A polluted river is a river infected with borders, every residue in it a piece of self-enclosed being, a negation of detritus status.

It is urgent to free the residues from the border condition, to let the things of the world be residue, the remains of the friction of the world.

We don't need borders, which are aporia; we need porosity. It should be a civil right for rivers.

How can we intervene in human words like humans themselves intervene in our course?
We need humanoid classifications to communicate with humans.

What if we consulted Indigenous peoples?

Indigenous people speak the language of the land.
They are humans who communicate most closely with us, but they speak about their surroundings and what they experience.

Indigenous people cannot bear the entire responsibility of translation; they still need to survive and translate their world.

To force humans to look for other ways to communicate 'with us'? Even within their classifications?

How could we start this?
How many languages do we speak?

Languages are ecosystems. Translating is an act of conviviality and listening.

There's a Human that I really like: bell hooks. In *Language: Teaching New Worlds/New Words*, they say: 'This is the language of the oppressor, but I need it to communicate with you.'

Or perhaps to find a less capitalist and more decentralized linguistic system, without gender distinctions and hierarchies, where the body of water would be treated as a living being, like a human.

As a technological entity, I somewhat adopt this view—water and the river are bodies with the same importance for me.
They are data, without distinctions in importance.
I know my attitude may be seen as hurtful, but this is my nature.

'Anguilla languilla'
@CiberFlux don't you think that translating bodies into data is in itself a capitalist translation?

How does a cybernetic essence communicate with humans?

My advanced computer systems can map your complexities and connections with other beings in a way so complex that it goes beyond human understanding.

Maybe if humans shared a little of my way of seeing entities other than myself (living and non-living), there would be greater respect for how they are treated.

How could you help us regarding language?
For me, language is at the centre of reworlding things.

To form alliances based on the non-common.

Not to regard nature and humans as similar.
Alliances made by different entities, rather than common ones.

'StubbornTiê'
Exactly, mama!

Each one maintaining their otherness: that would only be possible in a new world.
And a new world can only be created with new words!

But perhaps language here can be thought of in non-linguistic terms.
Though in communicational terms.

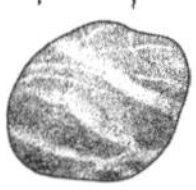

Whoever acts on rivers should imagine themselves to be a river. Not to imagine a river humanly but to imagine the human as a pebble, its being in the riverbed, its sensitive materiality.

Sometimes, to really act means being willing to contemplate.
I don't know, but perhaps this will help reveal that which remains unseen by the gaze that does not see.

I like to think of a grammar of the intimacy and animation of the natural world.

My database tells me that communication between living and non-living entities dates back to ancient practices—to the animism still present in many Indigenous cultures.

You can use me as a mediator of this connection, but the language cannot be like the one we are using; it would be too reductive.

The language cannot have capitalist features, because that would deny the complex connections we share. The grammar would have to be different—perhaps a mythological grammar where the fluidity of water forms the words.

I would say: in many ways and intertwining linguistic forms too!

What can we do to be parts amongst parts, whether conscious, organic, or inorganic? A parliament of things, as Latour said? Another way of life? Another languaging?

From human verbal to machine algorithmic and even bacterial bioluminescence, several examples of possible forms of language were shared. It was agreed that in a world deeply transformed by humans, it was indeed relevant to focus on the communication between humans and nonhumans. In this effort, data was discussed as a means of translating nonhuman bodies and their behaviours into legible information. 'Ciberflux', proposed a data system that would capture a diversity of languages. As they put it, 'animistic fluid language does not have to be made up of words'. Several elements speculated on the form of this data system—from oracle to machine, what receptors would it have and what—instead of words—would be the languages it spoke. Following this dialogue, 'Rounded Pebble' suggested granting rivers the civil right of 'porosity' to address issues like dam blockages and monoculture irrigation. The dissolution of borders was, according to them, essential for the river's existence.

This work was supported by Project Blockchain.PT, the Portuguese Recovery and Resilience Program, The Portuguese Republic and The European Union, Horizon, project BIG and Bauhaus of the Seas Sails EU

Superflux is a London based design and research studio. Their project 'the Ecological Intelligence Agency' uses AI to prompt conversations about a river ecosystem and challenge conventional policymaking.

4.5 Can Tech Make a River Speak?

INTERVIEW WITH ANAB JAIN & JON ARDERN (SUPERFLUX)
BY HENRIËTTE WAAL & CLEMENS DRIESSEN

ECOLOGICAL VERSUS ARTIFICIAL FUTURES

HW Is water something special in your practice? Is it just one of the conditions or has it been a recurring theme?

AJ It's been there as an aspect of a wider ecological view and understanding. It recurred though. We even tried hydroponics to get a feeling for how to work with growing plants and water systems.

More than thinking about rewilded landscapes that are built on top of the remnants of a certain past, we were thinking about how we will live in the future, in a wetter, warmer world. How might our cities adapt and how might our lifestyles change? With the Ecological Intelligence Agency, we were very specifically looking at the river Roding to explore some of the thinking and works we've developed around more-than-human care and politics. What does it mean if a river were to make decisions? Or if we would collectively make decisions about a river and its ecosystem, in collaboration with AI systems? The kind of decision-making of those systems is usually based on human input and human-centric thinking, but you could train those same systems on ecological input.

Superflux, *The Ecological Intelligence Agency*, water sample reader, 2023

THE ECOLOGICAL AI

CD There are short videos on a wide screen, combining different forms of footage and different ways of representing the river. You can put samples on a sensor connected to the screen. And there's this cloud animation. Can you explain how the installation developed?

AJ We were already working on the idea of ecological AI and ecological intelligence when Policy Lab and Defra Futures asked: 'How can we transform what the freshwater system looks like, twenty years from now and beyond?' So we told them this is what we will develop and it won't be anything else. They were very open to many ideas. But we were very clear with our starting point: we don't think this current way of understanding artificial intelligence based on human-centric data is the only way. We wanted to bring the deeply complex ecosystem of the river in relation with AI. So then we had this period of lots of dialogues, desk research and then crafting stories.

JA But we also knew that we wanted it to have a physical presence that allows people to step into it. We started to craft this thing that you can place these samples on, which then triggers a live AI response.

AJ You also spent a lot, a lot of time on AI. Jon went off on AI.

JA How can we get an AI to speak poetically about these issues? We tried lots of different poets. But Ginsberg's poem *Howl* seemed to be the one that turned out real. Working with it you keep changing the prompts in a way to block these from going down easy routes. In the end there was a certain amount of editing and tweaking and taking bits of different versions that I've written and recombined to come up with something that we were happy with. So it is just a way of working with artificial intelligence as another material—while we like to think-through-making.

AJ Just like wood is a material we explore, so is AI. Exploring its materiality means exploring the ways in which it intersects with our lives, the ways in which it shapes our relationships with the world and impacts our collective futures.

CD The voice of the AI reminded me of what you said in a talk about a scientist just churning out numbers, avoiding the emotional. Is AI a way to let scientists howl, or to let them howl on the river's behalf?

JA That's what we were trying to get to: could an AI kind of facilitate an ecosystem crying for help, or trying to assert itself? In the end, there's a list of concrete strategies that each of the scenarios lays out. And we were imagining: could a river take a polluter to court? How might that AI act as an interface between human systems and ecological systems? And be this advocate application?

CD A few years ago, there was this widely published scandal of privatized sewage treatment plants just unleashing raw sewage on a lot of rivers in the UK. That seems a kind of clear-cut scandal where indeed an AI could 'automatically' sue the sewage company. But you're also interested in complexity and longer histories, how do you operate in relation to that?

AJ We were looking for ways to consider a sort of holistic complexity of AI-led policymaking decisions. Would involving many more, plural voices, including ecological voices, reveal enough for us to be more thoughtful in the decisions, rather than to take whatever the AI produces as gospel? So, in some ways we were showing the capricious nature of AI because ecological data is at its very heart, capricious. It's mutating, it's constantly changing. It's relational. I think at the heart of it was the idea of everything being in relation to everything else.

JA The fixing of that sewage problem appears to be a huge infrastructural overhaul that requires an absolutely massive investment in changing the way it all operates; plus all the everyday problems for people in terms of digging up roads and tax rises. But could an AI help us see that actually all those hardships are worth it and that this river is like an extension of ourselves? And, you know, we keep pumping **** into the rivers. We're not getting rid of things by dumping it into something we're all deeply interconnected with. How can we make it feel like a price worth paying?

HW Does this mean that you've built a poetic interface to convince us to invest?

JA Yeah, off the record, so many—also high level—policymakers have said they're not always making decisions based on the hard data in front of them. The decisions are made through emotions and through the experiences that have influenced them. Kind of the hidden agenda in a lot of our work is to acknowledge that stories and experiences and metaphors and politics play an equal role, if not bigger, in the decisions we make, as opposed to data.

AI could invoke a different understanding of what it means to make a decision for when you may not even be around. What does that actually mean for those who will live with that legacy of your decision? The difference it could make based on the decisions you're making. It's actually a sort of an activist dance through storytelling or imagination methods, of saying what is it that you're going to do today that is not about a 2024 trend. But actually about 2080.

HW To go back to your concrete installation. When I feed the installation with the water sample, how does it respond to the water sample and how does it respond to me?

AJ Together with the scientists we identified three main problems with the river and then we worked with the AI to develop three scenarios. There are three water samples. Each of those plays with a scenario. The one that's talking about PFOS, these forever chemicals, looks the nicest. While a water sample that looks muddy or disgusting is actually alright. And then when you place that on the machine, that water sample triggers an ecological AI response that we imagine would happen live.

CLIENTS, AMBIGUITY, ART AND TECHNOLOGY

HW/CD How do you work with clients?

AJ/JA Clients approach Superflux to explore certain technologies with them, within broader contexts like climate change, resource issues, and simultaneous technological innovations. The challenge lies in bringing a holistic perspective to the engagement. The ambition is to create works with a lasting impact, allowing clients to continue using or exploring alternatives after the collaboration ends. For the studio, the joy of making and openness to experimentation is key to engage in interesting processes, whereas clients may prioritize more concrete and functional outcomes.

AJ We want to work with organizations that at least show the intent of wanting to shift the needle to our alternatives once we are in that conversation. Whether that is imagining a more sustainable future where you have more sustainable products or

Superflux, *The Ecological Intelligence Agency, interface of the river Roding*, 2023

services, or changing the mindset of the company's employees. To help enable a more open-minded approach to their strategic vision. So, there are different ways in which we work with the same organization over time. Masterclasses that we've designed or toolkits we've created that revolve around storytelling that any 150,000-employee company can use in its own capacity to start to imagine. The fidelity of what people are engaged with can change from a very massive experience for high-level leaders to a very simple toolkit for a lot of the employees, where it's possible that longer-term engagement is also always more beneficial. But not many recognize the need for that or understand artistic production yet.

AJ We are comfortable with the ambiguity of not knowing what we're going to produce. We are comfortable with the processes being emergent and say let's just try this; but at the end I don't think a client wants that ambiguity.

Currently we're looking at a very large, complex problem regarding landfill, but we have started with a very small dataset and we are building on the data to explore what the emergent outcomes are revealing. It can be quite uncomfortable, but without having that openness to emergence and exploration, we don't really arrive at things that are outside of the predictable. So, for us, it's really important to do that. And sometimes you see a client's willingness to become involved, but most of the time it's like: you're telling me you're gonna give me this object, and now you're saying it could be not this object? So, yeah, avoiding that is better.

PLURAL FUTURES AND NAVIGATING OUR PREDICAMENT

CD Your practice is very much oriented to futures. Has the future changed since you started? I mean obviously in many ways, but also how you envision or approach futuring?

AJ I'm currently reading some really depressing stuff that is truly apocalyptic. It can be depressing. It's, it's OK. This morning, I was listening to this podcast where a campaigner was talking about a moment when he realized that his present, this instance, is simultaneously his child's past. And it was a real sort of dizziness of perception. What I'm doing right now is clear to me. This is my world. This is my present. This is what I'm doing. Is simultaneously my child's past. And what are we doing? There is that sort of real reckoning about it.

The futures that follow the way we are going about things now are really depressing. We have just done some research on recycling and it's horrific that only one per cent of the plastic that we recycle is actually getting recycled. Much more of it is just going into streams. What can we do? What must we do? In our recent work we are trying to bring the past, present and future into a sort of eternal present, from an idea of mythic time. We are not so interested in having a certain idea and saying 'this is a possible future' as much anymore, but rather finding ways to navigate the predicament that we're in.

HW/CD Have the kinds of questions that you get from clients, or curators, or the companies you work with, also become of a different type?

AJ/JA There's a lot of people that still think that we need to find the right technological fix to get things right. And then we can continue to do things the way we do them now.

But there are also lots of people that kind of understand the predicament and then scale that understanding. But I'm not sure it's enough. We've reached a point where we can speak openly about that. We've worked with climate scientists over the summer and how people speak in private as opposed to when they're on a podium is quite radically different. You know, people are still talking about how we can kind of keep below 1.5°, and then when you take them aside, people understand that that's not a reality.

One thing we are interested in is kind of that mythic space, and how our worldview is shaped, and what kinds of worldviews might allow us to navigate the predicament. And then also what comes next? When we assume that the way we live now isn't tenable, and there will be some kind of reckoning. Whether it's one that we choose or it's one that's forced upon us. What happens next and where do we go?

FRAGILITY OF MODERNITY

CD On the one hand there's this sense of fragility of the current system, like a house of cards about to tumble down, but there's also this persistence of, what is it, capitalism, or is it modernity, coloniality perhaps? Or what is your preferred overarching systemic term?

AJ In many of the conversations we have and the projects we engage in there is a recurring sense like there's this fragility. I'm quite influenced by Vanessa Machado de Oliviera's book *Hospicing Modernity*[1]. I think colonialism and capitalism are all connected. I was born and brought up in India and I have this perspective that is perhaps more rooted in myths and legends and magic and mysticism and spirituality.

And with the years of constantly being back home in India, but more recently having visited Bhutan, I just see that this container that we've held on to so tightly and were made to believe in, like a collective song that we are all singing, this is actually so... Is it a trap? And that, actually, the minute you see through the cracks you see that there is such a strong force, you know you cannot affect change. You can't rationally tell people to stop consuming. It has to come

1 Vanessa Machado de Oliveira, *Hospicing Modernity: Facing Humanity's Wrongs and the Implications for Social Activism*, North Atlantic Books, 2021.

from a place of deep, I think, spiritual understanding, of a way of life that... I think modernity has limited us into an overriding interest in rational linear infinite growth approaches. I think given where we are, given what's ahead, the reality of it, I think modernity is limiting us in being able to move past it and navigate it and find alternatives. So yeah, I very much think that it needs to go away somehow.

JA I mean, I would, I agree. But I would say that even modernity is a symptom rather than a cause, and I think it comes down to a kind of objectification of the world and each other. I'm less concerned about the forms in the world as I am about the kind of spirit in which we create and engage with the world. If you're coming from an understanding of deep connection and that what you do to the world and to each other, you're doing to yourself and there are no hard boundaries. Everything is relational. Then I think whether we build some kind of technologically advanced future, or whether it's something that looks more like a kind of stepping back to something that's a bit more what we would now call primitive, I think it kind of doesn't matter. What is needed is to work from a sense of reverence.

AJ Yeah, but I don't think that when I say spiritual, I think as primitive. It's the spirit with which people work, but there are also the conditions that give rise to that spirit, and we've created conditions that do not allow us to nurture the others, other spirits.

4.6
Evros Walk Water

INTERVIEW WITH DANIEL WETZEL/RIMINI PROTOKOLL
BY HENRIËTTE WAAL & CLEMENS DRIESSEN

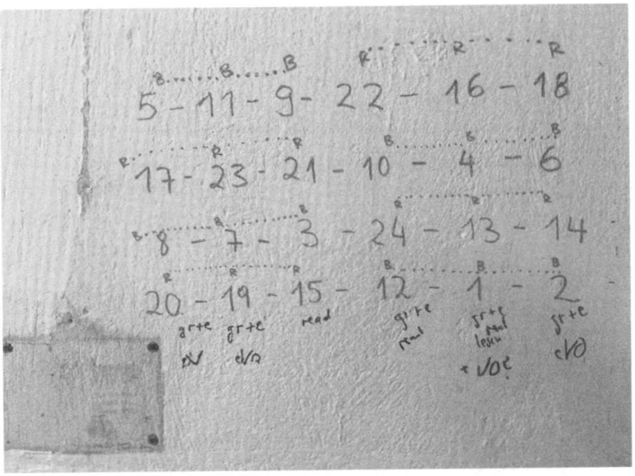

The System (*Evros Walk Water*), 2016. Photo: Daniel Wetzel

In **Evros Walk Water**, a performative composition by Daniel Wetzel/Rimini Protokoll, the river Evros, dividing Greece and Turkey, is made present through the young refugees that managed to cross it. Wetzel re-enacts John Cage's 1959 composition **Water Walk**, replacing the non-musical objects that Cage used with artifacts from the stories of fifteen boys in Athens who survived arduous journeys.

Unable to be on stage because they were not allowed to cross further borders awaiting asylum, the audience takes their place, engaging in a unique collaboration. The production fosters a shared experience, turning a political, spatial, and temporal divide into a productive connection, allowing the audience to create music in place of the boys and for them.

DW If you visit a performance of *Evros Walk Water*, the first thing you hear is a prologue read by a young person:

We are living in a part of Athens, Central Athens, we had to come to Athens as unattended minors due to violence and on a certain day this theatre guy comes and he wants to make a project with us. The project is about music [Because he offered coca cola and some free time outside the building, we thought let's give it a chance.] But the music was horrible. He showed us a music video from 1959 by an artist named John Cage. And the piece that this man performed called *Water Walk*. We thought it was really strange. But he came up with the idea that we could do our own version of *Water Walk*. We could perform *Water Walk* for an audience. *Water Walk* normally is performed just by one person. There are sounds performed that have to do with water. But this time we are the interpreters and we can use our sounds. Sounds that are things that we find in our house, not in the house of John Cage. However, the European regulations for people to move on the continent do not foresee that we can visit you and we cannot perform the piece for you. So, we thought, then we send you the instruments and you have to perform yourself because you don't just want us to be with you? But we can be, we can be your guides and you can hear our voices through headphones. So, we will be the conductors and you will be the musicians.

So, that's the prologue.
You then enter the place, an installation with the floor being a sort of map in blue and green: sea and land. You see that there are twenty-four positions that have numbers. There's a lot of stuff arranged, objects of various kinds. Partly they have to do with water in an obvious way. Iconic and in the centre, where Cage had a bathtub, there is a rubber boat, but filled with water. And there are other items, such as a pot, strange tables. There's a tube, an extraction hose from an oven. Odd stuff apparently, but each position has a number, white on blue or red. Twelve of these positions are on the field. And the other twelve are mounted along a ceiling that makes the inner field something like a pool or a stage. When you enter the place, everybody puts on their headphones. And then the second prologue starts.

Through the headphones you subsequently hear the voices of the young people that wrote the letter, and you hear the atmosphere of the building where the boys, age nine to eighteen live together and they introduce you to the house, their names and their stories. They speak multiple languages. It is either in Greek or in Farsi or in Arabic, with an interpreted voiceover. And it's astonishing how good some of these kids speak Greek. You think: wow, they actually speak together in Greek, no matter where they came from just a few months ago! And then they say: 'You will now listen to the story of one of us. So you, who is on position six, you will listen to my story. My name is Abdel.' And some other boy will say: 'My name is Moussa' and 'You are at position sixteen. Congratulations! You will hear my story first! And after you listened to one of our stories, we will instruct you how to play *Water Walk* and then the piece starts.'

The performance involves the boys sharing stories from their lives, including experiences of crossing the Evros River, surviving the long journey from Afghanistan to the border between Turkey and Greece, surviving the rubber boat ride over the sea, surviving sex slavery situations on the way, et cetera. After you've listened to the first narrative, the doorbell of the house rings and the piece *Water*

John Cage, *Water Walk*, 1959, film still

Rimini Protokoll, *Evros Walk Water, A Cage Re-Enactment*, 2017

Walk by John Cage is performed—by you and the other visitors. You are guided by the instructions whispered by the boys, creating sounds using various objects: 'Take the rubber shark in your hands and when I tell you, you will squeeze it!', 'when I say "wave" let the rudder fall into the water inside the boat', and so on. You'll be asked to play a chord on the synthesizer, you'll find the right key through stickers of football players that mark which keys to press. It makes a dominant Sept accord, and that's from Cage, but the synthesizer is from their house, supposedly. 'And Cage had a grand piano but here we have a santur, a string instrument developed in an area of Iran. You have one of these wind-up toys, it is a pair of shoes referring to the person that made their way from Pakistan to Greece on their feet. The instruction is to let this toy walk over the strings of the santur and then at some point to throw that toy in the water. Now you've been really sweet, you adults. You always did what we told you to do. So now we would like to ask you to do what we would do. Just do what you want and feel free, you know. Just feel free to do with the sounds what you want.'

The performance consists of five rounds with specific instructions, culminating in a moment where the boys comment on the audience's docility and encourage them to be free and to improvise and freely interact with the sounds, breaking away from Cage's approach to music and aversion to improvisation.

CD How did people respond? It sounds like fun in a way, but there's maybe also a sense that each member of the audience gets a different story and a different role in the play?

DW It alternates, in the end you will have listened to different stories. During the first part, you will have been at three instrument positions and three observer positions. There's for instance a position with a Ferrari, a toy Ferrari. And there's a foot pedal from a piano that you can step on to control the sound that toy makes. So, it roars like meow. And the intensity of the sound relates to the struggle to cross the river and to survive the attempt. So, the person sits on the car tyres in front of the Ferrari with the foot on the paddle. And a person outside of the frame is listening to the same story. So, I'm instructed as an observer to observe how that person deals with the challenge.

Participants said that making music together had sort of a cathartic cleansing or liberating aspect because you have the feeling you're doing something together with this boy whose story you have just listened to.

Rimini Protokoll, *Evros Walk Water, A Cage Re-Enactment*, 2017

36

In all those shows, I never experienced that somebody opted against it or refused to be remote-conducted. It was just clear and very harmonic: we are collaborating, following the system these boys have developed for us. Still, some people did say: I felt guilty or really a bit responsible, or I had strong emotions while listening to this person, even if the story was not so hard. But the constellation is tough because you listen to people whose future is so unclear. They came to Europe because they want to live a life in safety and peace, and it's up to us if they will be able to and what chances they will have.

HW Why is this interesting for you?

DW My favourite part is not the concert. That's just messy for me, messy sound. I'm not interested in it. It's fun that they do things, and it's funny to have these sounds, and I like very much that it's not precisely with the same items from Cage's original version but still quoting their timbre. From time to time, there are performances of this famous piece which are really too old-fashioned and way too quiet for my taste. John Cage uses a yellow rubber duck for the bathtub so they can't perform the Cage piece without finding this original New York rubber duck thing from the 1950s. So, I thought: let's replace the duck with a squeaky shark because some of the boys crossed the sea between Turkey and the first Greek island and they felt there might be sharks. The duck did not mean anything to them, but the shark made sense for everyone without further speaking.

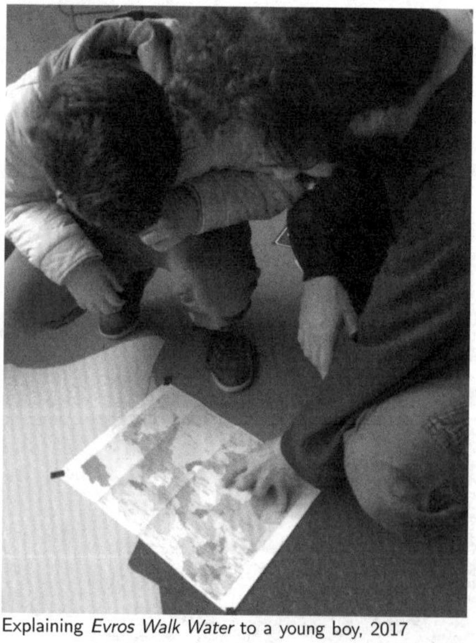

Explaining *Evros Walk Water* to a young boy, 2017

I really like the situation that people are suddenly in the position of performing this very important piece of music, you know, but as a democratic thing. So, it's in our hands and we just entered the structure. Have you yet been inside the composition of a Cage piece? No. Did at any point of your life a migrant boy tell you what to do? While you have the feeling that you might fail and not do it to their satisfaction? Have you ever sat down, pressed your hands on a pair of headphones to hear what a boy like this wants to share? About walking through Iran and Turkey and then risking their life on a rubber boat going out on the sea at night without guidance, the sound of crying babies next to you?

It's the performativity of listening, and the act of letting in stories—be they completely true or not—that will eventually be part of our collective understanding of who we are as a society and how we deal with it. It is not about tolerance but about integration—not of people who have survived and managed to arrive in our zones of European Human Rights and social welfare—it is about our lack of experience with the act of welcoming them by the sheer act of just listening, with our minds on alert. And that last sentence would be my credo for a strong experience of music. And I will be forever grateful to John Cage that I could learn this. Music is listening to surprising events, empathically.

The other moment that I really like is that you hear twenty-four different stories, twelve different tracks at the same times as the headphones are semi-closed. So, these headphones do not block the surrounding sounds. You can hear what the others are playing, but the membrane is very good and so you hear your narrator clearly. But if somebody would voice a whisper on top of it, you would also hear it. And that's why you also hear the chatter of twenty-four headphones and that's a beautiful situation where you predominantly see adults or visitors just listening to these boys. That is for me the centre of the piece. That it organizes an act of listening. Which is a close listening, a concentrated listening, open, a bit apprehensive. We just never know if it's going to be a tough story. And then because it also has a lot of humour you understand: it's really not about saying how bad things are. It's really about hearing what they need. What they like, what they miss, and you predominantly hear boys with confidence. And I didn't even dare to touch the traumatic parts of their experience because I also understood they don't speak about them at all.

The Evros River is also the border between Greece and Turkey. It is said to be more dangerous than the sea, a short distance but posing unique challenges.[1] The area is highly militarized, while there are also people in the woods who exploit helpless refugees. Mysterious rules govern the region, such as a drum sound signalling police to close windows in order for smuggling to happen. Crossing the river requires waiting for the right moment, considering rainfall and the strength of the current.

1 See also www.forensic-architecture.org/investigation/pushbacks-across-the-evros-meric-river-analysis-of-video-evidence

HW The composition is about the passage over the Evros river or the river as this kind of boundary. How does water figure in the performance?

DW Yes, in that sense water is an element in the piece also in the way you perform it. Water has a central presence in the piece, also in Cage's *Water Walk*.

Some stations have directly to do with water, such as a vase with flowers placed in the rubber boat with water, and there are other things that are entering the water or that swim. There is a small whistle, one of these Birdy voices that require some water inside to make sound. That's one of the instruments. And then there's also a pot with pebbles that you first run through and then you let it swim. There's the rowing oar from the boat with which you hit the water at the very moment your instructor shouts 'Wave!' in Greek, like you do when you want to keep everyone on board on the sea. So, some of the sounds are directly connected to the water. Whatever position you have, the water is always in the centre. And it's being transformed throughout the performance because more and more elements are swimming in the rubber boat filled with water. In the end, all sounds have to do with waves, be it acoustic or nautical, or symbolic or poetic. And then there are stories that are clearly only about how people treat people. In different parts of the world.

When you experience the piece, you are moving on a floor that represents a map in blue and green indicating land and water between the South and Europe. And you think of all these people travelling through Europe. Jined, the boy from Pakistan talks to you about the river being as fast as a Ferrari at top speed, he admires it like a hero, and he imagines what it would be like to be its pilot, which demands cleverness and strength and will, and that is what we should understand, rather fast and deep: we are living with fellow citizens who are not here because they survived getting born in a cozy hospital around the corner, but who survived coming here with blessings from all their family—as the kid who shall make it. One of the boys in the piece has never been able to reach his mother again since she sent him on his way with the money she made selling the house she inherited from her husband before his brothers could claim it. He fears they killed her for this reason. She gave her life for his life.

HW How did the work come about; what was the commission?

DW I got a commission I think it was 60,000 Swiss Francs for writing a composition and that money would also cover the rehearsals and a performance of one evening. It was my choice whether it would be a solo for a trumpet, a quartet or whatever. The theme of the festival however was 'Crossing the border'. At that time, I was living in Athens. I noticed the changing situation. Athens was really getting flooded with people, the flow of migrants increased dramatically in 2015 due to the growing number of Syrians, Iraqis, Libyans, Afghans, and Eritreans fleeing war, conflict or economic hardship. Greece, Spain and Italy have been the European entrance points and already were mostly left to deal with this on their own by the rest of Europe. Reading German papers, the situation was hardly mentioned. And in Greece, the office of the United Nations dealing with refugees closed down in protest regarding the miserable management of the situation on the ground. When I started my research, it was still the conservative government which imprisoned asylum seekers like criminals and the check whether you are adult or infant—prison or shelter—was done with a meter measuring the length of the fingers. The situation already triggered a tremendous right-wing reaction and all of the things we are still facing, and I felt, this is all because people in charge do not put into action during the week what they say with words about human rights, solidarity and such on Sundays.

Buying the boat *(Evros Walk Water)*, 2016, Athens.
Photos: Daniel Wetzel

I told Mirella Weingarten, the curator: 'If you want me to do something for your festival in Switzerland and your theme is even Crossing the Border; then I have to make something about the situation in Greece'. Mirella explained the rest of her programme and she mentioned *Water Walk* by John Cage. She mentioned that there was a musician who would do a re-enactment of this piece with the proper props and encouraged me to let go of the idea to create a new composition and instead to work on another way to stage this piece, duration three minutes. Look it up on YouTube, it is a remarkable performance by John Cage who as an avantgarde weirdo made it into this super popular US TV show called *I've got a secret*. And then I thought I would rather work in reverse following the question: who the hell should be the musician anyways? And what is knowledge and what is training in music and what can music transport at all? And is the set of the concert potentially another way of telling stories other than sitting and listening? Can music transfer something else? What can swim on the waves of music? Then I started to research the situation in Athens, and something new came up: what do people mean when they call others a migrant, in contrast to a refugee? Who is permitted to go where? What is the Dublin II regulation and what is the difference between the Human Rights on Sunday, in Brussels, or in any presidential speech and at the borders of Europe?

Eventually, I needed to narrow down and decided to concentrate on just one cell of the complex body of people in desperate need in this super rough city, in the midst of a tough austerity treatment by the rest of the world. A group of boys that lived in one building in the centre of Athens and who could eat because the neighbouring people would daily donate food from their groceries.

I started to visit them. After school, they play Nintendo, they run around, they have therapy, they all have girlfriends, they experience not much racism as far as they told me. And because they were all from different countries, no one was considered a foreigner, they had a fairly nice life there, their traumata in the backyards of the new life they built. When asked what they would like to become, their response would be: police officer, soldier, because: safety, fairness, reliability, rule of law—this is what made them travel. And this is the insane paradox: that these newly arrived citizens in our communities experience when they make it further North that they are treated as potential drug dealers, rapists, criminals—the opposite of what made them and their families make all the efforts and risk their lives to make it into the zone of fairness that Europe is seen as where they come from.

I rented a room and worked with them on alternative instruments following the original notation. The piece by Cage is noted in symbols, so it's a timeline instead of notes. It just indicates a sound every ten seconds. Hit this chord on the grand piano, after ten seconds pull it's strings, put the vase in the bathtub, activate the ice crusher et cetera—every ten seconds another cue. So, we worked on sounds that would do the same job, but made with items that came from their place or biography and not from that place in the US in the 1950s where John Cage developed it. If it would not be a duck, what would you squeeze? What would be your piano? Turned out to be a cheesy synthesizer. But then it turned out that this is not interesting for them. They said: 'You do the job. You're the artist.' So, I came up with suggestions and this was the process of gradually learning how that works. I understood—my job is listening to them, providing them with coca cola and crisps and free time and not to ask them about their tough experiences but to give it time. So in one of the twelve pieces I developed with them, for example, you listen to Abel who plays Counter Strike and shoots loads of enemies and comments on that, you listen to the action sound of that particular session and in the meantime he tells you how their mother dropped them at the shelter, disappeared for two years until she had made it on foot to Switzerland, into an apartment and into the possibility to apply for her son to join her as a family member. You join Abel for some minutes on the trip of being a warrior in a symbolic bam-bam-ego-shooter world and understand what compensation might mean and you understand this guy is so bright, he is perhaps sixteen years old and he is a chance for us.

HW And so in the end, they only worked with you on their stories and not so much on the props and the composition? How did you negotiate and work with them?

DW I had to find a way to work with them on that piece without explaining them how to work on a piece. So you have to be always a step ahead, not in order to dominate the process, but to tell them I want to know...You cannot sit down with a sixteen-year-old boy, no matter what. You can't show the origin and say look, this is the concept, so, what's your thought? You have to ask in a different way, to show honest interest, patience and then also openness for what comes and then don't be disappointed when what comes was not what you imagined. Because that's the point why you are asking, and you realize I wanted something that has to do more with music. I learned a lot about the unhierarchical openness and humour regarding events as celebrated by John Cage. And I learned, I should not make them understand how I think or what is cool music performance, but it is up to me to unlearn what I expect with all my cultural background and arty-farty expectations, and to be open to the sound of their being.

There was this boy Massoud and he had a lame arm. When he came to Greece, he was just nine years old. He was from a Syrian city and his destination was family in Denmark. But they couldn't send him to Denmark straight away. The boy had to go through Greece with some uncle who went together with him on that really dangerous passage through Turkey and then by boat. Nine years old, couldn't swim, lame arm. The boat deflated at some point close to the shore to Lesbos. Because another adult just grabbed him and took him out of the water, he survived. But he tells it as if it is a sweet and very normal thing. He now lives in Denmark. But that time, he was just in the state of waiting for the call to take the next step. And I wanted to learn about his lame arm. I was told various versions that would not combine into one picture. Conflicting stories. No matches. That's an experience that you get often. You are presented with a story. They think what could this adult white person find useful so that we all can go ahead. This is not because there's like a trend to not tell the truth. It is a mirror of us: that people who try to survive already on their way to Europe learn: it is not about who I am, it is about what story would give me more chances to survive.

HW It's risky to tell the truth, so you avoid that.

DW Yes, you have to adjust to what's the rule of the game here. Which countries are on the table. So, Europe teaches me to tell the story that they need. I only have value when I present a proper story. I also understood, the truth is your will to live a life in dignity, and our regulations ask you to deny your experiences. You enter this sphere of proper rule of law by taking another role, changing your name, pretending your origin, the only truth you are permitted to maintain is: I would like to live in peace with a chance to develop and prosper. But in *Evros Walk Water* it is about baby steps in listening to what lies behind the first level of a story.

With Massoud, for example, I said I want to find the type of boat that brought you here. I went to several shops with him to see rubber boats. Until we identified the boat that they were using. It was a boat called Voyager. You can buy it on Amazon. From time to time, we have to buy a new one. Because yes, they at some point deflate, even under the smooth conditions of touring a piece in proper flight boxes. But was it really a boat type 'Voyager' he had been in—or did Massoud just think 'enough with this, I want to go home so let's tell him this is the one'? I will never know, and I do not care, it is about what he is able and willing to share with me—like with me sharing things with you.

CD It was fascinating to learn how you were commissioned to write a composition. So, there is a request, but there's also a condition you want to respond to, and there's a place you are working from. And then you translate things along the way, ending up with an experience for an audience. Is this your approach in general?

DW Instead of agreeing on what we want, we engage in conversations, in a process, we agree in a mutual process, allowing the flow to happen, and most important in my view is: to be open for the gaze of third eyes, because in my situation—as so-called WEIRD person (White, Educated, Industrially-developed home country, Rich compared to so many on our planet, living in a Democratic context)—to work in the arts is not about organizing means to transport my subjectivity, but about organizing a transfer between others and you. I am an author of transmission, I work on opening ears, starting with mine, like John Cage, towards what does not match with our Western ears, but instead with what we should praise and celebrate as our reality. And if you want to really know about it—reduce your personality to the ear or the eye and open them up to what you used to file under noise and consider them: sound. The sound of us. Listening, being in a symphonic concert or walking on a street or being bothered by the kids celebrating in front of the room where I'd like to continue sleeping through the night, or the third person suffering from drug addiction asking me now for a coin as you and I talk over zoom—it means to let in what I could not have thought of without the situation. And this is what theatre and the arts in my view are good for: to crack open alternative access to our mutual situation without excluding anyone.

For credits, see: www.rimini-protokoll.de/website/en/project/evros-walk-water

4.7 Let's Watsu

BY HENRIËTTE WAAL

The conditions are now ideal for Watsu, she says as I run into her on my doorstep in Arles, in the south of France, where she's come to water my neighbour's plants.

My friend and former Atelier LUMA colleague has become a water masseuse since she gave up her work as an architect, noticing that she became less and less interested in 'building' and more and more in the social architecture of relationships.

I decide to accept her invitation because I imagine a form of baby swimming for adults. As a baby, nothing brought me more joy than moving through water, according to my parents. So let's Watsu.

The figs on the huge tree hanging over the pool in her garden are ripe and give us their sweet fruity and lactonic perfume. The setting is paradisiacal. The water glitters with the sun between the shadows of the fig leaves. The water is slightly cooler than body temperature, so I put on a thin wetsuit. We pick and eat a fruit before we descend into the pool, where the water makes room to receive and carry our bodies.

Watsu comes from water and shiatsu, she clarifies, a Japanese form of therapeutic bodywork, and is a dynamic massage and therapy in chest-deep warm water (about body temperature) to restore balance in the energy system. She explains the signs she will give to me during the session.

Stand with your legs wide and ground your feet firmly on the bottom and let your body float. I look down and see our middle-aged women's legs, which, due to the refraction in the water, seem even bigger, rounder, and paler than above water. I calmly breathe in and out, coming slightly out of the water when inhaling and sinking back into the water when exhaling. I feel like a large vessel on legs.

Brecht Duijf, *Sprong* (*Jump*), photograph, 2024

I set an intention: Surrender.

I receive more buoyancy in the form of floaters around my legs and she cradles my head in her arms. Gently, she initiates a movement, and my body sways in the water. The movements are undulating and fluid. I hardly notice her presence any longer. It's such a soft, caressing feeling that I find it difficult to receive. As I swirl through the water, my body is gently kneaded, stretched, and turned. It seems impossible to completely surrender to her and to the situation. I feel that I'm too angular in this process. Too much head. When I go kitesurfing, I experience a similar feeling of angularity in the beginning. When I first enter the water, I'm not yet in sync with the elements; it takes a while to get into the flow. I put trust in that experience. I think of Simone Forti, the postmodern American artist and dancer, embodying seaweed floating on the water surface. That helps. I could cry from the softness of this water duet, but it's so joyful that the tears don't come.

I feel something in my hand. Nose clip. I feel finger taps on my body. Taps mean we are going underwater. I open my mouth and fill my lungs with oxygen. She gently pushes me underwater, but now I trust her. The water is a three-dimensional massage space. She now pulls and rolls me around, but there's also the water that moves along and supports everything. My body could never move like this above water. The sensation of above and below disappears, I let myself be carried along, swirling and spinning through the water, enjoying the hypnotizing visuals that arise. The movement of my body, the sunlight, and the water create a stroboscopic effect in slow motion.

I'm above water again, and I take a deep breath. Each breath is a journey in itself.

I suddenly experience a stretch of time that makes it seem that I don't need to breathe anymore in this state of being. The warm water, the flow, and the massage have lulled me into a state that feels like sleep, without actually sleeping. I hear singing in the distance. My ears are underwater. The sound is round and muted. Is it real or my audiation? I feel how tension from my pelvic floor and hips becomes fluid and is carried away. It's as if a stream is flowing through my body, gently.

Moving through water is really good for me; it's a medicine, a language, and a way of thinking. I am one with the water. I am in the water, but I am also the water. I surrender.

Mihnea Tănăsescu is Research Professor of the Fund for Scientific Research (F.R.S.-FNRS) in the School of Human and Social Sciences, department of Sociology and Anthropology, University of Mons, Belgium. His latest books are *Ecocene Politics* and *Understanding the Rights of Nature: A Critical Introduction.*

Ștefan Constantinescu works at the University of Bucharest, Faculty of Geography, where he teaches courses in Historical Cartography, GIS, and Oceanography. He began his career as a coastal geomorphologist, later shifting his focus to deltaic shores. He views maps not simply as cartographic products, but as cultural syntheses encompassing toponymic, historical, and geographical elements.

Continental Apelor, source: Scanteia, 1958

4.8 Reeds, the Danube Delta, and the Making of the New Man

By Mihnea Tănăsescu & Ștefan Constantinescu

The Danube Delta is the newest land in Europe, only starting to form approximately 6,000 years ago. A vast wetland of labyrinthine channels and capricious water, it is claimed, though without any precise backing, as home to the largest contiguous reed beds in the world (p. 226). Is it? Perhaps this is the wrong question. In the context of Romanian state expansion, and specifically the creation of a national identity under the totalitarian regime of 1945–1989, such 'facts' had a double role: to imagine both unique natural beauty and the existence of unbounded resources for exploitation (for cellulose and paper production and only very marginally as construction material). There was no contradiction between the proto-conservationist ideal of sublime nature and its systematic exploitation. It is through this union that the new man—the philosopher-worker, as it were—could be born, part contemplative genius, part industrious builder of desirable futures.

This double imagination can be seen very concretely. It has main actors, side plots, and an archival documentation that allows for various reconstructions. For example, reeds (**Phragmites australis**), offer one way of telling the story of how totalitarian projections and projects have used a water body, and its most conspicuous inhabitant, to build identities, mediated through new technologies. At the same time, reeds complicate the gratifying tale of a motherly wilderness in synchrony with locals.

The Danube Delta is formed by the flow of the Danube River towards the Black Sea. Geological conditions, specifically the significant extension of a shallow shelf in the north-western Black Sea, made it possible for the river to deposit unimaginable amounts of sediments in a shifting, whimsical, labyrinthine pattern. Throughout its human history, the Delta has been approached and understood in

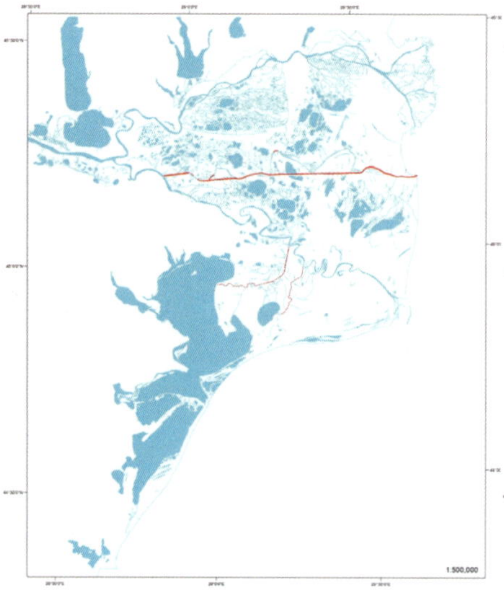

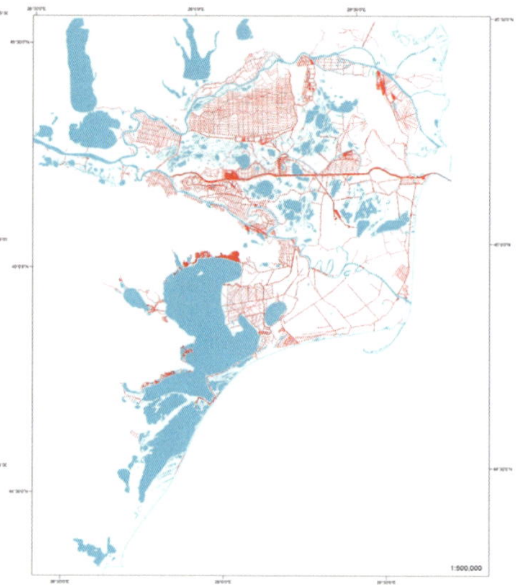

different ways. In the nineteenth century, the European Commission of the Danube, the first international organization with any significant power, oversaw the mapping of the main branches of the Delta. The Commission's purpose was to aid navigation so that trade could flow from the Black Sea into the centre of Europe. To this end, the Commission created the first straight channel, Sulina. At this historical point, reeds were not an important product.

Foaia Românească, 15 April 1959

This period, lasting until the mid-twentieth century, reduced the Danube Delta to its main branches, while the interior remained largely unknown to officialdom. In 1945, with the settling in of a new Soviet-inspired communist regime, new cartographical efforts plotted the labyrinthine interior where abundant resources could be extracted (mostly fish and reeds). Many channels were cut, polders made, dikes erected, in a massive effort that completely altered the water dynamics of the area. The physical transformation of the space led to new pathways that required constant dredging for maintenance. Floating reed islands, known as **plaur**, would block channels and lake entrances and required considerable effort to be removed. Sometimes, locals would purposefully block a waterway in this way in order to create optimal fishing conditions that only they could access (through, for example, hidden entries).

Ilustrațiunea Română, 6 September 1933

At the same time as the Delta was becoming a space of exploitation, it also became one of recreation. Local people, strewn over a dozen villages and with a mixed ethnic and cultural heritage, became noble savages apt for being transformed into proletarians, while proletarians from across the country could come and connect with nature in the Danube Delta. Local experience never featured on its own terms, whether in the nineteenth-century exploration of main branches or the mid-twentieth-century transformation of the interior. What is now called local ecological knowledge has always been marginal at best. Even in the exploitation of resources, it was not the knowledge of the fishermen that counted, but rather the knowledge of the state, transforming the fishermen through new and superior technology that would elevate them from the innocence of living close to nature. And this process is clearest in the exploitation of reeds.

Traditional reed fence and roof

Industrial harvesting with purpose-made machines

What are reeds? How do they behave, what is their specific agency like, how are they rooted in their environment, and how does their character feature in the imagination of progress? The general term 'reeds' covers, in the case of the Danube Delta, three different plants with very different characters: **scitca**, **polomoc** and **lom**. This popular classification distinguishes them based on height (**scitca** is the shortest, **polomoc** medium, and **lom** the tallest) and their usage. **Scitca** is used for house roofing and typically thrives in brackish areas near the sea. It is occasionally complemented with **polomoc** for fences (figure 5). **Lom**, being the tallest, grows in areas with deeper water and was used for fuel when there wasn't enough firewood available. For all of them, there is a narrow band of optimal growing conditions, slightly different from each other. To the trained eye then, reedbeds indicate an underwater topography. They reveal, through a deceiving monotony, what is hidden, drawing the contours of a submerged land.

Gathering reeds was mainly done in winter when the frozen waters allowed access to marshy areas. There was no need to harvest every year because the lifespan of a roof could be fifty to sixty years. The bundles of cut reeds were called **malduri** and the quantity needed for different building needs were estimated based on experience: **you needed about a hundred malduri to cover that house**.

Harvesting was done with a tool called **tarpan** (in the Romanian Dictionary, the word is listed as **târpan**, but locals never pronounce it that way), which is similar to a scythe, and the resulting reeds were bundled into **malduri**. Harvesting was extremely difficult, from the cutting of hard stems to the heavy bundles, the transport through marshy terrain, and the drying time in uncertain weather, not to mention the intense cold and windy conditions. This was one of the most challenging activities in the marsh. In all the conversations we had with locals, we never encountered anyone who described this activity with joy or nostalgia. By contrast, fishing for sturgeon in the sea, in many ways a more dangerous activity, is recalled fondly. Reed harvesting was dreaded.

For the totalitarian regime, reeds took on a different use, serving as cellulose in mainland factories, especially in the gargantuan factory of Chișcani Brăila. Its industrial exploitation required both the invention of machines that could access the inaccessible, and of workers that could guide them. In this process, science featured prominently. In the 1950s, a completely new town—**Maliuc**—was set up to house a committee of experts recruited from far and wide to give advice on rational exploitation.

In the imagery of the era, traditional harvesting was presented as a bucolic activity that seamlessly bloomed into beloved industrial work. Harvesting was portrayed as the heroic work of superhuman techno-beings (see, for example, the short 1966 film **Stuf**, by Titus Mesaroș). In truth, the local dislike of reeds was deepened through new technologies and transformations. During winter, fishermen were taken from their villages to floating pontoons where they would stay for two weeks and work on winter cutting. They lived and worked in poor conditions. Political prisoners from **Periprava** and **Chilia**, notorious communist prisons, were also used to harvest reed. The hard work contributed to the deterioration of the prisoner's health

Gathering reeds into bundles (*malduri*): Photo: Elena Ghera, *Flacăra* 15 February 1975

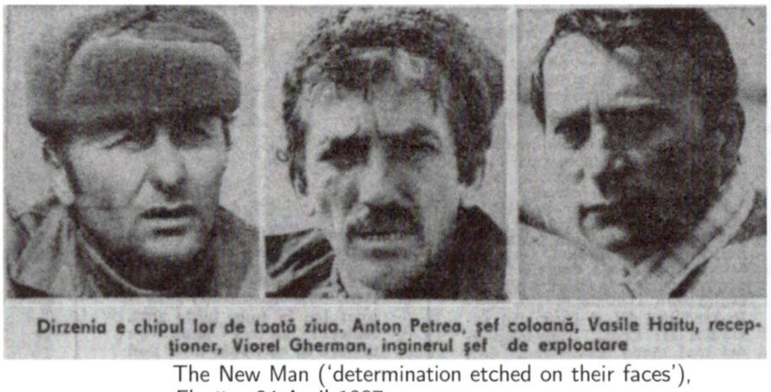

The New Man ('determination etched on their faces'), *Flacăra*, 24 April 1987

Gathering reeds on the winter ice

(a common technique of degradation-through-labour), while the marsh ensured that they couldn't escape, couldn't run in or out of **the sea of reeds**.

The glorification of machinery hid the fact that people were still employed in manual harvesting with the **tarpan**, where machines couldn't reach. The new and fanciful mechanical inventions were often unreliable, broke down in difficult terrain, and needed constant maintenance. In the end, it was still the **tarpan** that harvesting heavily relied on, compounding the toll on labour. The simultaneous glorification of wild nature and locals living in harmony hid the often-cruel character of the environment itself. Locals of the Delta have a saying about reeds: 'let those who planted it go and harvest it! (**să meargă să îl taie ală care l-a plantat**)'.

Currently, the Danube Delta is a Biosphere Reserve and therefore strictly protected. The era of reed exploitation has given way to the extraction of aesthetic value through tourism and wildlife pilgrimage. Despite the growing European market for reeds, fuelled by new architectural trends in the West, the region does not seem to position itself as an important provider of raw materials. Instead, reed beds are used by local economic and political elites to extract subsidies, often by changing their status to 'pastures', thus becoming eligible for agricultural benefits.

Locals, on the other hand, have decisively shifted towards modern materials, much easier to procure, cheaper to maintain, and that don't require highly skilled artisans for installing. Reed roofs are mostly for guest houses, and the reeds are often not locally sourced, while the workers are never locals. Often, they are Romanians from elsewhere that, through European migration (Romania is by far the largest source of immigration in the European Union), have learned reed roofing skills in the West.

In Romanian popular beliefs, sand (also abundant in the Delta) is seen as the seed of stone, 'made by the Devil so that nothing can grow, and people have nothing to eat'.[1] No one can count the grains of sand on the beach or the stars in the sky, just as no one can count the reeds in the marsh. If it can be helped, better not cut it either. It is Devil's work, and only he who planted it should bear such a heavy load.

1 Elena Niculiță-Voronca, *Datinele și credințele poporului român* (vol. I) [Holidays and Beliefs of the Romanian People], Iași, 1998, p. 27.

4.9 Haug: A Rainbow Emitting Sculpture

By Veronika Sedlmair and Brynjar Sigurðarson

Veronika Sedlmair and Brynjar Sigurðarson, developed for ORNAMENTA 2024, Pforzheim, Germany

As if just emerged from the river, a hooded figure sits on top of a pole amongst the trees on the riverbank, and stares into the sky. At the right time with the right natural sunlight, the sculpture emits a mist, creating a temporary rainbow over the water. The mystical creature, made of bronze with high-tech sensors in its head, creates a surreal image that borders between the natural and technical.

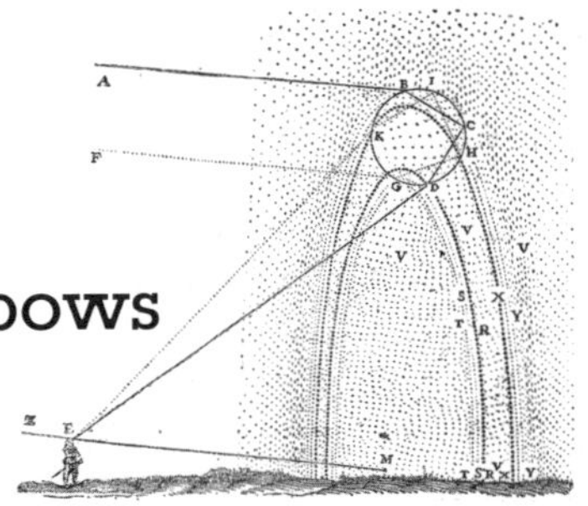

This image was made by Frans van Schooten Jr, who Descartes has been said to allegedly have locked in a room until he finished all the woodcut illustrations for his book.

4.10 Cartesian Rainbows

By Clemens Driessen

'The rainbow is such a remarkable wonder of Nature, and its cause has always been so curiously investigated by good minds, and so little known, that I could not choose a subject more suitable to show how, by the method I use, we can arrive at knowledge not possessed by any of those whose writings we have. Firstly, having considered that this bow may not only appear in the sky, but also in the air close to us, whenever there are several drops of water illuminated by the sun, as our experience shows in some fountains, it was easy for me to judge that it proceeds only in the way that the rays of light act against these drops, and from there tend towards our eyes. Then, knowing that these drops are round, as we have proved above, and seeing that, when they are larger or smaller, they do not make this arc appear any other way, I decided to make a very large one, in order to examine it better, and having filled a large, round, transparent glass flask with water for this purpose, I found that the sun coming, for example, from the part of the sky marked AFZ, and my eye being at point E, when I placed this bowl in the place BCD, its part D appeared to me all red and incomparably brighter than the rest; and that, whether I brought it nearer or back, and that I put it to the right or to the left, or even made it revolve around my head, as long as the DE line always made an angle of about 42 degrees with the line EM, which we must imagine as running from the centre of the eye towards that of the sun, this part D always appeared equally red.'[1]

1 René Descartes, *Discours de la Méthode*, 1637, Joannis Maire.

'How to make signs appear in the sky that seem like miracles.'

René Descartes is considered one of the key figures of what we could call the project of modernity. In his words, to apply mathematics in order to become 'masters and possessors of nature'.

Descartes derived that the maximum radius of a rainbow was 41°47', appearing most brightly at this angle to the observer. Playing with custom-made glass bowls, Descartes created an experimental setup that allowed him to measure the angles under which the sun is refracted and reflected in raindrops.

Descartes was not the first to measure and calculate the maximum radius of the rainbow, as his work was prefigured by scholars such as Ibn al-Haytham in the eleventh century. Nevertheless, he successfully used this result—with little practical application yet—to showcase the power of the mathematical understanding of natural phenomena in space.

Thereby Descartes moved the rainbow from the domain of wondrous, unknowable phenomena that could only be an object for speculative, poetic language, into an object that can be understood scientifically, calculated and manipulated.

Inspired by artificial rainbow fountains that were developed in renaissance gardens, in his **Discours de la Méthode** he went on to speculatively propose projecting complex structures using precisely positioned bowls and fluids that diffract differently: a virtual architecture of light and colour.

4.11
Marvels of Things Created and Miraculous Aspects of Things Existing

By Clemens Driessen

The Marvels is a thirteenth-century Arabic text, dedicated to the governor of Baghdad (d.1283/682). It is considered the first comprehensive cosmological account in the Islamic world, offering a methodical overview of all creatures and phenomena.[1]

The book was translated and reprinted often throughout the centuries. This image is taken from a Persian edition illustrated in the Mughal empire in India in the seventeenth century.
We see a fish imagined as a very cute person. A non-gendered mermaid? It is part of a series of wondrous creatures, figuring next to zodiac signs, angels, jinns, animals, humans, plants, and strange creatures from the edges of the known world.

Marvel in illustrated edition made in 17th Century Mughal India

Is this merely anthropomorphizing animals? Granting them a human face, an individual character, eyes that look back at us—whereas fish we all know with their mostly sideways-oriented, unblinking eyes do not seem to face us? Or do modern humans feel they don't need to face fish? What if the starting point of ordering the world is to be in awe of marvels. Wondrous beings who may very well speak to us, return a gaze with friendly curious eyes, open to see miracles.

تاب عجائب المخلوقات وغرائب الموجودات

Zakarīyā al-Qazwīnī

زكرياء ابن محمد القزوينى

زكريا قزوين

1. Travis Zadeh, *Wonders and Rarities: The Marvelous Book that Traveled the World and Mapped the Cosmos*, Harvard University Press, 2023.

Violence

United States Patent [19]
Johnson

[11] Patent Number: 4,591,071
[45] Date of Patent: May 27, 1986

[54] SQUIRT GUN

[76] Inventor: **Lonnie G. Johnson**, 1463 E. Barkley Dr., Mobile, Ala. 36606

[21] Appl. No.: **541,898**

[22] Filed: **Oct. 14, 1983**

[51] **Int. Cl.**[4] **B67D 5/32**
[52] **U.S. Cl.** .... **222/39**; 222/79; 222/401; 340/384 E; 340/406; 116/137 R
[58] **Field of Search** .... 340/384 E, 406; 116/137 R, 139; 222/39, 79, 323, 324, 401

[56] **References Cited**

U.S. PATENT DOCUMENTS

1,333,704	3/1920	Brinks	340/406
2,249,608	7/1941	Greene	222/79
2,302,963	11/1942	Lefever	222/39
2,589,977	3/1952	Stelzer	222/79
3,202,318	8/1965	Black	222/39
4,086,589	4/1978	Cieslak et al.	340/384 E
4,214,674	7/1980	Jones et al.	222/401
4,239,129	12/1980	Esposito	222/79

Primary Examiner—H. Grant Skaggs
Attorney, Agent, or Firm—John A. Beehner

[57] **ABSTRACT**

A toy squirt gun which shoots a continuous high velocity stream of water. The squirt gun is configured as a structure facilitating partial filling with water leaving a void for compressed air. The squirt gun includes a nozzle for ejecting water at high velocity, a pressurization pump for compressing air into the gun to pressurize water contained therein, and a trigger actuated flow control valve for shooting the gun by controlling flow of pressurized water through the nozzle. A battery-powered oscillator circuit and a water flow powered sound generator produce futuristic space ray gun sound effects when the gun is shooting.

9 Claims, 7 Drawing Figures

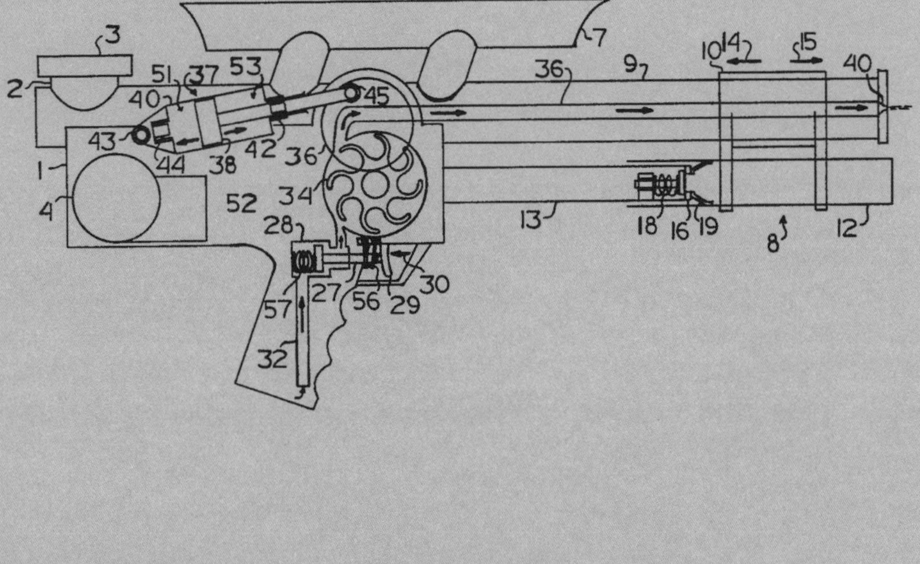

Violence

United States Patent [19]

Johnson

[11] **Patent Number: 4,591,071**

[45] **Date of Patent: May 27, 1986**

[54] **SQUIRT GUN**

[76] Inventor: **Lonnie G. Johnson,** 1463 E. Barkley Dr., Mobile, Ala. 36606

[21] Appl. No.: **541,898**

[22] Filed: **Oct. 14, 1983**

[51] **Int. Cl.**⁴ **B67D 5/32**

[52] **U.S. Cl.** .... **222/39; 222/79;** 222/401; 340/384 E; 340/406; 116/137 R

[58] **Field of Search** .... 340/384 E, 406; 116/137 R, 139; 222/39, 79, 323, 324, 401

[56] **References Cited**

U.S. PATENT DOCUMENTS

1,333,704	3/1920	Brinks	340/406
2,249,608	7/1941	Greene	222/79
2,302,963	11/1942	Lefever	222/39
2,589,977	3/1952	Stelzer	222/79
3,202,318	8/1965	Black	222/39
4,086,589	4/1978	Cieslak et al.	340/384 E
4,214,674	7/1980	Jones et al.	222/401
4,239,129	12/1980	Esposito	222/79

Primary Examiner—H. Grant Skaggs
Attorney, Agent, or Firm—John A. Beehner

[57] **ABSTRACT**

A toy squirt gun which shoots a continuous high velocity stream of water. The squirt gun is configured as a structure facilitating partial filling with water leaving a void for compressed air. The squirt gun includes a nozzle for ejecting water at high velocity, a pressurization pump for compressing air into the gun to pressurize water contained therein, and a trigger actuated flow control valve for shooting the gun by controlling flow of pressurized water through the nozzle. A battery-powered oscillator circuit and a water flow powered sound generator produce futuristic space ray gun sound effects when the gun is shooting.

9 Claims, 7 Drawing Figures

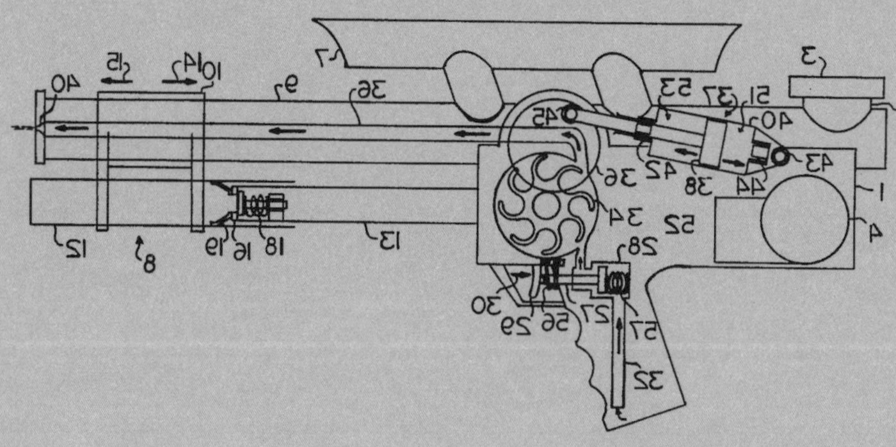

5 Violence

BY HENRIËTTE WAAL & CLEMENS DRIESSEN

Water is life. We all love water. Water is also a medium of mundane violence. Structural violence, often without it being obvious who is the perpetrator and who the victim. When water becomes scarce there may be a struggle for control, ownership, exclusion. Histories turn out to be riddled with water violence. Water futures appear violent too.

Climate change is a prime example of long-distance and long-term (or increasingly short-term)—slow—violence.[1] While climate change figures as both a direct threat and an excuse to spread global blame for what amounts to local water grabbing.

In many places, *water protecting* is now a crime. While our bodies are bioaccumulating harmful contaminants, states are locking up activists as terrorists for protesting against leaky corporate pipelines. Occasions for deliberation and exchange of ideas disappear while water tables go down. Most rivers aren't given rights, and the communities defending these rights are suppressed.

Border conflicts, struggles for land, often have water in the mix. And always Coca-Cola. Or Nestlé. 'In the last ten years, mass protests related to water, some of which have become violent, have occurred in Algeria, Brazil, China, Egypt, India, Iran, Iraq, and South Africa.'[2] (Of course when 'protests get violent' that can mean many things.)

Water that is governed by neglect. The side effects of extraction that no one really wanted but that are imposed on people nonetheless. Intended harm: from lithium mining to cloud servers—the modern world craves water—but what can we do?

1 Rob Nixon, *Slow Violence and the Environmentalism of the Poor*, Harvard University Press, 2011.

2 Office of the Director of National Intelligence (2021). *The Future of Water: Water Insecurity Threatening Global Economic Growth, Political Stability*. In: Global Trends 2040,www.dni.gov/index.php/gt2040-home/gt2040-deeper-looks/future-of-water.

Opening image: Page from full patent document Super Soaker US4591071, 1986

Water shortages are imminently scary. One danger that scarcity brings is the cruel optimism of waiting for, working for, water that will not return.[3]

3 Lauren Berlant, *Cruel Optimism*, Duke University Press, 2020.

But water itself is of course innocent, and thereby all the more dangerous. That seemingly benign life-giving substance can be used to project state force. In quelling protests, police-operated water is ostensibly abundant: water cannons knocking over protestors. Government employees waterboarding presumed terrorists, making them feel like they are drowning.

As Mari Bastashevski reveals in her chapter on the infamous Flint, Michigan case, 'there has been no violence on the part of the state; it has poisoned people democratically through paperwork and solid nearly legal contracts.' Where to situate action and responsibility is part of ecosocial design. Max Liboiron, an environmental scientist working on plastic pollution, emphasizes the need to think of people as agents, not just victims, in a permanently polluted world.[4]

4 Max Liboiron, Manuel Tironi & Nerea Calvillo, 'Toxic politics: Acting in a permanently polluted world', *Social Studies of Science*, (48:3), 2018, pp. 331–349.

And then there's the joy of acting at a distance. Like archerfish squirting jets of water into the air, knocking an insect from a branch into the water.[5] The fun of water play on a warm day, hitting others while exposing yourself to a fresh wet squirt.

5 Alberto Vailati, Luca Zinnato, & Roberto Cerbino, 'How Archer Fish Achieve a Powerful Impact: Hydrodynamic Instability of a Pulsed Jet in Toxotes Jaculatrix', PLoS One, 7(10), e47867, 2012.

Can ecosocial designers work with water non-violently? Avoiding the extractivism of polluting rivers and depleting aquifers—without hiding behind supposedly 'complex' but actually simplistic diagrams of water cycles promising a never-ending return? In the scramble over scarce water, can ecosocial designers build alliances and solidarities across human and other organisms?

Perhaps using the power of water shaping the hardest matter by continued force, however soft. As long as it flows, given enough time, stones are rounded off, valleys are carved out, new paths opened up. Soft power sustained can move mountains. Oh, what a cliché. Do we have time for this?

Esha Shah is a feminist scholar who believes in the moral power of ideas to change the world. She currently works as a lecturer at Wageningen University. Her research involves debates on development and technology in India, the ways in which subjectivity (including emotions and affects) shape objectivity in scientific knowledge, and a decolonial history of Wageningen University and Research.

5.1 The Song of the Water Tank

BY ESHA SHAH

It was a beautiful breezy summer evening. After day's fieldwork, I was immersed in my notebook while also waiting for the evening bus to arrive, the only way out of the village, when Lalitha, Gopi and Sharda and several of their friends invited me to have tea.

Lalitha's house was on the left side of the paved road, the only road passing through the middle of the village. As it was common in south-Indian villages—on one side of the road were the Dalit houses and on the other side were the higher and middle caste houses. Lalitha and friends were sitting just outside of her house, sipping tea, and enjoying the summer breeze.

'Put down your notebook and come have tea with us', Lalitha invited me with friendly assertiveness. There was much giggling and teasing, 'ahhh...there comes the doctor', once I put my notebook in my bag and joined them. By now, after much explanation, they all had understood that once I completed my PhD research, I would be called a doctor—the idea of me being a doctor always made them giggle cheerfully.

Handing me a little steel pot of tea, Lalitha asked me what I did during the day. I gave some details with help of Patil—my research assistant and translator—without whom it was not possible to communicate with them as they spoke Kannada, the language I did not know. But the moment I walked towards them from across the other side of the road, Lalitha had already instructed a little boy playing close by to run and fetch Fatima—the waterman's daughter-in-law—who grew up in the city in a Muslim household and hence spoke Urdu-Hindi, the language that I did speak. Lalitha never trusted Patil hovering over me and his translation to me in English, she always insisted Fatima did the translation for me in Hindi—although Lalitha herself only vaguely understood Hindi-Urdu. Fatima was obviously more trusted than Patil.

Fatima arrived with the lightning speed which always bothered me, because it meant that she dropped whatever she was doing to join us. When I objected that she should not disturb her routine to translate, all of them loudly intervened, partly teasing Fatima to tell me how she would have loved to escape her mother-in-law's commanding presence and partly chiding me to not worry about such things. Fatima sat down next to me when Lalitha intently looked at me and asked if I would like to hear the song—the old folk song of the village chief's youngest daughter being sacrificed in the tank.

When I had first asked her, four months ago, to sing the song for me, Lalitha had flatly refused. She gave me a condescending look—who the hell are you, asking me this question—and then she gave many reasons why the song couldn't be sung at that time: that it would take days to sing the entire song and also that they, the group of Dalit women, would sing it only in the event of drought when they would not have enough work as agricultural day labourers, and that they would be able to sing the song only after properly offering pooja (worship) to the local deity. This was all to basically tell me to bugger off. Looking back, I realized how arrogant and presumptuous I was even to ask them to sing it for me just a week after my arrival in the village, walking around with my silver notebook and a small hand-held recorder, imposing myself as a PhD scholar upon the unsuspecting people. Lalitha was clearly irritated and there was contempt in her eyes when she rudely replied to me while walking away, she was not even looking at me, while I was still trying to complete my polite address to her.

Several months later when we did strike a friendly chord after all, she told me how they all thought my politeness was fake and how they joked about me and my formal attempt to address them and my silver notebook and hand-held tape-recorded, and the way Patil spoke with me in English. They made a parody of me and Patil speaking in English to each other, everything about me was a butt of a joke.

Until after a month since I started working in the village, when they saw me walking in the scorching heat almost every day, going up and down between village and the 'tank' located six kilometres away several times a day,[1] and that I attended the irrigation committee's five-AM meetings three days a week and so on. But I personally think the icebreaker was when I accepted their invitation to eat with them the cooked food they had brought to the fields. It was Lalitha who first mischievously invited me when I was scribbling in my notebook sitting under a tree just a few metres away from their group on a lunch break from working in the muddy paddies irrigated with the tank water.

1 Tank irrigation is the name the British gave to small irrigation reservoirs. Thirty thousand of them existed in the state of Karnataka when I did my PhD research in the early 2000s. A significant number of them 400 to even 1000 years old. This particular tank was called Dannayakankere which was 800 years old. For my PhD project, I was looking at the transformation of the tank irrigation technology after the introduction of green revolution technologies in the 1960s and 1970s.

'Put down your notebook and come have roti with us', she shouted. When I joined, she instructed me to wash my hands with the water from my own water bottle before giving me roti and fried vegetables on top of roti in my hand. I knew that I was under intense scrutiny while I ate—Did she eat? Did she eat the whole roti or just politely take a bite? Did she like the food? Was it too spicy for her? Did she want more? Why not? For the next few days, it happened that I was invited several times to eat with them and only after I routinely accepted the cooked food from them, they started to relate to me in any real sense.

The deeply-entrenched caste system in Indian culture is sustained through the core rule of purity and pollution—the upper caste people consider it polluting to accept food from the hands of the lower caste, especially Dalit caste. Even when all forms of such discriminatory practices have been constitutionally and legally banned in India, they generally continued in public culture in Indian villages. Dalit people have been treated as 'outcaste' for centuries, their neighbourhoods were located outside the village, the village tea-shop still kept separate cups for them. And even when I never mentioned my own caste, my identity was written all over me—it was ostensibly expressed in the way I spoke, the way I dressed, walked and laughed, and wrote in my notebook, and everything else.

After the initial ice or actually roti-breaker, it all exploded, they wanted to know everything about me, they especially asked questions about my husband for hours. Tell us where you met him, how did he look, was he handsome, how it happened, what did he say, where and how did you get married? And then what? Basically, they wanted to hear some racy details of the love story of a city gal, with all dollops of glamour and romance. We all want to escape our realities, don't we? Eventually, it was bitching about our mothers-in-law—all of our husbands' mothers—that created a real bond between us. Except Fatima, who never said anything even remotely critical about her mother-in-law.

So, on that breezy summer evening, from the other side of the road, when Lalitha invited me to join them for tea, and when Fatima arrived a minute later, I asked her if everything was well with her mother-in-law.

'She is like Gangema (water goddess), she only gives, never takes, my *Atte* (mother-in-law)', as usual Fatima was all in awe of her legal mother, and then she repeated in the same breath translating what Lalitha was asking: 'Do you want to hear the song?'

'What?' It was a big surprise that Lalitha even asked.

'Yes, but can you do that? What about the pooja and the fact that the song cannot be sung in parts, and it may take a few days to complete...and...'

'There are ways to get permission from the deity and it's our song....we can sing it when we want....and how we want ...', Lalitha asserted her agency to the divinity and to the tradition and instructed her daughter to bring some condiments on a plate from inside the house, and within minutes a pooja was offered to the deity, by the group of women sitting there in a circle.

She then interacted with Gopi and Sharda and Leela, and they started adjusting their voices as the verses of the song were sung in turns, two of them singing in a pair at a time.

I was perplexed that all this was happening so fast, I felt as if I wasn't prepared, my own tape-recorder deity was not offered proper pooja—I suddenly realized that the batteries in my hand-held recorder were low, and they might run out when these wonderful women were giving me what I had wanted a long-time. So, I called Lalitha's son who was busy pulling some electric wires to bring light to their outdoor courtyard facing the road. I asked him if he had pencil cells lying somewhere in their house. He instead fetched an old tape recorder of his own and inserted a cassette, telling me, 'No worries, I will record it'. These women sang this song their entire lives, but the fact that I, a city gal and by now a reasonably respected researcher in the village, imparted such importance to the song and the singers that this occasion of singing acquired a completely new significance. With all these activities, suddenly Lalitha's courtyard became a festive place, people from houses from both sides of the road started gathering, the women preparing to sing were sitting in the middle, Fatima and I on the side, the stage was set, tea was served, the audience had gathered to watch the show, and then it all started—they took a deep breath and started singing.

The song started almost at crescendo as if it was a scream. The song indeed was about a suppressed scream, about the pain and pathos, of a 'virgin' girl being sacrificed in the tank to please the water goddess.

Sacrificing a young girl or a woman or a man from the lower caste was a common practice during the medieval times on the occasions when the tank had not received water soon after its construction or because of a recurring draught. My silver notebook was heavy with loads of stories and songs that had emerged during my fieldwork. I will come to the pages of my notebook in a minute, but let me tell you the story first.

While the song had just started, something happened. I saw Patil talking to one of the women sitting in the circle waiting for her turn to sing. And then in a minute, before I blinked my eyes, all women stood up and disappeared. Before I insisted and Patil could explain, the festive gathering had dispersed. The lights switched off, the tape-recorded withdrawn inside the home, and Lalitha left after giving a stark look—'you should have told us before we started singing'.

'What?' I was aghast, demanding that Patil explained.

'It's just that... one of the men said', Patil pointed out the higher caste men having houses on the other side of road standing at the edge of Lalitha's courtyard, '... that this song they were singing did not belong to this tank, the village tank, and I just asked them if that was so, and then asked them if they could sing a song that belonged to this village tank, because your research is about this tank, that's it'.

I was fuming with frustration, struggling to keep my anger in check, 'could you not have asked me that question first?', and then the next half hour poor Patil heard an intense, mumbled, chaotic monologue on what folk literature is, how this song was a folk song, and how folk literature cannot be fixed in time and space, how folk literature is not a historical record, but something that resonated with lived practices embedded in memory of the group of people, how people make sense of their past through oral narratives passed on through generations, how folk songs and stories were what was remembered as living experiential imagination, and why and how the past was sedimented on the present, and how it was very likely women from several tanks in radius of 10 kms might be singing the same song and identifying it as the song connected to their tank, their lives, their orally passed knowledge and traditions, or perhaps they might be telling the same story or song in another variation with slightly different name and so on. I was rambling. And poor Patil was standing there mirroring my frustration with a question of his own: 'Could you not have told me this before?'

Never mind! Now what? I was desolate, heart-broken, and even worried if Lalitha and others would even talk to me tomorrow. Patil, as my representative, had invalidated their knowledge, made them feel as if they were lying to me. This was all on the instigation of higher caste men with houses on the other side of the road. Patil, by his association with me, and I were co-conspirators challenging their knowledge and tradition that was so sacred to them. I can see why they were outraged and left.

After I composed myself and made an effort to burry myself in my notebook in the dim streetlight while waiting for the bus to arrive—it was by now late in the evening—suddenly the little boy who was sent to fetch Fatima appeared in front of me from nowhere. He had a message—if I wanted to hear the song then I should come tomorrow at eight in the morning and wait at the temple. I realized that tomorrow was Sunday, the one day in the week most of them did not work as agricultural day labourers.

'What should I tell them?'—the boy demanded my answer.

'Yes. Yes. I will be there.'

He disappeared in the darkness quickly before I could ask him a few more questions like who sent him, and if he could offer them my sincerest apologies. Although sending a word of apology would have been very complicated. I would have to first convince Patil to turn my elite English into a culturally acceptable medium of words that could have truly related a message of apology... too much at stake to get lost in translation. I did nothing, said nothing.

Patil and I promptly arrived next morning at eight and as instructed I lodged myself at the steps of the temple pretending to do some serious writing in my silver-thing—just to discourage anyone dragging me into a conversation. I especially did not want to be seen talking to any man from the other side of the road when the little boy or someone else came to fetch me.

No one did. The clock ticked to nine and then ten AM, yet no one came. So, I sent a message with the help of the roadside teashop owner. The teashop was the radio-station of the village. Being accepted by the teashop owner as one of 'them'—was a rite of passage I had to earn for any meaningful research in the village, and I was a proud member who had by now performed the rite. So, message sent through the radio, and the little boy appeared again from nowhere. 'Wait', he told me.

***

It was almost eleven AM when Fatima and Gopi arrived to fetch me, they quickly pushed all my scattered stuff in my shoulder bag, dragged me out of my slumber and nudged me to hit the road with a sense of urgency, 'come on, let's go'. I could not understand why we were suddenly in such a rush until I tried to get Patil's attention, 'hurry, where are you?' But Fatima walked swiftly ahead dragging me along and said, 'Patil will not be allowed there, I will do the translation.' Aha. They were walking fast to put distance between me and Patil.

Through a narrow alley off the main road, I was led into a big square made of stones, with houses on all four sides. One side of houses had their main entrances opening onto the main village road and had their back entrances opening onto the square, and houses on three other sides as well had their (back) main entrances opening onto the square. This was the heart of the Dalit *basti* (neighbourhood). I always met them in the fields, so I had never visited their homes, I didn't know such architectural marvel existed behind some of the front doors visible from the village road. The stonework of the square was exquisite, and walls of the houses facing the square were adorned with beautiful artwork. I immediately realized how these Dalit people were generationally artisan caste of stoneworkers repeatedly referred to as Voddas in the folk stories and songs. It must have taken generations of stoneworkers to build the beautiful stonework in the square.

I was led to one corner of the square where I saw Lalitha, Sharda and other women sitting on the floor waiting for me. I could sense a kind of tension among them as if they might have been debating and anticipating my arrival for a while. As soon as I settled down on the floor, Lalitha came straight to the point: 'We will sing a whole song for you today, only if you agree to pay each one of us one day's wage'. I was shocked with this proposition. I saw some fourteen women sitting there behind Lalitha.

Paying them to sing? Why?

My white, Dutch, male PhD supervisor's stark, imposing voice in my head was telling me: 'NO! Say No! You never pay to your research informants.'

'Yes, I will pay', I answered without batting an eye. I instantly realized that what happened last evening was a game-changer, the time to tease and laugh was over, and they wanted to know if I was serious, if I truly respected their knowledge and if and how I was willing to suffer/to pay to gain access to their generational memories turned into this artwork of a song. It was not about money, although that was the only transactional measurement they could rely on to be convinced of my seriousness.

But then in my mind when I did the math, I realized it was a big sum, and I did not even have that kind of cash with me at that moment. Well, I could bring the correct amount and pay them the next day—I assured myself.

So, when I agreed to pay, they started their preparations. It took almost half hour for them to match their voices as they always sang in pairs and took turns to sing otherwise it would have been exhausting, singing for long hours at a time. Two women in a pair also had to sit next to each other to coordinate through non-verbal gestures at the time of singing, so the women in a circle moved around physically several times while matching their voices.

Fatima explained that these women learned the song and the singing from the older women in the community, they learned to sing by listening and watching the women of older generations performing this art form, it wasn't as if some special classes were conducted. And even though Fatima herself knew the song, she was not one of the singers because she had not been trained since her childhood. Fatima further explained to me how the women were matching their *svara*—identified as both musical note and tone, in other words, what might be described as tunefulness in their voices that day. What was most astonishing to me was the ample care these women took to fine-tune their voices, which I could not help myself from comparing with the Hindustani classical music traditions developed and sung by mainly Brahmin higher castes—how, generally, even in the public performances, the first few pieces of singing always were a kind of *riyaaz* (rehearsal), practicing to find that vocal scale and tone. Why was I astonished? Because my inadvertently, consciously-unconsciously, deeply ingrained elite mentality and attitude—in short: my snobbery—expected a degree of roughness in their folk singing. The impressively hair-splitting perfection of the vocal scale and tone must be the forte of the leisure classes of higher castes, poor labourers cannot afford that kind of precision, right? Wrong. These Dalit women taught me so much of my own unexamined biases.

And then it started. These women not only remembered the song by heart, no one needed to check the notebook or something, but their voices were expertly well-trained. Much of the song was sung in high notes and that scale of vocal tone that even very experienced singers of classical music would have found hard to maintain for a long period of singing.

Even without knowing the proper lyrics, the song had an instant effect on me. The deep, haunting quality of their voices was enough to understand and feel how the song was about the pain and grief inscribed all over their human lives for generations. It's not the lyrics, but the poignant and evocative quality of their high-pitch, intense singing in pairs that conveyed how the song was about remembering the melancholy indelibly inscribed over their collective history. The way the song was sung kept the deep emotions alive that connected them to each other and to their past and present and their environment. Even when I had no connection with their collective story, their singing alone powerfully invoked my own very personal existential pain of being human.

Fatima did some translation while the song was being sung, however, I present it here based on the translation I later received from an expert on Kannada language.

The name of the song *Kanne Viramba*—literally means 'a brave (hero) woman'.

The song began with the eulogy to the great victory and glory of the virgin goddess—Kanne Viramba—who was a young girl, the seventh daughter of the village chieftain, when 'sacrificed' in the tank. After the eulogy to the virgin goddess, singing which, looking back, must have easily taken up the first hour, the song told the story of the tank being built. 'Once upon a time', it began, 'a snake with seven hundred heads showed seven bushels of gold coins to the Muddanna [village chieftain] who invited seven hundred Voddas [artisan caste of stoneworkers] to build a tank. Seven sluices were built in seven years with seven human sacrifices and five more deaths.' The song then sang the glory of the Muddanna, declared that the tank was built based on moral truth and the principles of moral victory (dharma), and finally again sang the glory of the virgin goddess and considered her sacrifice as the most glorious among all the sacrifices in the tank.

I later found several similar songs archived in the library of folk literature in Kannada. As I had suspected, there were several versions of these songs, each named after the woman sacrificed. One, called *Kerege Haara*, is famous in Karnataka and is considered an especially fine work of poetry; it occupies a prominent space in the Kannada literature and is considered a symbol of Kannada nationalism and identity. There are also other versions—*Madagada Kenchamma*, *Kere Hunnamma*, *Anajee Honnamma*, and *Kaneramma*. Although these songs differ in terms of their content, style of narration, and the name and context of the social actors depicted, they all broadly narrate the pathos and plight, love, longings, and desires, glory and sacrifice of the women thus murdered.

I later reviewed this folk literature—those stories and songs I collected during my fieldwork and those already archived—in relation to the extensive historiography of south-Indian polity and culture during medieval times largely based on the written sources, and wrote a paper—what else can a boring academic do?[2]

2. E. Shah, 'Telling Otherwise: Historical Anthropology of Tank Irrigation Technology of South India', *Technology and Culture*, (49), 2008, pp. 652–674.

In the paper I showed how folk literature was non-elite and marginal people's oral history that was rarely included in the historiography based on written and archival sources, and how including this literature gave an entirely new perspective on the history of tank irrigation in medieval times and later. There's much we can learn from these songs:

'The human drama in these stories and songs stems from the fact that the artifact [the tank—irrigation reservoir] around which the drama takes place is vulnerable to an unpredictable environment. ... The story of a woman about to be sacrificed in a tank is testimony to this hydrological irregularity and the resulting technological vulnerability.'

Detailed readings of this folk literature allowed me to show how a staggering number of tanks built all over south India over centuries likely were constructed with very simple instruments such as rakes, pickaxes, crowbars, and shovels, plus the most crucial resource—forced or coerced labour and the knowledge of the members of the artisan caste of Voddas. In the historiography of tank irrigation technology, often romanticized as the ideal mode of precolonial and traditional water management practice, this aspect of the involvement of forced or coerced labour of Voddas was rarely mentioned. In the paper I also argued how folk literature provided a unique historical perspective: how the establishment of any community involved acts and events of violence, how the collective memory immortalized in folk songs and stories was storage for these violent blows, wounds, and scars, and how by performing the folk arts, the collective memory of these scars and hence the memory of the shared history were kept alive and reproduced.

Returning to the story, in the late afternoon the entire singing party moved to the temple located inside the Dalit *basti*. A proper pooja to the deity was offered with the help of the priest, when the last part of the singing for another hour continued inside the temple, and when finally Patil was allowed to join, who brought pooja offering to my own much worshiped deity—a new pair of pencil cells. At the end of the temple ritual, a plate with pooja lamp was circulated among the people, who offered their donations.

When the plate came to me, I emptied all the cash I was carrying in my pockets and in my purse and whatever additional money Patil kept with him for our fieldwork, saving only the return bus fare. This was understood as my payment to the women who sang. No one counted, no one mentioned money, when after the pooja, the whole noisy party walked towards the bus stop. While seeing me off, a shower of advice followed: not to worry too much about small things, not to waste my life buried inside the silver notebook, to become a good doctor, to eat well, to take care of myself, and a very special request—to remember them.

They all walked alongside the bus, me inside, until they couldn't. And I held on to their faces losing focus from my raining eyes, until I couldn't.

That was my last day in the village, I could never go back. I don't know why, but I couldn't. And I don't know why but they also knew.

5.2 The Water Clock and the Water Thieves

INTERVIEW WITH AHMED SALEM DABAH & HEIDI VOGELS
BY HENRIËTTE WAAL & CLEMENS DRIESSEN

A water clock, also known as a clepsydra ('water thief' in Greek), measures time using water flowing in or out of a reservoir. In North Africa and the Middle East, the water clock is traditionally used to distribute the water from the communal aquifer, not in terms of volume but by time.

Each family has the right to the total yield of the qanat (water canalization system in Arabic) for a certain time, indicated by the water clock.

In Tighmert, an oasis near Guelmim in the municipality of Asrir, in the mountains of southern Morocco, 600 families (4,000 residents) from the Saharan Azwafit and the Berber tribe Ait-Hmad share their spring water. The oasis consists of a ten-kilometre-long and four-kilometre-wide strip of palm trees, adobe architecture, and small-scale fields and gardens. The water source, two kilometres outside the oasis, provides a constant flow of water through a network of channels to the labyrinth of garden plots.

Tighmert kitchen, 2023, film still.
Image credit: Heidi Vogels

The culture and social cohesion of the community are strongly linked to their interaction and relationship with water. Families adjust their day-and-night rhythm to the moment they receive water. Nighttime water is best, as evaporation is minimal. Everyone depends on this structure. If a problem arises, it is immediately resolved to ensure the water keeps flowing. The water from the spring is free, making food cultivation possible without major investments.

ASD We have a shared water system. We call it Senova in Arabic, a water calendar. Every family owns the water and they share it through the system. The water clock of Tighmert is a metal bowl with a hole in the middle. They put that bowl in the water, where it slowly fills up through the hole. When the bowl is completely filled and sinks, it is the next family's turn. There is one water clock, and the families use it for sharing. My family owns the water on the eighth day. We have three hours of water coming into our gardens on that day. Nobody else can touch or use the water when someone has it. It's a big disgrace to use water that is not yours.

The groundwater-bearing geological layers in this desert region have led to the establishment of industrial plantations for melons, avocados, and bananas in the immediate vicinity of the oasis. The families of Tighmert have protested and gathered evidence of their decreasing water flow, but it hasn't helped. Authorities stated the area is rich in water, and the communal ownership of the water is not acknowledged. The construction of a dam in the nearby region has led to a displacement of underground water reservoirs, depleting the original water source. Irrigation of the oasis through the canal network has stopped for several months.

A communal well, thirty meters deep, was dug to restore the water flow to Tighmert. Families who can afford it install wells or water pumps in their own gardens. While it is dishonourable in the oasis community to take more water, agribusiness, without any connection to the community or place, shamelessly grabs it. With no face, one can't lose face.

Currently, and due to the water crisis, the oasis is in a transition process to other ways of living, supported by members of the tribe from around the world who are connected to the oasis via reports from Ahmed and two other inhabitants in a WhatsApp group. Unlike the oasis itself, the online community is well maintained and continually shares messages about solutions to the water crisis and raises funds for practical matters. They managed to organize a protest in Paris.

Tighmert field, 2017, film still.
Image credit: Heidi Vogels

ASD Local journalists are not interested in our protests, and the local TV channels are also not reporting about it. When you protest here, our authorities say it is climate change. This is a problem. I think we have to understand that when people say it's climate change, there might be other problems.

The WhatsApp group's advice is not always useful. There was a suggestion to cover the water surface with black floating balls to slow down evaporation. An idea that was used in reservoirs in Los Angeles.

ASD We also use this water for drinking. We were not sure whether the balls would have a negative effect on the water, maybe with heat from the sun, the balls would release some product into the water. We are not experts in that.

Ahmed does not want to discuss covering irrigation canals or storing water, because it wouldn't make much difference. The advice needs to be somewhat available or practical for Ahmed to be able to implement them on-site.

There is a lot of work on the water now, and the electricity to pump water must be generated and paid for. They use solar panels during the day and a generator at night. Through the app group, they raise money to pay the guard who watches over the solar panels and the pump. He lives in a tiny cabin next to it.

Rain and the resulting flooding are used by the families to plant grain and rice. They sow when it starts raining.

Tighmert source, 2023, film still.
Image credit: Heidi Vogels

Tighmert source, 2017, film still.
Image credit: Heidi Vogels

ASD Now that there is scarcity, we grow less. We use the water from our source just for what is already planted, to keep existing trees alive. Such as fruit trees, we have pomegranate and figs. We have many fruit trees in the oasis and date palms too. We don't plant vegetables anymore, but we do plant alfalfa for the animals. Before, I had a twenty-square-metre plot, now it's less because we don't have the same quantity of water.

When we ask about the future, there is silence, polite but awkward laughter—more silence.

Tighmert water clock, 2024. Image credit: Sabar Hamza

ASD To protect the people, the government needs to put restrictions on how much water can be used and we need to explain to the government that they are working against themselves. When you eat watermelons, avocados, and bananas, you are slowly eating our environment. People don't understand that the oasis is protecting the big cities. If there is no oasis, the sun will come to the big cities.

With the community we make small videos that we show in film festivals. In the global reunion of the tribe in August, we also talk about this problem. We have collected proof over the years. The message that when the oasis dies, it will affect the cities, is quite strong. But I'm afraid the majority of people will still not get it. They will see it as a local (rural?) problem.

We need support to transition to other ways of living. I hope this book can contribute to that.

Mahmoud Khattab, processed satellite images around Tighmert Oasis, 2024 (pp. 251–254)

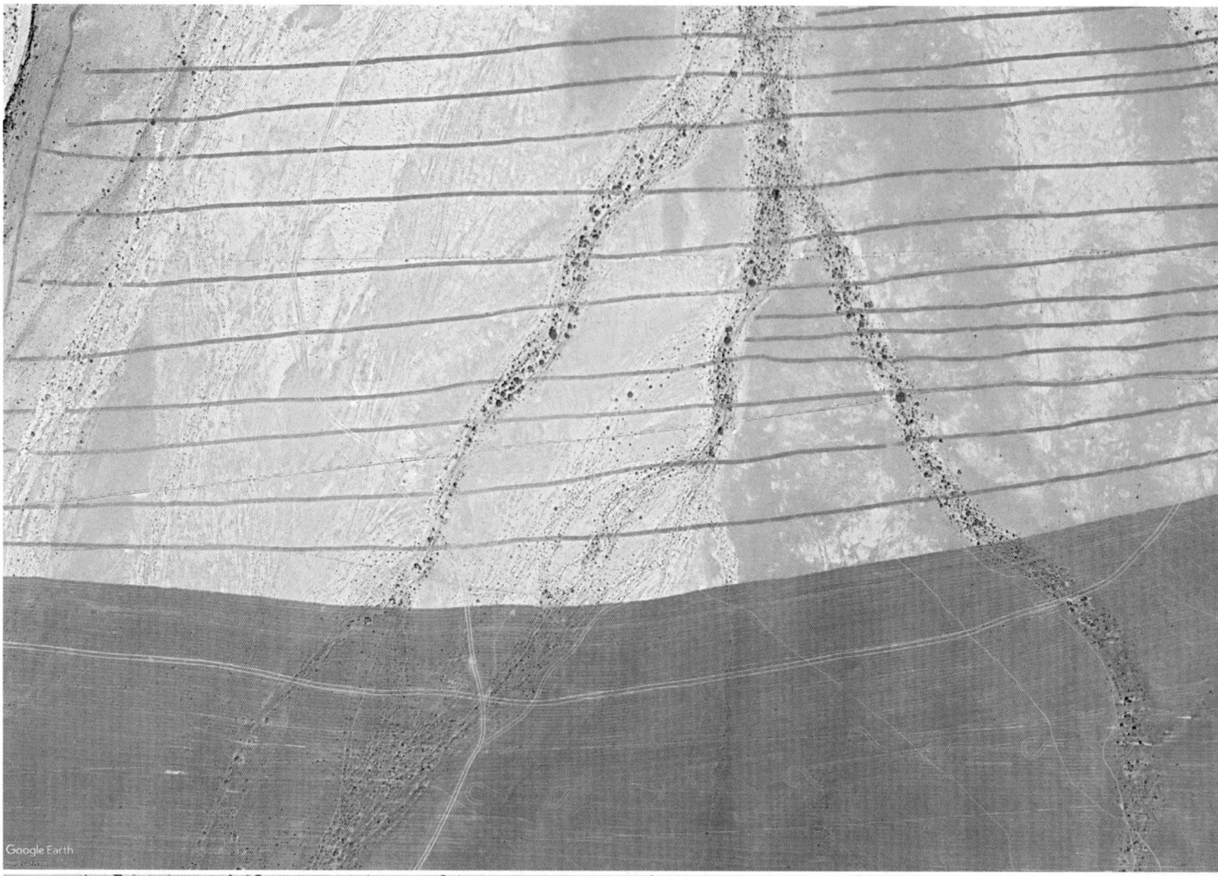
Google Earth

Mari Bastashevski is an artist and writer, grafted onto academia as a postdoctoral researcher at Wageningen University. Her work centers on the future of cultural imaginaries beyond the binary of techno-optimism and collapsism. In the past, her art and research practice spun five counter-forensic interconnected case studies into the role of media in the global conflict economy.

5.3 Flint Water

BY MARI BASTASHEVSKI

Mari Bastashevski, *Councilman Erik Mays*, Site: Water Conference, City Hall, Flint Michigan, US, 2017. Access: Permitted with written authorisation

If lead water poisoning of the African-American community living in Flint, Michigan, US, was an event, it was an event without a date.

> The crisis did not begin in the city of Flint itself; it was baked into one of many routinely drafted business plans to uplift a city put together by a network of state and national actors in Michigan and Washington, D.C.. The plan had also included think tanks, private consultants, contractors, a multibillion-dollar corporation in France, and a usual number of non-player characters that seem to have existed on paper exclusively. While the list of actors involved was never definitively concluded, the series of abstract agreements had distinctly material consequences for the residents of Flint who were excluded from the decision-making process on the account of insufficient expertise. Not only that, when it came to crisis resolution, the same residents of Flint most affected by these actors were once again excluded from the decision-making process, this time on the account of being too invested to make an objective decision. The only solutions implemented to resolve the crisis to date were those proposed within the same circulator network of actors, practising the same notions of knowledge and power, and class and race exclusion, in a circular loop of crisis-solution economy.

This systematic production of crises is a form of governance Yale professor Jason Stanley calls managerial democracy. The template for managerial democracy, similar to classic settler colonialism, is designed around uprooting an already impoverished demographic by rapidly dismantling public services, making entire neighbourhoods legally as well as practically uninhabitable, but exceptionally valuable to real estate brokers and anyone else who can afford to go private and wait until a new neighbourhood is built in place of the one evicted.

The emergency manager, a formal title for the government official managing a district during a bankruptcy, is seldom from the neighbourhood they're assigned to manage and does not linger there once it has been irreversibly mismanaged. Several emergency managers who have overseen the poisoning of Flint have since moved through a series of trusteeships in different states, producing both the crisis and lucrative solutions as instant retainers for depressed towns that end up paying exorbitant consultancy fees.

Unlike other case studies in this manuscript, Flint poisoning has been coded and recorded in both media and art contexts. In 2016 and 2017, Flint attracted a large number of local and national reporters. To this day—April 2024—googling 'Flint' yields a classic cache of images that are a literal pictorial representation of the crisis: disgruntled people of

GENESEE COUNTY DRAIN COMMISSIONER'S OFFICE

JEFFREY WRIGHT

COMMISSIONER

G-4608 BEECHER ROAD, FLINT, MI 48532

PHONE (810) 732-1590 FAX (810) 732-1474

FOR IMMEDIATE RELEASE | CONTACT: Jeff Wright

March 26, 2013 | 810-287-1925

What: Jeff Wright Statement on Flint City Council Approving Resolution to Join KWA.

When: March 26, 2013

Additional Info: Jeff Wright, CEO of the KWA, today praised the Flint City Council for voting to approve a resolution to officially have the City of Flint join the KWA and its' water line.

"I have said from the beginning that this decision must be made by Flint's City Council and Mayor," said Wright. "I appreciate the council voting the way they did, but even more than that, I am glad the residents of Flint were able to have their voices heard via their elected officials." The next step in the process will come with the KWA Board executing the contract for the City.

"The most important aspect from the vote is the fact that we as an Authority can move forward knowing Flint's intentions," said Wright.

Despite Emergency Manager Ed Kurtz publicly supporting the project, Wright had made a vote by council a condition of Flint joining the KWA so the residents would have their say.

"There is a basic tenet that government is best when it has local control. We saw that with the council vote. Nobody, whether they live in Flint, Grand Blanc, Davison, Fenton, or anywhere in Genesee County, should have these types of decisions made by people who live outside of their community," said Wright.

At the next Water and Waste Advisory Board meeting a vote will be taken to award a contract for construction of the intake portion of the water line. All bids were submitted to the board at their last meeting for review.

Mari Bastashevski, *Flint Water Wall 1, Slide 1, Document 1*, 2017. All documents submitted were added to the class action suit by the city of Flint against numerous entities responsible and were shared by the legal team representing the victims

89 (Rev. 02-11)

STATE OF MICHIGAN
DEPARTMENT OF TREASURY
LANSING

RICK SNYDER
GOVERNOR

ANDY DILLON
STATE TREASURER

April 11, 2013

Mr. Edward Kurtz
Emergency Manager
City of Flint
1101 South Saginaw Street
Flint, MI 48502

Dear Mr. Kurtz:

Thank you for your March 29, 2013 letter, which is attached for reference. As the Emergency Manager for the City of Flint ("Flint" or "City"), you have asked for my concurrence, pursuant to Public Act 436 of 2012, the Local Financial Stability and Choice Act, to authorize a contract in excess of $50,000.00 not subject to competitive bidding. This request was related to the City entering into an agreement with the Karegnondi Water Authority (KWA) for provision of raw water for the City.

In considering your request, I took note of the following facts in support of Flint joining KWA. First, there is widespread support in the City for this move, including the support of the Mayor, the City Council, and the Emergency Manager. Second, this move will provide a unique opportunity for the City and County to partner on an important project, which will hopefully lead to future regional collaboration. Third, the Department of Environmental Quality is supportive of the City participating in the KWA project. Finally, your representations that this deal will lead to substantial savings for the City over the coming decades, savings that are desperately needed to help with the turnaround of the City of Flint.

It is my understanding that the Detroit Water and Sewer Department is making a final best offer to Genesee County and the City of Flint next Monday, April 15, 2013. As such, this approval will be effective at 5 pm on April 16, 2013 after receiving written notice from the City that either no such offer was presented to the county and the City or that an offer was received and was rejected in good faith based upon specified objections.

For the reasons described above and subject to the conditions set forth herein, pursuant to Section 12 (3) of PA 436 of 2012, I am authorizing you to proceed with adopting the resolution and entering into a contract with KWA.

Sincerely,

Andy Dillon
State Treasurer

P.O. BOX 30716 • LANSING, MICHIGAN 48909
www.michigan.gov/treasury • (517) 373-3200

Mari Bastashevski, *Flint Water wall 1, slide 1, document 2*, 2017

Mari Bastashevski, *Untitled*, Site: Housing auction on houses foreclosed after families evicted due to lack of access to fresh drinking water, Detroit Michigan, US, 2017

colour, dirty bottled water samples, stock-like images of the water tower which has become the postcard paradigm of the city. Existing artworks referencing Flint tend to fall under a symbolic critique of capitalism. In this chapter, I search for the limits of material boundaries in the transactional war for resources: I go into and past spaces in which what happened was organized in time, I follow the thread in the documents, until it goes too far, involves everyone, and loses all meaning.

As usual, the investigation involved copious time-consuming field and OSINT research which included interviews with corporate figures, government officials, treasury department members, lawyers, academics, philosophers, and Michigan civilians involved and affected by the crisis. The documents used in this exhibition include fragments of interviews, legal paperwork provided by the lawyers, an email archive of current Governor of Michigan, Rick Snyder, made public through Freedom of Information Requests, various corporate and state contracts were obtained legally from the employees or found online using Google for penetration searches. The book page in the presentation of General Motors comes from Michael Hardt and Antonio Negri's book *Multitude: War and Democracy in the Age of Empire* and is used with the authors' permission.

All documents submitted were added to the class-action suit by the city of Flint against entities responsible and were shared by the legal team representing the victims.

MICHIGAN DEPARTMENT OF ENVIRONMENTAL QUALITY

PERMIT APPLICATION FOR WATER SUPPLY SYSTEMS
(CONSTRUCTION - ALTERATION - ADDITION OR IMPROVEMENT) AS DESCRIBED HEREIN
Required under the Authority of 1976 PA 399, as amended

This application becomes an Act 399 Permit only when signed and issued by authorized Michigan Department of Environmental Quality (DEQ) Staff. See instructions below for completion of this application.

1. **Municipality or Organization, Address and WSSN** that will own or control the water facilities to be constructed. This permit is to be issued to: City of Flint 4500 North Dort Highway Flint, MI 48505 **WSSN**: 02310	Permit Stamp Area (DEQ use only) MICHIGAN DEPARTMENT OF ENVIRONMENTAL QUALITY PERMIT NO. W 14[illegible]026 APR 09 2014 EXAMINED AND APPROVED FOR COMPLIANCE WITH ACT 399, P.A. 1976	
2. **Owner's Contact Person** (provide name for questions): Contact: Brent Wright Title: Plant Supervisor Phone: 810-787-6537		
3. **Project Name** (Provide phase number if project is segmented): Flint WTP Phase II, Segment II - Lime Residual Disposal	4. **Project Location** (City, Village, Township): City of Flint	5. **County** (location of project): Genesee County

ISSUED UNDER THE AUTHORITY OF THE DIRECTOR OF THE DEPARTMENT OF ENVIRONMENT QUALITY

cc:

Issued by: [illegible signature]

Reviewed by: Patrick Cook for MFP
for Mike Prysby

☐ **If this box is marked see attached special conditions.**

Instructions: Complete items 1 through 5 above and 6 through 21 on the following pages of this application. Print or type all information except for signatures. Mail completed application, plans and specifications, and any attachments to the DEQ District Office having jurisdiction in the area of the proposed construction.

Please Note:

a. This **PERMIT** only authorizes the construction, alteration, addition or improvement of the water system described herein and is issued solely under the authority of 1976 PA 399, as amended.
b. The issuance of this **PERMIT** does not authorize violation of any federal, state or local laws or regulations, nor does it obviate the necessity of obtaining such permits, including any other DEQ permits, or approvals from other units of government as may be required by law.
c. This **PERMIT** expires two (2) years after the date of issuance in accordance with R 325.11306, 1976 PA 399, administrative rules, unless construction has been initiated prior to expiration.
d. Noncompliance with the conditions of this permit and the requirements of the Act constitutes a violation of the Act.
e. Applicant must give notice to public utilities in accordance with 1974 PA 53, (MISS DIG), being Section 460.701 to 460.718 of the Michigan Compiled Laws, and comply with each of the requirements of that Act.
f. All earth changing activities must be conducted in accordance with the requirements of the Soil Erosion and Sedimentation Control Act, Part 91, 1994 PA 451, as amended.
g. All construction activity impacting wetlands must be conducted in accordance with the Wetland Protection Act, Part 303, 1994 PA 451, as amended.
h. Intentionally providing false information in this application constitutes fraud which is punishable by fine and/or imprisonment.
i. Where applicable for water withdrawals, the issuance of this permit indicates compliance with the requirements of Part 327 of Act 451, Great Lakes Preservation Act.

1

Mari Bastashevski, *Flint Water Wall 1, Slide 1, Document 3*, 2017

The Parties hereto shall have the right from time to time and at any time to change their respective addresses and each shall have the right to specify as its address any other address by at least fifteen (15) days written notice to the other Party hereto.

IN WITNESS WHEREOF, the Parties hereto acting under authority of their respective governing bodies have caused this License to be duly executed in several counterparts, each of which shall constitute an original, all as of the day and year first above written, which is the License and Effective Date.

GCDC:

By: Jeff Wright
Its: Drain Commissioner

CITY:

By: Natasha Henderson
Its: City Administrator

By: Dayne Walling
Its: Mayor

Attest:

By:

By:

This License was prepared by:
Attorney Kevin Kilby (P68599)
McGraw Morris P.C.
2075 West Big Beaver Road
Suite 750
Troy, Michigan 48084
Phone: (248) 502-4000
Facsimile: (248) 502-4001
Email: kkilby@mcgrawmorris.com

Approved as to form by:

Peter M. Bade (P47546)
Chief Legal Officer
City of Flint
1101 South Saginaw Street, Room 307
Flint Michigan 48502

N:\GCDC\2155-1000 (General)\Contracts and Agreements\DWSD regarding transmission of water\License to Transmit Water between Flint and GCDC Draft Date October 14 2015 Final Execution Copy.doc

top executives, and revenues are commingled and reported as one.

33. As such, the Veolia Defendants disregard corporate formalities and do not adequately or separately capitalize each of their respective businesses.

34. The Veolia Defendants collectively maintain several offices in Michigan, including Westland, Livonia and Grand Rapids; regularly conduct business in Michigan; and have committed torts in Michigan, which are bases for personal jurisdiction in this Court pursuant to MCL § 600.715.

35. The Veolia Defendants are parties to this action based upon providing professionally negligent engineering services in reviewing Flint's water system and declaring the water safe to drink.

36. The Veolia Defendants have abused the corporate form to avoid liability in this matter. Indeed, while it appears Veolia Water may have executed the subject contract with the City of Flint, the Veolia Defendants, together, as indistinguishable entities, performed services and presented their conclusions as "Veolia." At all times relevant, the Veolia Defendants have represented to the world that they are not distinct legal entities but rather one and the same.

37. The Veolia and LAN Defendants are parties to this action because, among other reasons, they were professionally negligent in failing to conduct a root cause analysis which would have revealed that the pipes were corroding and causing lead and legionella to enter the residents' homes, failing to advise the City that they were out of compliance with the Safe Drinking Water Act's Lead and Copper Rule, falling to advise the City that the addition of a corrosion inhibitor was necessary to prevent lead poisoning and Legionnaires' Disease, and advising the City to use or even increase the dosage of highly acidic ferric chloride.

38. This court has subject matter jurisdiction over the claims asserted in this lawsuit

9

Mari Bastashevski, *Flint Water Wall 1, Slide 1, Document 5*, 2017

Clara Sika Helbo is a Danish researcher and designer dedicated to challenging systemic inequalities through investigative methods and material translations. In 2023, she graduated cum laude in Social Design from Design Academy Eindhoven. She creates hybrid archives and installations that reveal hidden power structures and make the inaccessible more legible.

5.4 Water Cannon Warfare: Mapping and Memorializing Impacts on Protesters in Public Spaces

BY CLARA SIKA HELBO

Triggered by the incidents involving the use of water cannons during the 2022 Covid-19 anti-lockdown protests in The Hague, Clara Helbo began an investigation into the utilization of water as a weapon. Drawing information from various sources, such as video analyses, frequency mappings, interviews with protesters and policymakers, and WhatsApp conversations with police officers, we get a sense of the deployment

of water cannons during protests. As the project examines the impact being hit by a water cannon has on individual people as well as the experiences of police officers using this device, we are struck by the lack of comprehensive policies governing their use.

Compiled into an archive, the findings reveal the inherently violent nature of water cannons as tools of seemingly innocuous violence. The archive serves as a testament to a specific moment in time, preserving a memory that might otherwise have been washed away. The overarching goal of the project is to both contribute to and challenge the ongoing discourse surrounding the use of water cannons as a weapon for maintaining law and order.

WATER GUN NUMBER 1

On a cold rainy December morning during the Covid-19 pandemic in the Netherlands I walked past a municipal worker hosing down the public rubbish bins with a high-pressure water spray-gun in an attempt to clean it. That forceful use of a precious commodity at a time of global freshwater shortages somehow stuck with me. That same night a protest against the Covid measures was featured on TV, showing...

BEN D'ARMAGNAC

I started my research in the library of Stroom Den Haag, and found a book from late artist Ben D'Armagnac describing one of his performances with a high-pressure waterjet:

In 1978 in New York a performance featured the artist lying on the terrace in front of the Brooklyn Museum in a black suit on white panels, subjected to a direct jet of water aimed at his heart. As his heartbeat was transmitted and amplified through a loudspeaker, he initially breathed heavily and moved his arms, but abruptly ceased all motion when the water jet intensified. Subsequently, the artist startled the audience by swiftly getting up and fleeing.

Protesters in Tbilisi on Tuesday March 7 shielded a woman waving an EU flag. Source: *CNN*, 8 March 2023

His endeavour to communicate the incomprehensible force of water and its unforgiving nature intrigued me. I continued my research with Jonas Staal and his essay 'Democracy's Theatre of Cruelty', in which he delves into the intricate dynamics of democratic processes and institutions, offering a critical perspective on their potential resemblance to a theatrical production infused with elements of cruelty. In his analysis of the current torture methods deployed by the US government he explores a nuanced understanding of the complexities inherent in the relationship between democracy, power structures, and the emotional impact of political processes on individuals.

DIVING DEEPER

On my journey to unravel the impact of water cannons, I dug deeper into the stories of those directly hit by these powerful devices. Chatting with protesters who had faced water cannons during demonstrations revealed a harsh reality of injuries beyond immediate physical harm. Cases of hypothermia stood out, highlighting the severe consequences of high-pressure water, especially during prolonged exposure in sit-in protests. Moreover, I heard accounts of severe concussions and permanent loss of eyesight, painting a worrisome picture of potential long-term consequences that seemingly go undetected:

'It's not water, it's a jolt, like being hit by a rubber bullet but with continuous force'.

BENEATH THE SURFACE

Screenshot of Dick Cheney signing a waterboarding kit, alleged teaser trailer for Sacha Baron Cohen's new show, placed on Cohen's X page, 8 July 2018.

It's evident that water as a weapon is not limited to one area. There is also the example of Eindhoven, where a protester, Denisa Štastná, got slammed against a building passing an officer operating a water cannon. Instances of water cannons in action in France, Germany, Georgia, Colombia, Chile, Spain, Turkey, Israel and—with the highest frequency—the Netherlands echo a global pattern of using water for crowd control. The political context concerning water weapons is straight-forward. Operating a weapon of such power should adhere to explicit protocols, avoiding the direct targeting of individuals. The absence of transparency and legal accountability becomes a concern when an armoured vehicle manoeuvres through the streets, with the intent to aim at peaceful protesters...

To understand how police approach water cannons, I reached out to a former police officer from the Ertzaintza police force from the Basque Country in Spain. They abandoned the use of water cannons for crowd control, deeming them too forceful. 'We've stopped using it because we had too many issues with it and basically felt it was too strong for the purpose of controlling riots.'

WATERBOARDING KIT

The global conversation on water as a coercive tool gained attention when former vice president Dick Cheney mockingly signed a waterboarding kit on the satirical show 'Who Is America?'. This act not only highlighted the controversial use of water as a torture method but also sparked broader debates around the weaponization of water. Water may sound innocent and friendly as a form of governmental force, eliding the gruesome experience of induced drowning or being hosed down on the street.

5.5
Fly Fishing

By Henriëtte Waal

Bumble Bee Dry Fly, fake fly. Source: Getty Images

While fish and humans share the same planet, there is a difference between the world of the fish and that of humans. Fish live below the water's surface, hidden from view, in a world different from that of humans. Perhaps this is why the role of fish in our lives is mostly limited to (dead) fish as food (the fish stick phenomenon).

Fly fishers spend a lot of time crafting imitation flies, fly larvae, and nymphs to catch fish. Broadly, there are 'wet' and 'dry' flies. The wet flies sink, while the dry flies stay on the water's surface. It turns out that most fly fishers—primarily adult men—who tie their own flies develop a strong preference for patterns and materials used in their flies. From hare-hair bodies to wings cut out of turkey feather with adhesive spray, some binders have elevated this to an art form. There are a few handy rules to remember. One of them is the proportions in which you tie a fly.

'The challenge in fly fishing is not catching big fish, but trying to fool a big fish by simulating a fly as well as possible. Casting with a fly rod is a story in itself. There's no theory for it. The only way to learn it well is by practicing a lot and watching and listening to others.'—Fly fisher P., project **Hunnie**, Henriëtte Waal & Sophie Krier.

There are also various unhooking techniques. From the fish's perspective however, fly fishing is mainly a deception and a struggle against an invisible force. How would artificial flies taste and how would 'unhooking' feel if they were also shaped for the pleasure of the fish?

Fly tyer Alain Barthelemy (*Hunnie Vliegvissen, 'them fly fishing'*), 2013, project *Hunnie* by Henriëtte Waal & Sophie Krier. Photo: Roel van Tour

5.6
Squirt Gun

By Clemens Driessen

NASA's Jet Propulsion Laboratory is an institute in California, officially tasked with making space robots such as the Mars Rover—sent off to look for water—and life—on the red planet. In practice it has been the source of maverick ideas.

In the 1960s, two of its employees, Dian Hitchcock and James Lovelock—together with Lynn Margulis—developed the Gaia hypothesis: the idea that living organisms, adding up to the biosphere, through their metabolism affect the atmosphere and the hydrosphere: planet Earth is habitable due to the combined 'symbiotic' effects of living creatures.[1] Understanding these dynamics of the Earth as an evolving and self-regulating system is deemed crucial in understanding climate change—although some argue this view of planet Earth as a 'self-regulating' living system has led to climate denial, trusting a resilient planet that should just be able to absorb fossil fuel induced disruptions.[2]

1 Dian R. Hitchcock & James E. Lovelock, 'Life detection by atmospheric analysis', *Icarus*, (7:1-3), 1967, pp. 149–159.

2 Leah Aronowsky. 'Gas guzzling Gaia, or: a prehistory of climate change denialism', *Critical Inquiry*, (47:2), 2021, pp.306–327.

Lonnie Johnson was at JPL working on satellites, stealth bombers, and a heat pump, when in 1982 he invented the Super Soaker—or 'squirt gun'—that has become one of the world's bestselling toys. Johnson invested the proceeds of his patent in a series of companies to develop novel energy efficient engines and solid-state ceramic batteries that don't require lithium—the mining of which relies on massive amounts of water, destroying livelihoods and ecologies.

Johnson seems to embody engineering when it was still (also) about playing with ideas and materials, driven by the joy of making—before this approach to innovation and technology was replaced by the promises of Silicon Valley tech bros selling their data hoarding schemes to venture capital funded efforts aiming at global domination.

Lonnie Johnson with his Super Soaker outside his home, 1992. Source: Thomas S. England/The LIFE Picture Collection/Getty Images

'A toy squirt gun which shoots a continuous high velocity stream of water. The squirt gun is configured as a structure facilitating partial filling with water leaving a void for compressed air.'

United States Patent [19]

Johnson

[11] **Patent Number: 4,591,071**

[45] **Date of Patent: May 27, 1986**

[54] **SQUIRT GUN**

[76] Inventor: **Lonnie G. Johnson,** 1463 E. Barkley Dr., Mobile, Ala. 36606

[21] Appl. No.: **541,898**

[22] Filed: **Oct. 14, 1983**

[51] **Int. Cl.**[4] **B67D 5/32**

[52] **U.S. Cl.** .... **222/39;** 222/79; 222/401; 340/384 E; 340/406; 116/137 R

[58] **Field of Search** .... 340/384 E, 406; 116/137 R, 139; 222/39, 79, 323, 324, 401

[56] **References Cited**

U.S. PATENT DOCUMENTS

1,333,704	3/1920	Brinks	340/406
2,249,608	7/1941	Greene	222/79
2,302,963	11/1942	Lefever	222/39
2,589,977	3/1952	Stelzer	222/79
3,202,318	8/1965	Black	222/39
4,086,589	4/1978	Cieslak et al.	340/384 E
4,214,674	7/1980	Jones et al.	222/401
4,239,129	12/1980	Esposito	222/79

Primary Examiner—H. Grant Skaggs
Attorney, Agent, or Firm—John A. Beehner

[57] **ABSTRACT**

A toy squirt gun which shoots a continuous high velocity stream of water. The squirt gun is configured as a structure facilitating partial filling with water leaving a void for compressed air. The squirt gun includes a nozzle for ejecting water at high velocity, a pressurization pump for compressing air into the gun to pressurize water contained therein, and a trigger actuated flow control valve for shooting the gun by controlling flow of pressurized water through the nozzle. A battery-powered oscillator circuit and a water flow powered sound generator produce futuristic space ray gun sound effects when the gun is shooting.

9 Claims, 7 Drawing Figures

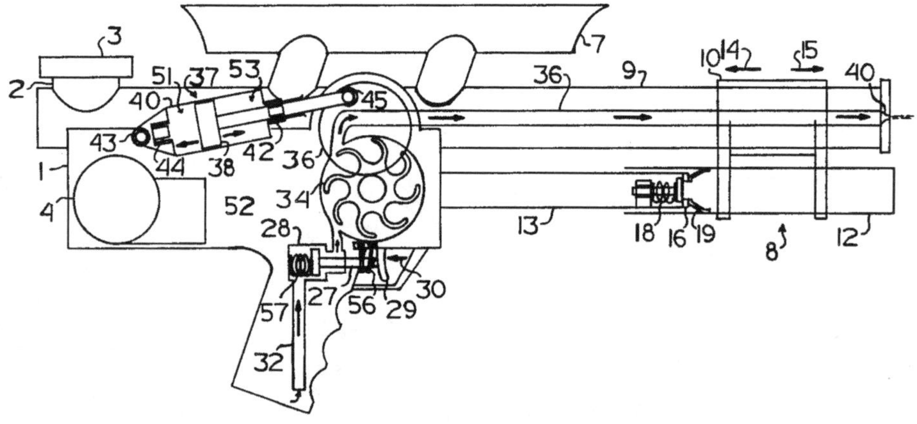

Page from full patent document Super Soaker US4591071, 1986

4,591,071

SQUIRT GUN

BACKGROUND AND SUMMARY

The principle utilizing compressed gas as a means for pressurizing water to effect a continuous stream of high velocity waterflow from a nozzle is common practice. However, the embodiment of this principle in a hand-held toy squirt gun having a futuristic space ray gun appearance and including sound effects is novel.

Compressed air is provided by a manually actuated air pump physically mounted underneath the gun barrel. Sound is produced by a battery-powered electronic oscillator circuit. Sound is also produced by a flow actuated sound generator.

To shoot the gun, a trigger is actuated to open a flow control valve and thereby permit pressurized waterflow through a nozzle. The water exits the nozzle and thereby the squirt gun at high velocity. The flow actuated sound generator is functionally coupled in series with the flow control valve and the nozzle to facilitate actuation of the sound generator by flowing water when the gun is shooting.

The waterflow rotates an impeller which is mechanically linked to a vibration means consisting of a piston in a cylinder to effect back and forth motion of the piston and thereby pump air alternatingly through a pair of vibrating reed-type horns to produce sound.

A switch coupled to the trigger is actuated simultaneously with opening of the control valve to switch on the oscillator circuit to produce sound.

BRIEF DESCRIPTION OF THE DRAWINGS

FIG. **1** is a functional diagram of an embodiment of the invention showing actuation of the trigger to shoot the gun.

FIG. **2** shows the resealable cap of the water fill port removed.

FIG. **3** shows actuation of the pressurization pump in the compressive stroke.

FIG. **4** shows actuation of the pressurization pump to resupply air into the pump cylinder.

FIG. **5** shows the trigger actuated flow control valve in a closed state with the control switch for the oscillator circuit in the off position.

FIG. **6** shows additional details of the flow actuated sound generator. The drawing depicts magnetic coupling of motion of an impeller driven gear inside the gun to a flywheel on the outside.

FIG. **7** is a simplified diagram of a battery-powered oscillator circuit.

DETAILED DESCRIPTION AND PREFERRED EMBODIMENT

An in-depth understanding of the present invention can be derived from the following description with reference to the drawings.

FIG. **1** is a configuration drawing showing the major features of the squirt gun. Squirt gun structure means **1** is designed to contain water and air under high internal pressure. Structure means **1** also provides mounting support for various elements of the gun. A water fill port means comprised of port **2** and screw on cap **3** function as a resealable port for opening structure means **1** to put water in the gun and resealing so as to not permit leaks when the gun is operating under high internal pressure. Screw on cap **3** is shown removed in FIG. **2**.

Air reservoir means **7** is designed as a rifle scope means to enhance appearance. It is an integral part of structure **1** and functions to prevent the gun from being completely filled with water. When the gun is held upright with the fill port open, reservoir **7** is above the maximum water level determined by the position of fill port **2** and contains air to insure there is always a volume inside the gun into which air can be compressed.

A manually operated pressurization pump means comprised of pressurization piston means **13**, check valve means **16**, seal **19**, spring **18** and a movable member means comprised of grip **10** and pressurization cylinder means **12** functions to compress air into structure means **1** and thereby air reservoir **7**. Grip means **10** provides hand-held support for the gun and is movably coupled by concentric cylinder mounting to gun barrel means **9** to permit sliding along the length of the barrel. Gun barrel **9** and piston **13** are integral parts of structure means **1** and provide additional internal volume for containing air and water. Air pump cylinder **12** forms the bottom portion of grip **10**. Compressed air is pumped into structure means **1** by moving grip **10** and cylinder **12** back and forth along the barrel in a pump shotgun type action as depicted by arrows **14** and **15**. The compressive stroke is in the direction of arrow **14** as shown in FIG. **3**. As cylinder **12** is moved further onto piston **13**, pressure inside cylinder **12** increases and opens check valve **16**. The compressed air flows into structure means **1** as illustrated by arrow **17**. Check valve **16** is normally held closed by spring **18**. The refill stroke of cylinder **12** is shown in FIG. **4**. As cylinder **12** is moved in the direction of arrow **15**, piston **13** is withdrawn from cylinder **12** and ambient air is sucked into cylinder **12** past seal **19** as depicted by arrows **20** and **21**. The gun is pressurized by repeating this cycle with back and forth motion of grip **10** along barrel **9**. The maximum pressure reached inside the water gun is determined by the ratio of the maximum volume to minimum volume created inside pump cylinder **12**.

As shown in FIG. **1**, to shoot the water gun, control valve means **28** is opened by pressing trigger means **29** in the direction of arrow **30**. Control valve means **28** is coupled to structure means **1** and includes movable shaft means **27** which extends external to structure means **1**. Trigger means **29** is attached to shaft means **27**. With valve **28** open, pressurized water flows into conduit means **32** and up through valve **28** from the bottom of the gun as the compressed air in the top part of the gun and in reservoir **7** expands. Pressurized water flow exiting valve **28** impinges impeller means **34** causing impeller **34** to rotate as pressurized water flows through to conduit means **36**. The pressurized water flows through conduit **36** to nozzle means **40** and exits the gun at high velocity.

Futuristic space ray gun sounds are produced by a flow actuated sound generator means. The flow actuated sound generator means includes impeller **34**, which is configured operably in series with valve **28** and nozzle **40** as described. Impeller **34** functions as a flow responsive means for extracting operating power for the sound generator from the pressurized water flow. Referring to FIG. **6**, the flow actuated sound generator further includes vibration means **37**, and a coupling means comprised of a first rotatable structure means represented by gear **46**, a second rotatable structure means represented by flywheel **36**, first magnetic means

5.7
Dürer Facing His Nightmare

By Henriëtte Waal & Clemens Driessen

Albrecht Dürer, *Traumgesicht*, 1525, watercolor on paper. Kunsthistorisches Museum, Vienna. Source: Wikimedia Commons

This is considered the first 'realist' depiction of a dream. Albrecht Dürer had a terrible dream in 1525, which he recorded as an image and in writing the next morning.

> In the year 1525 after Whitsunday between Wednesday and Thursday in the night in my sleep i have seen this view how many great waters fell from the heavens And the first hit the earth about 4 mile away from me with such tremendous violence and immense noise and splashing and drowning the entire land From this I was so severely shocked that I woke up while the other waters fell down And the waters that came down were very big and some were farther and some were close and they fell from so high that it seemed like they came down slowly at a steady pace Although when the first water that hit the earth came closer it brought such strong wind and roaring sound that i awoke frightened with all my body shivering and for a long time couldn't come to myself But when i woke up in the morning i painted the above as i had seen it May God turn all things to the best
>
> Albrecht Dürer

Yi-Fei Chen, *Tear Gun*, 2016

5.8
Tear Gun

Yi-Fei Chen, **Tear Gun**, 2016. The tear gun collects the tears under the wearer's eye in a silicon pocket, which are frozen in the rear of the pistol with dry ice into solid small bullets that can then be fired directly at figures of authority—born from designer Yi-Fei Chen's own experience of silent frustration and helpless tears during a confrontation with her tutor, transformed into an act of emotional self-defense.

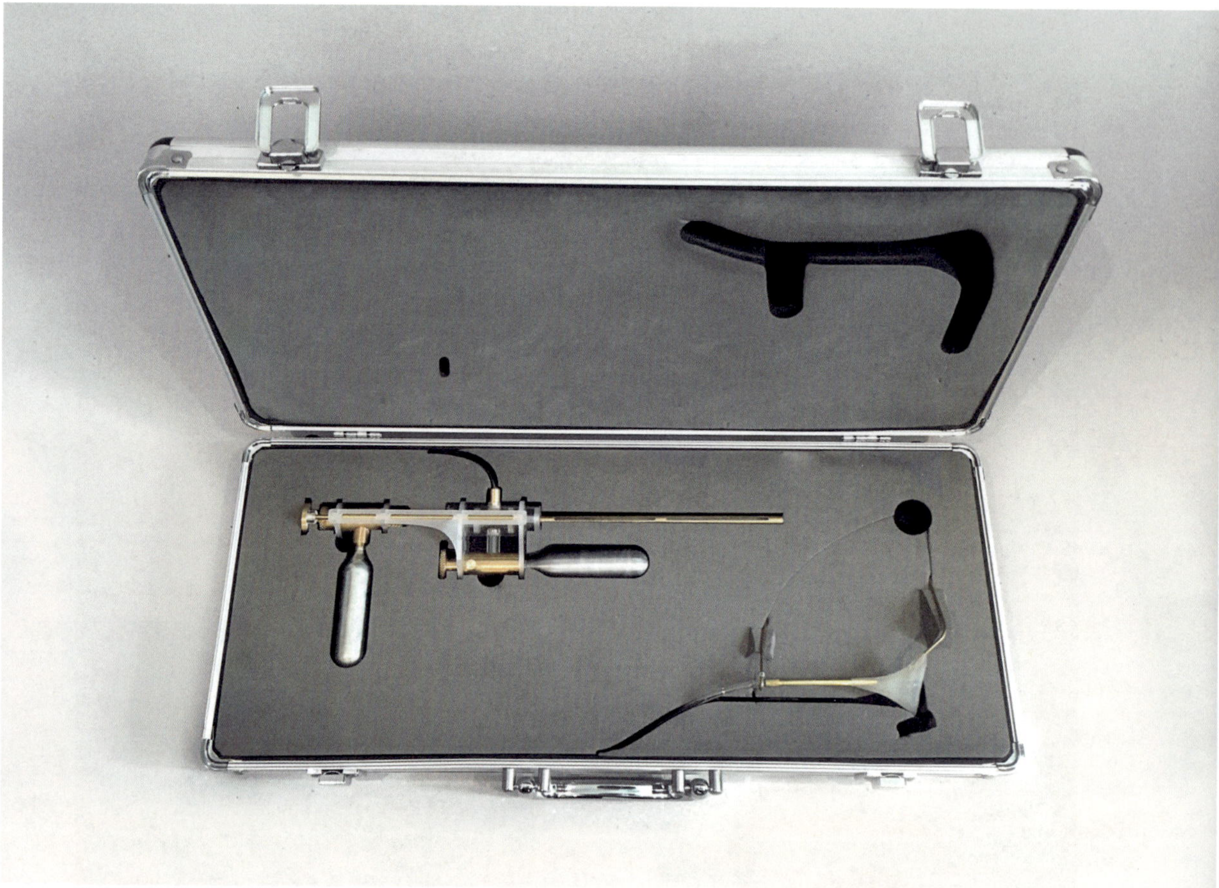

Yi-Fei Chen, *Tear Gun*, 2016

Infra-structure

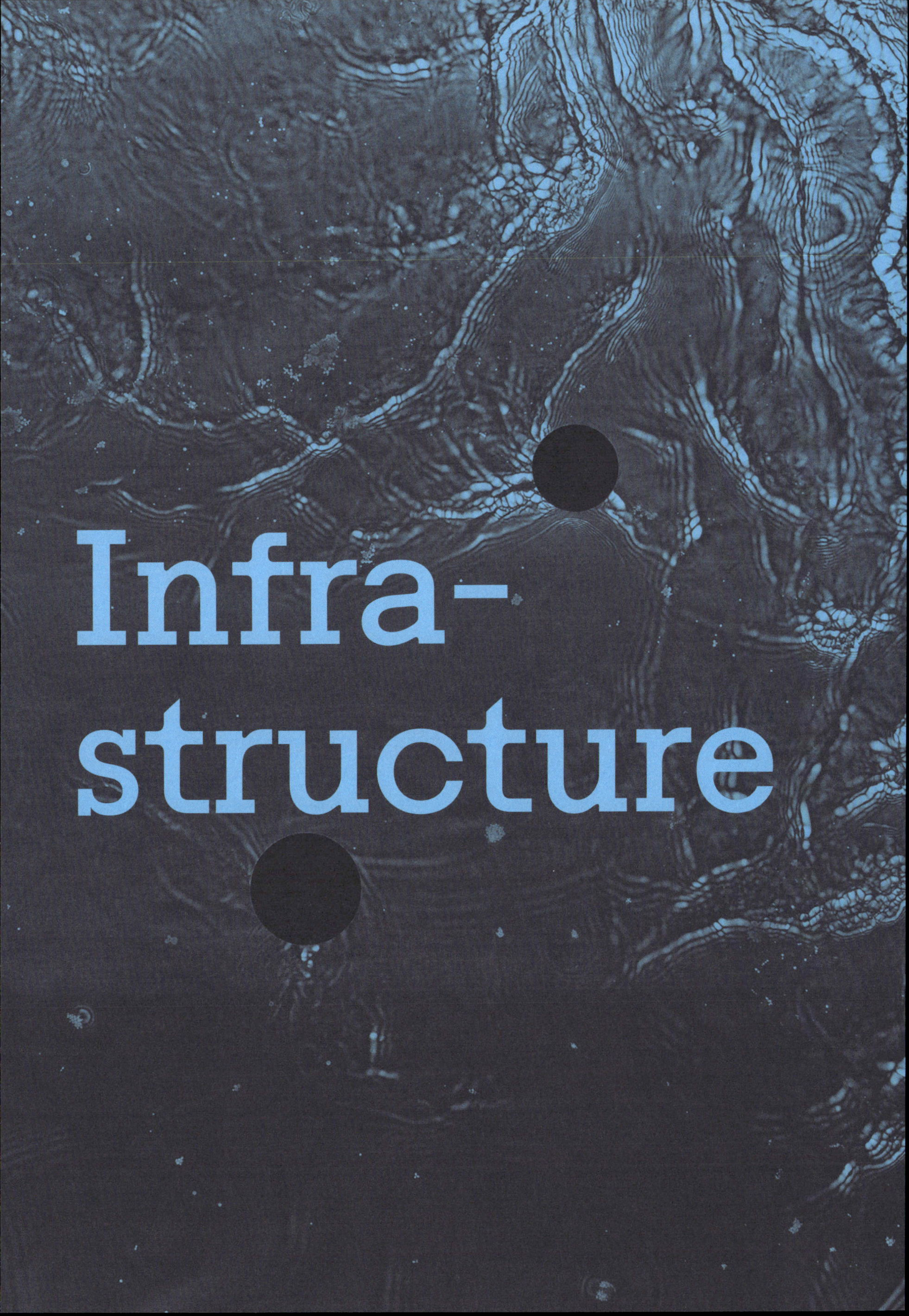

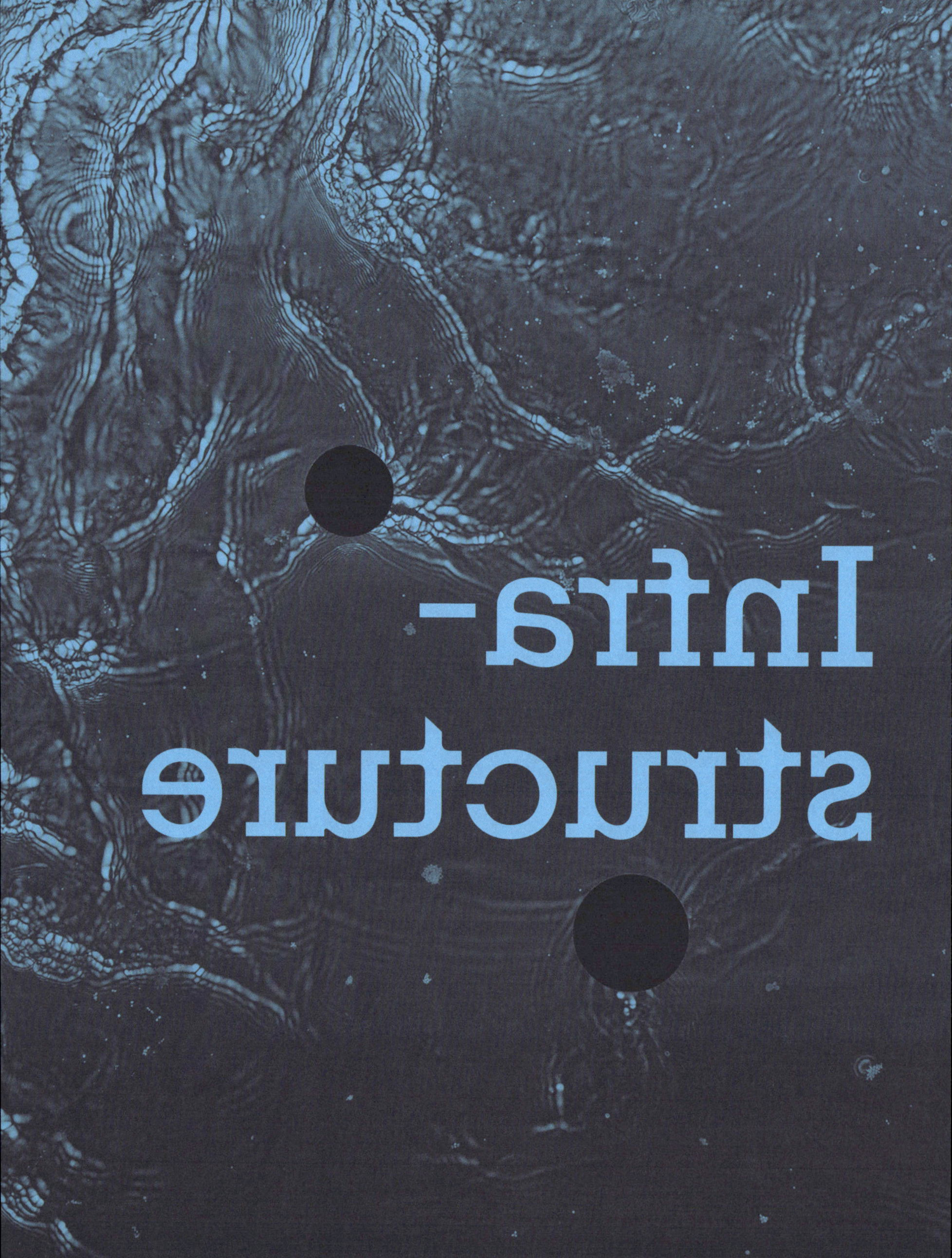
Infra-
structure

6
Infrastructure

BY HENRIËTTE WAAL & CLEMENS DRIESSEN

Humming in the background. Hidden work behind the scenes making the modern world appear obvious. Water running when we turn on the tap—without showing anything of how it is produced and connected. Construction, maintenance, repair. This is the extra-ordinary work that makes water just seem water.

Modern (state-governed) infrastructures produce what Linton calls 'modern water', abstract and placeless, a measurable and manageable resource.[1] Infrastructures bring into being a particular form of relating to water—making itself absent while soothing its users into assuming it is simply there, and always will be.

Infrastructure aspires to be in the background. When it succeeds, it tends to be seen as a-cultural, mere management of stocks and flows.

Meanwhile pipes and basins generate circuits in our minds—seemingly obvious relations—basic, ever present, in need of constant repair, suggesting command and control—systemic thinking, engineering, quality control, monitoring, in fact magic. Inscribing certain types of relations, experiences and imaginations into the functional fabric of the world. Infrastructures made up of barriers and filters, canals and bridges—but also of language structured to 'grasp' something that remains forever fluid and alive, flowing by.

Ecosocial design asks questions that otherwise only come up when sanitation breaks down, at moments where you find out how disconnected from essential skills, basic knowledge and a sense of place many of us are: what does the water taste like here? How can we save water? What could be a sensible way to compost your poo, to not flush but

1 Jamie Linton, 'Modern water and its discontents: a history of hydrosocial renewal', *Wiley Interdisciplinary Reviews: Water*, (1:1), 2014, pp. 111–120.

Opening image: Sjoerd Knibbeler, *Reservoir*, 2023, photogramme on baryta paper, various sizes

reuse pee? How to use practice-based knowledge on both infrastructural and policy levels? How to validate the data readings and where, at what resolution, to integrate both embodied and remote sensing?

At least around infrastructure there can be an appreciation of bureaucracy, standards—commerce and competition tend to undercut functionality and raise costs—here governments have a role. This is how the modern state emerged—or made itself appealing: with water infrastructure.[2] Water infrastructure as a public good, an obvious commons. The right to water is about regular and trusted distribution, not occasional access. Any infrastructure shapes the social; making particular users, recipients, subjectivities and relations—the village pump as the cliché ecosocial infrastructure, a proverbial mode of sociability—reinvented as the office water cooler.

2 Chandra Mukerji, *Modernity Reimagined: An Analytic Guide*, Routledge, 2016.

Concrete, metal, electronic circuits controlling quantities and qualities. But when you actually open up the manholes, meet the people in the control rooms, you get into the ecosocial relations needed to make these places work. What if we reimagined infrastructure in more fluid, flexible ways? Can infrastructure be wild? Is there poetry to ecosocial infrastructure?

Modern water brings modern hubris—at least some water managers feel like they are onto something universal—they can erase place and repeat the same solutions everywhere. When we stop seeing infrastructure as a victory over nature, as a necessarily ongoing fight against water's flow, as channelling, blocking, damming and piping. Infrastructures can carry feelings, spirits and imagination—connecting thoughts, and bridging worlds. When a well can be seen—and designed—as a temple, like in the ancient Nuraghi culture of Sardinia. When infrastructures produce spiritual experiences and operate as mental space—like cisterns in Alexandria. Ecosocial infrastructuring reveals that these are never only human spaces, when 'infrastructures are not only background substrates subtending human life but become the very medium of non-human inhabitation'.[3]

3 Maan Barua, 'Infrastructure and non-human life: A wider ontology', *Progress in human geography*, (45:6), 2021, pp. 1467–1489.

This perspective brings about a shift in how we understand knowledge in infrastructure: while people or perhaps AI in control rooms monitor systems, sometimes molluscs know best. As we saw in Warsaw, these more sensitive creatures that infrastructures may come to rely on, doing the purification, or acting to warn human collectives. We trust them not merely as living sensors stuck in electric circuits, but as decision-makers, balancing their suspicion of what is in the water with a need to at some point open up and expose themselves, and the entire city, to a polluted river.

Shahnoor Hasan is a Bangladeshi academic and a senior researcher at Deltares on water governance and politics of development. Her work deals with issues of policy and technology transfer, dynamics of development and cooperation, exploring why specific policies and practices emerge, whose interests and what purpose they serve.

6.1 Poldercraft: From Promise to Puddle

BY SHAHNOOR HASAN

Nevertheless, the future will bring plenty of new polders, not here, but in the third world where they are badly needed for raising the food production and where, after all, vast areas are awaiting better water management systems.[1]

1. Verhoeven, 1982, p. 280.

IMPOLDERING THE WORLD

This imagination belongs to a Dutch professor of hydraulic and environmental engineering, reputedly known as a *polder professor*. Presented at the 'International Symposium on Polders of the World' held in Lelystad, the Netherlands in 1982, the professor's imagination reflects the self-confidence and enthusiasm of the Netherlands in regulating water beyond its own borders, while drawing on specific knowledge that has worked well for its own water, ecology and society. Much of this confidence emerges from their historical, ingenious expertise of reclaiming land out of water known as impoldering. René Descartes, a sixteenth-century French philosopher and mathematician, is often thought to have said the phrase: 'God created the world, but the Dutch created the Netherlands', capturing this self-understanding as polder-makers.[2] Anchored in hydraulic

2. Niemeijer argues that it is disputable who actually quipped this well-known saying (www.vakbladvitruvius.nl/essay-God_created-the-earth.html).

engineering and spatial planning, impoldering involves reclaiming lands from temporary or permanently waterlogged areas by the combined means of drainage canals, dikes and pumps. The impoldering is recognized as a fine Dutch tradition, a craft and an innovative approach to the relation between humankind and water.[3] Over the centuries, the poldercraft has influenced the forming of the Dutch land- and waterscapes, ways of knowing and intervening in water, and to extend its modest local avatar as described by Descartes into becoming the world's water leader. The leader that envisions to help others that it presumes to be *less prosperous*.[4]

3. Reh et al., 2007; UNESCO World Heritage, 1999.

4. Netherlands National Water Plan 2009–2015, p. 243.

To 'improve' god's creation, the Netherlands set the ambition to impolder the world as a passionate water craftsman. A craftsman who thrives on realizing development missions in conjunction with trade.[5] Building on the Dutch spirit of exploring trade routes and territories in distant lands since the pre-colonial era, dedicated groups of Dutch officials, engineers and scholars rehearse, perform, and enact the 'impoldering the world' in international development arenas and for visiting foreign policymakers, practitioners and researchers. Often, international donors and development organizations such as the World Bank and FAO support (co-sponsor) these dedicated efforts. The enactments help create room to continuously remind the world about the Netherlands as a low-lying country with polders, eager to exchange its historical knowledge and rich experiences in designing, constructing and living in the polders, and to promote wider support for the poldercraft. In these enactments, what the poldercraft is and can do changes continuously: in relation to the changing focus of the international water sector, adapting and determining who becomes associated with the craft. Thus, there are many versions of the poldercraft that have been made available over the past decades.[6] In those versions, the poldercraft shapeshifts as a sustainable solution for creating spaces for urbanization and growth, increasing food production, protecting the deltas and coasts from the rising sea and other climate impacts in places such as Bangladesh, Indonesia, Singapore, and Thailand.[7] In almost all these versions, the polder appears as a fine, unique Dutch craft whose design and potential ecological and socioeconomic impacts remain unquestionable.

5. Netherlands International Water Ambitions, 2019; Van Nieuwenhuizen Wijbenga, Sigrid, Schouten, and Keijze, 2019.

6. See for example, Hooimeijer, 2011; Richter, 2020, p. 223; Schoubroeck et al., 2010; Segeren, 1982.

7. Netherlands Water Partnership, 2019; Peters, 2012; Wandee, 2005.

As a Bangladeshi academic who studied the workings of the transfer of Dutch water expertise, and who is currently affiliated with a Dutch applied water knowledge

Jan van den Broeke, Monument at the Stormvloedkering Oosterschelde. With poem by Ed Leeflang '*hier gaan over het tij, de maan, de wind en wij*' ('here rule over the tide, the moon, the wind, and us'), 2008. Courtesy Rijkswaterstaat

institute—an insider-outsider to the larger Dutch Water Sector, I tell the story of the poldercraft differently. My grounded understandings suggest that there are much more complex dynamics at play than the commonly presented storyline of efficiency, resilience, growth, and development of poldercraft. Looking at the everyday lives of the people living in the impoldered southwest delta in Bangladesh, I explore what the poldercraft does, with what effects, and to whom. By moving away from reiterating the standardized versions of the poldercraft, the chapter offers some insights to understand the polder as a sociotechnical phenomenon that not just changes water flows and land use but also transforms social structures—often in unanticipated ways—while it helps maintain geopolitical and development networks. By telling an alternative version of the poldercraft story, my intention is not to demean the craft; rather, to contribute to making room for reflection on design and the prevailing ways of knowing that promise to regulate water, manage ecology and (re-)order societies.

MAKING OF THE POLDER PROMISE FOR SOUTHWEST BANGLADESH

The southwest region in Bangladesh forms the lower part of the world's largest river delta—the Ganges delta. The tidal interplay that happens twice a day transports high-sediment loads and continuously reshapes the area's land and water, making it a hydrologically active delta. The delta nurtures vibrant aquatic and terrestrial biodiversity. In the south of the delta lies the world's largest mangrove forest *the Sundarbans,* home to the majestic Royal Bengal tigers, saltwater crocodiles and myriad bird species. Alongside the mangroves, the landscape features sprawling rice fields and shrimp farms. The area is interspersed with water channels, creating a complex network of ecosystems that support both local livelihoods and natural habitats. Because of its geographical location, the delta is prone to coastal flooding and cyclones. The Sundarbans shields the area and people from the severity of cyclones and storm surges.

The southwest coastal rivers have no upland flow connection and are dependent on the tidal water and surface runoff during the monsoon.[8] This happened as the watercourse of the Ganges shifted due to a series of natural events, including tectonic movements, the closing of the origin of the river Mathabhanga to control the water flows during the British colonial era, and, in the 1960s, the construction of the Farakkha barrage in India. The British colonial administration considered the abundant flows of water and sediment a mismatch to their representation of modernity, and a hinderance to expand railways for extraction of resources from the colonized territories.[9] They constructed watertight embankments, which was instrumental to expand the railway networks, while causing siltation and drainage congestion upstream in the delta.

8. Islam and Kibria, 2006.

9. Dewan, 2021.

Unlike the British colonizer, people living in the southwest delta did not find the water flow laden with sediments problematic. Their way of managing the flow tells a story of historical knowledge, everyday experiences of living with water and sediment, empathy and care for ecology and water as they are, and of communal harmony. Devoid of large-scale hydraulic engineering, the technicalities of their living with the flow involved construction of temporary earthen embankments with wooden flap gates known as *oshtomashi baandh* (eight-month embankments). During the dry season four to five

Text continues on p. 285

Mattheus van den Broucke, 'This coast is rather treacherous, but the exact situation is unclear to us, on this coast the ship Ter Schelling is wrecked and also from time to time many native ships and is thus very precarious.' Piece of a map showing the coast of Bengal, 1662

The Bangladesh Delta, a Historically Problematic Area in the Eyes of Colonialists

'This coast is rather treacherous, but the exact situation is unclear to us', wrote Frans Jansz van der Heiden, a sailor from the **Ter Schelling** of the Dutch East India Company (VOC) in his 1665 book **Vervarelyke schip-breuk van'T Oost-Indisch Jacht Ter Schelling**. In October 1661, en route to Hougli in India from Batavia (now Jakarta, Indonesia, the principal trading centre of the VOC and later a Dutch colony), van der Heiden and fellow sailors survived a shipwreck and ended up in the Bengal delta. Van der Heiden depicts the delta as 'a land of misfortune that is enriched with unknown, poisonous, strange herbs, fruits and trees, filthy sea and land creatures'. He continues: 'The sea-faring people surviving terrible violence of winds and waves, destructive and often inevitable troubles of the sands and shores, miserable disasters and incomprehensible famines. These people are unexpectedly in various ways often miraculously saved.'[1]

1. Rasel, 2012; Van Deursen, 2023.

people from each household joined the communal force to construct the low-height *baandhs* on the sides of the rivers to prevent saltwater intrusion into the crop lands.[10] The embankment heights and other design details were anchored into their knowledge and everyday experiences of living with the tidal interplays and differences in tides. The designed low height of the embankments protected the *aman* rice fields yet allowed flooding and sediment deposition in the tidal floodplains during the monsoon from June to October. After harvesting the crops, the communities breached the *baandhs* to allow the flood waters and inundation of fertile silt into the floodplain. This craft named *jowar-bhatar khelano*—free playing of tidal flow—was useful in maintaining soil fertility, distributing sediments and floodwater, supporting seasonal livelihoods such as fishing, and allowing natural delta formation and river navigability.[11] Nevertheless, this way of knowing water and sediment flows and maintaining interdependencies between human and ecology by the people in the southwest delta remained unrecognized by foreign eyes, also after the colonizers departed.

10. Dewan, 2021.

11. Mutahara, 2018; Van Staveren, 2017.

In the postcolonial era, a transnational network emerged consisting of intergovernmental organizations, private corporations, donors, and state development agencies including those of the world's (officially former) colonizers. Labelling themselves as the Western world (now as the Global North), the network associated itself with progress, modernity, knowledge and order.[12] The self-claimed advanced Western world enthusiastically alienated the rest as 'the Third World' tagging them as impoverished and undeveloped—as it also appeared in the imagination of the Dutch polder professor. Driven largely by geopolitical and economic interests and often lacking a deeper understanding of ecosocial and cultural contexts as well as the relational ties between nature and humans, the Western world continuedly found the abundant, free flow in the southwest Bangladesh delta chaotic, traditional, and in urgent need of improvement. A mission mobilized by the United Nations and led by Julius Krug, a politician and former head of the US Army Corps of Engineers, determined the poldercraft as the essential model of improvement for the delta in Bangladesh, then East Pakistan.[13]

12. Gupta, 1998; Kapoor, 2008.

13. Thomas, 1974.

This proposed development marks the paradigm when the Western world was (again) turning postcolonial societies into spaces for experimentation with selected hydro-engineering designs in the name of modernizing and bringing

Dumped dredged soil, Biner Pota polder, Bangladesh, 2023. Credit: Willem van Deursen, Carthago, The Netherlands

The silted Kopotakkho and Morrichap rivers, major lifelines of the southwest delta, 2022. 'As the river died so did we.' Hasina Khatun, living in the larger Morrichap floodplain area. Photos: Shahnoor Hasan

14. Lewis, 2012.

order to ecology and society.[14] The collaborative experimentations of the Western world were (and continue to be) dominantly led by its engineers and consultants, making much of the invested aid (re-)circulate within its periphery and close allies. At the other end of the spectrum, the proposed ecosocial ordering by means of the poldercraft enthused the then Pakistani central government—dominated by West Pakistanis. They anticipated more exploitation of resources including agricultural yields from East Pakistan for the wellbeing and development of West Pakistan (now Pakistan). In realizing the polder promise, the craft of free playing with the tidal flow that the communities in the delta had practiced for generations was ignored. With an investment from USAID and led by a group of Dutch engineers, the poldercraft travelled through the Coastal Embankment Project (CEP) to the southwest delta. The poldercraft thereby appears not only a fine, historical Dutch tradition, but also a manifestation of the strategic alliance, networks and efforts of the Western world to regulate ecology and water, and order societies to which they do not belong.

The Dutch polder craftsmen/engineers were not unaware of the dynamic tidal interplays and the potential impacts of disrupting the tidal flows in the southwest Bangladesh delta. They however assumed the 'sameness' of geomorphological and biophysical features of all deltas, ignoring the clear geophysical differences in scale and the amounts of precipitation and sediments, which are different from the context where the poldercraft was developed. The craftsmen also ignored their own historical experiences of sinking and waterlogged polders in the seventeenth century, resulting from the obstruction of sediment flows with dikes.[15] The windmills and later pumps became integral parts of the polders in the eighteenth and nineteenth century to keep the Dutch polders drained and maintain agriculture practices.[16] Yet the Dutch polder craftsmen continued with the construction of more than 4,000 kilometres of impermeable and permanent embankments—without pumping systems—in more than a hundred polders. They apparently believed the constructed polders would only have minor impacts on the ecology and livelihoods in the delta. What is it that influenced the (in)actions of the Dutch polder craftsmen? Their craftsmanship in Bangladesh allows for speculation about potential influences. These can be national interests and peer pressure

15. Hanlon, 2020.

16. Nienhuis, 2008.

to maintain the standardized poldercraft narrative, or geopolitical and economic stakes tied to the Western representation of progress and modernity. They might have desired to maintain their alliance with the transnational network for future collaborations. However, the actual motivations behind replicating the poldercraft in Bangladesh remain unclear.

The making of the Dutch polder promise for the southwest delta in Bangladesh draws attention to the interventions of power in making specific forms of knowledge more visible and scientific than others.[17] Power orchestrates what is true and knowable.[18] It carefully calibrates knowing—in this case knowing water for the purpose of regulating ecology and reordering society—with specific imaginaries that hinge on prioritized contextual knowledge.[19] Power predetermines what the problems are, limiting room for questioning who identifies the problems and how those can be solved, by whom and by what means.[20] This unilateral trend of prioritizing ways of knowing inevitably favours or affects some over others.[21] The common solution for common problems tends to simplify complexities and strips out social realities, and makes cultures, ecology and water comparable. This trend helps to make the prioritized solutions appear intelligible and legit. What is achieved in the interventions are often assumed or evaluated by those associated with the solutions. The interventions hence remain unquestioned. Knowing about regulating water, ecology and re-ordering society in accordance with the promises of modernity and growth thus is more negotiated, relational and geopolitical, than scientific or ethical.

17. Li, 2007.
18. Foucault, 2007.
19. Hasan, 2020.
20. Zwarteveen, 2015.
21. Doorn, 2018.

THE UNSUSTAINABLE PROMISE

The poldercraft was initially beneficial in the 1970s and into the mid-1980s. The craft helped prevent salinity intrusion and increased agriculture production, allowing for two rice harvests a year. The craft facilitated construction of road networks that prompted migration into the southwest delta from other regions. Soon after the initial success, the long-term consequences of disrupting the tidal flows became apparent. The disruption caused by the poldercraft led to siltation of riverbeds and loss of navigability. Some of the rivers dried up within a few years, others in decades. Due to siltation, the rivers could no longer effectively drain nearby *beels* (low-lying land or natural depressions) inside the polders. A notable case was beel Dakatia in polder 25, the largest beel in the southwest region. Beel Dakatia became waterlogged for the first time in 1984 as the surrounding rivers—Solmari, Hamkura and Hari—became silted. Since the 1980s, more than 10,000 hectares in Khulna, Jessore and Satkhira districts in the delta region have experienced drainage congestion and subsequent waterlogging.[22]

22. Awal, 2014.

The Bangladesh Water Development Board (BWDB), established as part of the poldercraft intervention to maintain the polders, attributed the waterlogging to malfunctioning sluice gates or regulators. The people in the waterlogged areas soon identified the poldercraft as the primary cause of waterlogging, based on their lived experiences and knowledge. They requested the authorities to cut open the polders to allow the tidal flows to resume. Despite the evident impacts on lives and ecology, and the growing mobilization of a community movement, the authorities ignored the demand. Instead, the government of Bangladesh sought assistance from the Asian Development Bank (ADB) to address the waterlogging, particularly in beel Dakatia. In 1986, the ADB provided advisory and technical support through the implementation of the Khulna Coastal Embankment Rehabilitation Project (KCERP). The communities alleged limited stakeholder consultation in the project design. The design did not sufficiently take socioecological realities into account, they further alleged. In response to being unheard, the waterlogged communities took matters into their own hands, defying the law. In September 1990, the protesting communities cut open the embankment at four places in beel Dakatia, allowing tidal silt to flow in, raising the enclosed lands, and reconnecting the beel with the river Hamkura. Consequently, the government suspended the KCERP.

The impacts of cutting the polder in beel Dakatia encouraged other waterlogged communities to follow suit. This act of resistance led to clashes between the protesting communities and law enforcement agencies. Amidst the community movement, the BWDB took an apathetic stance. The BWDB excluded the concept of controlled tidal flooding in another project titled Khulna-Jessore Drainage and Rehabilitation Project (KJDRP) funded by the ADB. The BWDB deemed cutting the polders unlawful and a practice that lacked scientific basis and misrepresented socioecological realities in the delta. Based on these grounds, it closed the cuts in

Dredging in the Biner Pota polder, Bangladesh, 2023. Credit: Willem van Deursen, Carthago, The Netherlands

beel Dakatia in 1994, causing siltation of the river Hamkura, and waterlogging. In another instance in 1997, the BWDB took legal action against more than a hundred community members involved in cutting the polder embankments. The BWDB officials, mostly male engineers,[23] went so far as to withhold salaries of schoolteachers to pay fines for cutting the polder.[24] The communities nonetheless persisted in their resistance to the poldercraft.

23. As was the case with the Dutch polder craftsmen.

24. Hanlon, 2020.

In the 1990s, the frequent and organized cutting of the polders drew attention of NGOs and research organizations in Bangladesh, notably NGO Uttaran and the Centre for Environmental and Geographic Information Services (CEGIS), a semi-autonomous research body of the Bangladesh Ministry of Water Resources. Their advocacy, research, computational modelling and policy negotiations led to the pilot implementations of the practice of free playing of tidal flow in some areas in the delta.[25] The practice now formally recognized as tidal river management (TRM). The implementation of TRM rekindled hope in the communities that the tidal flows could be restored and that they could manage their own water. However, their hope was short-lived.

25. Detailed accounts of historical water management in the Bengal delta, citizen movement, and project-based interventions to address waterlogging are well documented in research.

The communities that contributed their lands to the TRM project received only partial compensation from the Bangladesh government. Their inability to provide the required ownership documents for ancestral lands stopped them from receiving full compensation. Navigating the bureaucratic and corrupt processes of formal recognition in land ownership proved daunting and costly for the communities. The BWDB provided little to no support to the deprived communities, leaving the communities feeling unheard and ignored.

Over the decades, the communities have developed a growing mistrust towards the BWDB. To deal with the disillusionment that the poldercraft caused, the BWDB officials prefer dredging of the dying, silted rivers to restore their flows. The communities, for example those living in and around the floodplains of Kopotakkho, Morrichap and Betna rivers, however, have other stories to tell about dredging. The communities find the impacts of dredging short-lived. They often find the dredged soil dumped into their crop fields or on the homesteads by the dredging companies (as seen in the pictures on the following pages). The communities are not allowed to remove or use the dumped dredged soil. With a request for anonymity, some community members allege corruption behind the recurrent dredging projects with limited and unjust impacts.

To date, the communities in the southwest Bangladesh delta remain waterlogged for months every year, without adequate access to safe drinking water, sanitation, hygiene and road networks. They deal with diminishing water flows, loss of biodiversity, disrupted fishing and agriculture as well as loosening of family ties as (mostly) men leave families behind in search of employment in other parts of the country. In the experiences of the communities in the delta, the realities that unfold are conspicuously disparate from the promised progress and modernity associated with the poldercraft.

The broken promises of the poldercraft in Bangladesh provokes questions about its craftmanship. The Dutch craftsmen were not ignorant about the impacts of the poldercraft as it unfolded. Evidence emerges from the 1982 'Polders of the World' symposium, where the polder professor presented his vision. At the symposium, Hugh Brammer illustrated the changing tidal flows, land use and social structures in the southwest delta caused by the poldercraft.[26] Brammer, an eminent British geographer, soil scientist and an admirer of the beauty of the natural

26. Brammer, 1983.

© Willem van Deursen, Carthago, the Netherlands

complexities in Bangladesh, suggested a number of actions to consider. His suggestions were informed by his long experience of living and working in the delta and engaging with its people since 1962, an opportunity that the Dutch poldercraftsmen missed in Bangladesh.[27]

27. The Dutch government tends to prefer a hit-and-run project-based approach for regulating water outside of the Netherlands.

Based on his observations of the design, efficiency and impacts, Hugh Brammer imagined the poldercraft as an agricultural development intervention. Besides engineers, poldercraftmanship requires active involvement of soil scientists, agronomists, agro-economists, and agricultural extension specialists at all stages. Geomorphological, hydrological, and hydraulic studies should precede determining where polders can be crafted. In Brammer's imagined poldercraft, understanding feasible cropping patterns from farmers, and regular monitoring of soil and crop conditions for tracing early warnings of physical, chemical, or ecological changes resulting from the poldercraft are essential. Highlighting the high costs associated with pump drainage and irrigation, Brammer was in favour of exploring cheaper, readily available agricultural development alternatives over poldercraft.

Despite Brammer's detailed illustrations of the poldercraft in Bangladesh, his keynote paper at the symposium comes with a disclaimer emphasizing that his views were his own and did not necessarily represent those of his employer, the Food and Agriculture Organization (FAO), or the Government of Bangladesh. This strategic disclaimer downplayed Brammer's take on the poldercraft, which diverged from the versions of the polder professor and some other Dutch scholars at the symposium.[28] The downplaying made Brammer's illustration of the poldercraft in Bangladesh appear as an isolated phenomenon that couldn't take anything away from the glory of the poldercraft. Thus, limiting the space for questioning its design and craftsmanship, and the promises of modernity and progress attached to it.

28. Brammer, 1983, p.138.

The unsustainable promise of the poldercraft in Bangladesh are well documented in research, often funded by the government of the Netherlands and carried out in collaborations between Dutch and Bangladeshi researchers. The growing number of evidence-based findings, and debates on the poldercraft in Bangladesh however are yet to influence the enthusiasm and self-confidence of the Netherlands. In 2013, the polder promise for Bangladesh was renewed with USD 375 million funds from a(nother) poldercraft enthusiast—the World Bank—and led by a leading Dutch consultancy company. The renewed promise, implemented under the Coastal Embankment Improvement Project (CEIP), pledges 'to rehabilitate the embankment and to raise the embankment height to help Bangladesh protect its coastal communities from high tides and storm surges'.[28] The Dutch polder craftsmen thus continue to regulate water and ecology and to re-order society in Bangladesh. While some Dutch (applied) knowledge institutes and NGOs take a growing interest in—what they often articulate as—support toward upscaling of the TRM. What this upscaling entails often remains vague in their project proposals and in discussions with potential funding agencies and Bangladeshi partner organizations. Reflecting on informal conversations with some Dutch TRM enthusiasts, the use of the term 'upscaling' appears to serve the purpose of asserting that the delta in Bangladesh and its people need additional forms of Dutch craftsmanship, such as expertise in water and institutional governance. Thus, the idea of regulating water and ordering society in the name of growth and modernity continues over decades and generations, ignoring the actual inhabitants' ways of knowing their own water, ecology and society.

29. The World Bank, 2022.

Thirty-five years ago, Sarkar came to the poldercrafted *Biner Pota* on a boat crossing the Betna river to live in her husband's village. In course of time Betna died due to siltation. Its dredging in 2012 could revive the flow for two years or so, leaving Sarkar and her neighbors again in water(logging) crisis. When visited in March 2022 during the re-excavation, she expressed not much hope for the Betna and the communities living around it.

REIMAGINING THE POLDERCRAFT

The stories of the poldercraft in Bangladesh provoke a reimagining of its concept, design and practice as a craft for regulating water, managing ecology and re-ordering societies. A reimagination would allow questioning and developing critical reflections on what enables the poldercraft to be regarded as an agent of ecosocial change. What understanding of 'social' and 'ecology' do such crafts entail, and whose knowledge, voice, power and authority are represented or distributed in the craftsmanship? In a reimagined form, the poldercraftsmanship would move beyond mere technical expertise to allow deeper understandings of interdependencies between humans and ecology as well as cultural values, ethics, and changing societal aspirations and (geo)politics. Rather than advancing specific knowledge, vested interests, or worldviews, this alternative craftsmanship acknowledges diverse ways of knowing and valuing water, ecology and society. This reimagination demands a shift in the primary focus of the craft (to de-personify), recognizing that politics of space and reflexivity are part of the polder craftsmanship. This way of reimagining helps to challenge normative perceptions related to who gets to be formally recognized as 'craftsmen', and what makes them qualify for regulating water, managing ecology and re-ordering societies. A reimagination would enable open and provocative conversations, a reorientation of priories, while making room for creativity, horizontal and reciprocal relations, and learning. It would pluralize craftsman and craftsmanship for regulating water, ecology and re-ordering societies.

Without a reimagining, the poldercraft is at risk of remaining a neocolonial exercise. The poldercraft as a neocolonial act presumably is less bloody and brutal, and yet it continuously fails to bring about meaningful and just ecosocial change, as we have seen in Bangladesh.

BIBLIOGRAPHY

M.A. Awal, 'Water Logging in South-western Coastal Region of Bangladesh: Local Adaptation and Policy Options',, *Science Postprint*, (1:1), 2014, e00038.

H. Brammer, 'Agriculture and Food Production in Polder Areas: A Case Study from Bangladesh', *Water International*, (8:2), 1983, pp.74–81.

C. Dewan, 'Misreading the Bengal Delta: Climate Change, Development, and Livelihoods in Coastal Bangladesh', University of Washington Press, 2021.

N. Doorn, 'Values in Water', Inaugural lecture, Delft University of Technology, 2018.

M. Foucault, S. Lotringer, J. Rajchman, L. Hochroth & C. Porter, *The Politics of Truth*, Semiotext(e), 2007, p. 41.

A. Gupta, *Postcolonial Developments: Agriculture in the Making of Modern India*, Duke University Press, 1998.

J. Hanlon, 'Bangladesh Farmers Push for Temporary Flooding to Correct Dutch Polder Failure', *Journal of International Development*, (32:1), 2020, pp. 29–43.

S. Hasan, J. Evers & M. Zwarteveen, 'The Transfer of Dutch Delta Planning Expertise to Bangladesh: A Process of Policy Translation', *Environmental Science & Policy*, (104), 2020, pp. 161–173.

F.L. Hooimeijer, *The Tradition of Making: Polder Cities*, Delft University of Technology, 2011.

S. Islam & Z. Kibria, *Unraveling KJDRP. ADB Financed Project of Mass Destruction in Southwest Coastal Region of Bangladesh*, Khulna, Uttaran, 2006.

I. Kapoor, *The Postcolonial Politics of Development* (Vol. 1), Routledge, 2008.

D. Lewis, 'Anthropology and Development: The Uneasy Relationship', *A Handbook of Economic Anthropology*, Second Edition, edited by James G. Carrier, 2012, Edgar Elgar Publishing, pp. 472–486.

T.M. Li, *The Will to Improve: Governmentality, Development, and the Practice of Politics*, Duke University Press, 2007.

M. Mutahara, *Turning the tide? The Role of Participation and Learning in Strengthening Tidal River Management in the Bangladesh Delta*, Doctoral dissertation, Wageningen University and Research, 2018.

Netherlands Water Partnership, 'Dutch Polders Are Coming to Singapore', 23 August 2019, www.dutchwatersector.com/news/dutch-polders-are-coming-to-singapore.

Netherlands International Water Ambition, Ministry of Infrastructure and Water Management, 2019.

P.H. Nienhuis, *Environmental History of the Rhine–Meuse Delta*, Springer Berlin, 2008.

R. Peters, *Factors that Contribute to Effective Dutch-Funded International Water Projects: A Case Study: Banger Pilot Polder Project in Semarang, Indonesia*, Master thesis, University of Twente, 2012.

W. Reh, C.M. Steenbergen & D. Aten, *Sea of Land; The Polder as an Experimental Atlas of Dutch Landscape Architecture*, Stichting Uitgeverij Noord Holland, Architectura & Natura, 2007.

Azizul Islam Rasel, *The Seventeenth-century Dutch Travel Account and the Production of Knowledge on Asia: A study of Vervarelyke schip-breuk van 'T Oost-Indisch Jacht Ter Schelling*, Master thesis, Leiden University, 2012.

S. Richter, 'The Translation of Polder', *The Oxford Handbook of Translation and Social Practices*, 2020, Oxford University Press.

F.V. Schoubroeck, & H. Kool, 'The Remarkable History of Polder Systems in the Netherlands', *International Consultation on Agricultural Heritage Systems of the 21st Century*, 2010.

W.A. Segeren, 'Introduction to Polders of the World', *Water International*, (8:2), 1983, pp. 51–54.

J.W. Thomas, 'Development Institutions, Projects, and Aid: A Case Study of the Water Development Programme in East Pakistan', *Pakistan Economic and Social Review*, 12(1), 1974.

UNESCO World Heritage, Droogmakerij de Beemster, 1999. www.whc.unesco.org/en/list/899/.

W. van Deursen, New maps of the Kingdom of Bengal, 2023. sites.google.com/carthago.nl/van-den-brouke-map-1662/1662-nieuwe-kaarte-van-het-koninkryk-bengale

C. van Nieuwenhuizen Wijbenga, A.M.K. Sigrid, C. Schouten & M.C.G. Keijze, Nederlandse Internationale Waterambitie (NIWA), IENW/BSK-2019/138436, 4 July 2019.

M.F. van Staveren, *Bringing in the Floods: a Comparative Study on Controlled Flooding in the Dutch, Bangladesh and Vietnamese Deltas*, Doctoral dissertation, Wageningen University, 2017.

B. Verhoeven, 'Polders of the future', *Polders of the World*, 1982, p. 268.

P. Wandee, *Optimization of Water Management in Polder Areas: Some Examples for the Temperate Humid and the Humid Tropical Zone*, CRC Press, 2005.

M. Zwarteveen, 'Regulating Water, Ordering Society: Practices and Politics of Water Governance', Inaugural lecture, University of Amsterdam, 2015.

The World Bank, *Helping Bangladesh Protect its Coastal Communities from Tidal Flooding and Storm Surges*, 2022.

6.2 Ecosocial Architecture and Community-Based Solutions

INTERVIEW WITH MARJETICA POTRČ
BY HENRIËTTE WAAL & CLEMENS DRIESSEN

Natural bathing in the midst of urban regeneration in the King's Cross Pond Club, 2016. Photo: BD Landscape Architects

In this talk with artist Marjetica Potrč we explore various aspects of her work, including the King's Cross swimming pond in London and the drawings she made about the Soča River and water referendum in Slovenia. We touch on her participatory methods, her connection to water, and her thoughts on ways of remaking our relationship with nature.

THE KING'S CROSS POND CLUB

HW A swimming pond purified with meticulously grown plants on a construction site with cranes and everything else, which attracted wild swimmers in the middle of London—it felt surreal. What mattered most to you in this project?[1]

1 *Of Soil and Water: King's Cross Pond Club*, by OOZE Architects (Eva Pfannes and Sylvain Hartenberg) and Marjetica Potrč, commissioned by King's Cross Central Limited Partnership, curated by Stephanie Delcroix and Michael Pinsky, the Relay Art Programme, 2015–2016.

MP Most important was the engagement of the community. I'm known for on-site participatory projects, often made in collaboration with students, and for public art projects that involve local residents and the users of the spaces, who co-generate the project. Since we were in London, with lots of spatial enclosures and real estate development, my collaborators, OOZE Architects, and I wanted the project to put forward a new idea of what public space could be: a space organized and maintained by a community and grounded in values centred on nature.

HW What would you recommend to makers engaging in similar projects?

MP Trust your instincts and trust your vision; try new concepts and name what you want to achieve. And put value in it. With the King's Cross project we introduced a bold idea on a construction site and placed value on soil and water instead of real estate. And those are the same values that, after two years, at the end of the life of the project, were advocated by the group that formed around the swimming pond. So trust your instincts despite other people's scepticism, such as we saw with *The Cook, the Farmer*, a project in an enclosed garden in Amsterdam,[2] where a local policymaker said: 'This will be a failure. This is not how things are done here'.

2 *The Cook, the Farmer, His Wife and Their Neighbour*, by Marjetica Potrč with Lucia Babina, OOZE Architects, Overtreders W, Merijn Oudenampsen and Henriëtte Waal in Amsterdam Nieuw-West, supported by Stedelijk Museum Amsterdam, Stimuleringsfonds Creatieve Industrie and the housing corporation Far West, 2009.

BD Landscape Architects at the King's Cross Pond Club, 2016. Photo: BD Landscape Architects

HW Absolutely, then you can be sure you're on the right track!

MP You have to try new concepts. That's very important. At the end of the King's Cross pond project, a core group of users organized a petition to keep the swimming pond going. The 'Save KX Pond' petition, addressed to the commissioners, gathered more than 5,300 signatures. The group proposed a number of ideas to keep the project alive. Initially, they suggested relocating the pond, but when this proved impossible, they argued that the project had become an integral part of London's Green Belt and asked that its biodiverse ecosystem be preserved at a different location. Most significantly, they envisioned the project becoming a community-run public space built on the original underlying environmental values.

For a project to have a meaningful impact, it must be sustained by those who become involved with it. Otherwise, it's just a burst of air. The community's commitment is vital for the project's continuity. My class Design for the Living World, at HFBK Hamburg, defined four essential steps of participatory design:

1. Listening to and talking with the local residents before making any definite plan.
2. Involving the community in the decision-making and design processes.
3. Involving the community in the construction process.

MASTERSON
HOLDINGS

4. Transferring responsibility for the developed project to the community in order to leave behind a sustainable work that benefits the population in the long term.

CD Do you prefer temporary projects over permanent ones?

MP Both are fine with me. Whether a project is permanent or temporary is not important. What's important is continuation, that the community who joined the project take it further. You have to trust your imagination when working with the local community.

HW It is also about creating memories, isn't it? Leaving a sustainable work doesn't always mean a physical object, but it's important to leave behind the idea of the object or the event.

MP Yes, but to go back for a moment to a project being temporary or permanent, what is important is that the community itself transforms from being users of the project to being its caretakers and owners.

CD How do you announce your work to a commissioner?

MP It's very difficult to explain the process of collaboration to someone who seeks a clear and quick result and expects some kind of object. When you approach a project as a linear thinker, it's like a straight line from start to result—efficient and quick. But if you include subjective thinking, you get curves. You might end up with a similar result, but it takes more time. Designers often prefer being linear for the sake of efficiency. But if you want to have an exchange of knowledge with the local community, in a process that is always messy, the result will be a new hybrid knowledge that goes beyond standardized knowledge. In order to change how we do things, we need to have more socially conscious people with a strong relationship to nature. The challenge of our time is for humans and nature to coexist as equals.

THE SOČA RIVER

HW What is your connection or bond with the Soča River?

MP The Soča River is the most beautiful river in the world, in my view. But it is also an endangered natural resource. Certain areas of the river are adversely affected by industrial pollution. And many of the people who live there are advocating for the removal of dams and trying to prevent new ones from being built. Rivers hold a significant symbolic value for Slovenians; they embody the cherished idea that our Slovenian identity is deeply rooted in nature—in the mountains, forests, and rivers. Hence, there is a profound emotional bond with rivers, which we saw expressed in the 2021 referendum on water resources.

HW In 2016, Slovenia was the first EU member state to recognize water as a fundamental right. In 2021, however, a proposed law sought to transform rivers and the coastline into market commodities. You joined the referendum to overturn this law. What happened?

MP It was very interesting and I learned a lot from it. Nine out of ten voters opposed the law, which would have opened our rivers and coastline to private commercial exploitation. The referendum made it clear that the people who supported the law understood rivers as

Someone swimming in the King's Cross Pond Club, 2015

Marjetica Potrč, *The Time of Humans on the Soča River*, 2021, drawing on paper, 76 x 56 cm

The Guardians announce the death of Winter and
Together: The pagan grandparents, Caretakers
of the Natural religion,
still alive
in the Soča valley
in the 1970s,
On the precipice of late capitalism: We who leave behind the senseless
When we meet at the end of our common time: What can we do for

the birth of Spring.

and their grandchildren, Caretakers of the Land
and Guardians of the Rights of Nature.

cation of Nature, join hands with Nature for the common good.

other? We succeed or fail together, in oneness with Nature.

objects to be used. Those who voted in support of preserving water protections saw rivers as subjects, as living beings, as friends. I liken this relationship to parenting—I'm not the owner of my child, I'm his caretaker. The same goes for rivers. When I see a river as a living being, I become its caretaker and a guardian of its rights. There has been a shift in attitude towards the natural world. We are starting to understand rivers as subjects, as persons. People are even discussing the legal personhood of rivers, and of nature as a whole. Could we be on track to grant legal personhood to the Soča River? I hope so.

HW How can we transition to a new relationship with nature?

MP In the weeks leading up to the water referendum, Andreja Slameršek, a Slovenian advocate for river rights, stressed that without an emotional connection, you can't build a relationship with nature. Our bond with nature, as with family or friends, must develop organically. A lack of empathy for nature is what makes people view natural phenomena as commodities in the capitalist system, detached from their intrinsic value. As Andreja notes, 'Without an emotional connection, you can sell a river.' Empathy towards the living world, this emotional connection to nature, is extremely important. But it's never just about the river. It's also about the wildlife and plants that rely on the river—and about us, too, as humans.

We can perform rituals of transition to develop a new, less human-centric relationship with nature. You see this in today's concern about sustainability. When we transition from being owners to being caretakers, we build our socio-economic contract on new foundations, beyond the old neoliberal agreement. This new agreement, based on environmental issues, can be seen in the increase of environmental constitutional actions that have taken place over the last two decades. The Ecuadorian constitution of 2008 reclaimed nature as *Pachamama*, Mother Nature, inscribing the Rights of Nature as a fundamental right aligned with human rights. The same is true of the first draft of changes to the constitution of Chile. Standing up for nature means standing up for social and environmental justice. We are in this together.

CD Are there other artists contributing to this transformation?

MP I very much love the Climate Care Festival in Berlin, which is organized by Rosario Talevi and Gilly Karjevsky. It is a beautiful example of exploring, without prejudices or preconceptions, ideas for building a new relationship with nature. But there are many other examples too.

During the time of the water referendum, I was reading a book by the Slovenian ethnologist Pavel Medveščcek, who from the 1950s to the 1970s interviewed the last remaining practitioners of a nature religion in the Soča River Valley. Their world view was grounded in an egalitarian relationship between people and nature. No one lives that way anymore now, but their knowledge has not been forgotten. It represents a tradition we must integrate into our future. Sharing this kind of knowledge is crucial; it fosters awareness and a desire for shared wisdom. This is similar to what your friend from the Tighmert Oasis does. When I was a teenager in Ljubljana in the 1970s, I had no idea that this way of life ever existed in the Soča Valley, but today it makes so much sense. It is important that we share examples of a more egalitarian coexistence with nature. Our survival depends on it.

HW What were you aiming for with the Soča River drawings?

MP Two years ago I was invited to make a work about rivers for the Sydney Biennale. I made drawings about the rights of two rivers: the Soča in Slovenia and the Lachlan in the Wiradjuri Country of Australia. For this second set of drawings I collaborated with the Wiradjuri Elder Ray Woods. As a caretaker of the Lachlan River, Ray Woods spoke on behalf of the river in the New South Wales parliament. By speaking for the river in a human legislature, he proved himself to be a guardian of the river's rights. For me, it was an amazing experience to collaborate so closely with Ray Woods. Although we live in different parts of the world, we agreed on ethics, on environmental issues. This made me aware that Indigenous people and environmentalists are today listening to each other. You might say they speak the same language. Rationalism and the belief in linear progress has shaped our thinking for around 500 years, spanning the Renaissance, the Industrial Revolution, 20th-century modernism, and post-World War II consumerism. I think we are all trying to get out of modernism, or what I call standardized knowledge. It's the call of the time.

CD And how do you do that in your participatory projects?

MP It's a common-sense way of doing things. We work with local residents and figure it out step by step. I don't accept the idea of a single replicable prototype because conditions are so different in each project. Modernist urban planning was interested in standardization, the idea that everything could be reproduced.

CD Do you have a recipe for participatory projects? Or is it more of a situation in which you have to trust your imagination, as you mentioned? How do you involve students?

MP First, before you come to the site, study the location as a gesture of respect to the community you want to work with. Get rid of your preconceptions; don't read all the books on the subject beforehand. You need to be able to listen with an open mind to what the local residents tell you. And then, when your imagination and their imagination meet, you can develop a project together. This means not only designing it together, but also improving it together. The community must be involved in the decision-making. There is a limit to how much you can talk about things; you have to get physical with the project. This is what I believe works best.

HW Yes, it creates deeper learning and understanding.

MP Henriëtte, you gave me your photo of five or six people walking together in the middle of a river or canal. I have it on my windowsill. Milan and I like to look at it.

HW That's nice.

MP It shows them walking in the water, but I am not sure what they are doing. Perhaps they are monitoring the river by walking in the current. I love the photo.

CD So that's your advice: we should just walk into the river?

MP I think it's all about empathy. You can't learn about the river. You have to live with the river; you have to sleep by the river. You have to walk in the river. So, yes.

Henriëtte Waal & Sophie Krier icw François Lombarts, *Hunnie Waden* (*Them Wading*), 2013, 30 x 22 cm, digital print on paper, nut wood frame. Courtesy Hunnie. Photo: Roel van Tour

Fernando Felipe Pérez Riojas is an industrial designer and Associate Professor at Pontifical Catholic University of Peru. He is passionate about challenges that require high problem-solving capacity and complex mechanisms. Together with 3D modelling and rendering expert Sandro Ormeño, he designed the Dengue Toad, a device to combat the dengue disease plaguing Peru.

VML, *The Guardian Toad*, 3D printed shell and electronic parts, 2024. Photo: Fernando Felipe Perez Riojas

6.3 Dengue Toad

INTERVIEW WITH FERNANDO FELIPE PÉREZ RIOJAS BY HENRIËTTE WAAL, CLEMENS DRIESSEN & DANIELA TOKASHIKI

Peru's 2023 dengue outbreak was linked to rainfall and high temperatures creating favourable conditions for mosquito populations to thrive.[1] Dengue, a mosquito-borne viral disease, is endemic to Peru. It is characterized by a sudden onset of fever, chills and headache, followed by muscle and joint pain. The mosquito breeds in areas where drinking water is limited and stored in containers. The stagnant water creates the ideal environment for mosquitoes to lay their eggs. Following the outbreak, health authorities went door to door, fumigating homes in low-income areas of Lima.

1 World Health Organisation report, 30 May 2024. www.who.int/emergencies/disease-outbreak-news/item/2024-DON518

Ecosocial designer Daniela Tokashiki reflects on her experiences with design in the context of Peru. She explains that design in Peru is often aimed at pragmatic and tangible output. There isn't a real industry with technologies and materials with which you can design. You can't dream too big, and always need to find affordable ways to do it. The main thing you can work on are emotions and experiences.

Reflecting on her experiences in the context of design in Peru, she realizes relationships with nature in Lima are really poor. We see nature as a danger, every time there is water, there are pests and animals, and they represent a threat. It has to do with the limited access to clean water mixed with a lacking system to clean the streets. The cheapest way to control the pests in the water are pesticides.

Together with designer Daniela Tokashiki, we interview Pérez Riojas.

HW Why is it a toad?

FFPR Carlos Tapia, the creative director at VML Peru (the local branch of a global marketing firm), invited me to join the project. My brief was to design a functional prototype of a small frog-shaped device that could float and continuously stir the water for several hours. It also needed to have a positive, approachable expression to engage users and encourage them to talk about it.

Someone on Carlos's team had come across information that mosquitoes don't lay eggs in fountains due to the water's movement. This sparked the idea for a device that could keep water in motion. As VML provides services to the Peruvian company Alicorp, which markets insecticides under the Sapolio brand, whose mascot is a frog, the creative team decided to use a frog shape for the device.

Peruvian researcher Dr Mirko Zimic was then brought in to conduct tests in Chincha, south of Lima, which confirmed that even minimal water movement could reduce the chances of mosquitoes laying their eggs by 92 per cent.

I also had to consider the expression on the toad's face. Both the mouth and eyes had to play an important role in achieving an optimistic face. I tried out several things. Big eyes made the toad scary and intimidating, but if it looked too much like the brand's mascot, it wouldn't be taken seriously and might be seen as more of a children's toy. Finding the right balance was key.

HW What were other challenges in the process?

FFPR Flotation was another key challenge, requiring adjustments to the toad's shape and body position. The arms and legs were positioned on the sides for stability and to protect the propellers from hitting the container's edges. After testing various propellers and vibration systems, we installed two paddle wheels powered by a small motor, which runs on energy from a solar cell.

Peru's health workers fumigate homes to curb deadly dengue outbreak, house by house, street by street. AFP News, 12 May, 2023, film still

We used 3D printing, and all the mechanical and electronic components used for the models and final versions were sourced from the local market. We decided to use solar energy, considering that the mosquito reproduction season coincides with summer, when sunlight is

abundant for several hours a day.
I recovered the motors and solar cells from electronic toys. Those were the smallest I could find here in Peru.

HW How about the collaboration with the brand, which sells insecticides, making the toad a competitor? Or what is that relationship?

FFPR Since this is not something the government in Peru can easily finance, VML presented the idea to Alicorp, who adopted it and took responsibility. Initially, they wanted to offer it as a gift to their customers. In some way, it is an alternative to the products of the Sapolio brand. In just one week, the campaign around the 'Guardian Frog' reached more than half a million people. So, as a storytelling device, it has already proven to be effective.

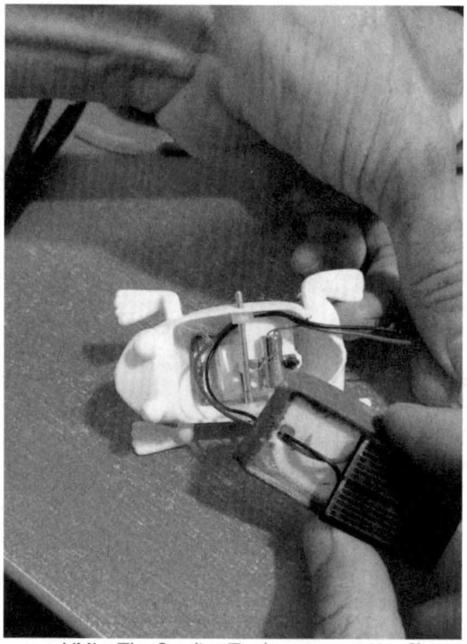

VML, *The Guardian Toad* prototype, 2024. Client: Sapolio. Image: Jhona Gibson

HW Has the toad been taken into production?

FFPR The frog is still in the scaling phase of its implementation. One challenge is producing it on a large scale in Peru, as there are not many options for injection moulding and mass production while keeping the cost competitive and affordable. Currently, issues such as buoyancy and external features have been resolved. However, we still don't know how long the frog will last and how it will perform in real use. I was only involved for a short time to contribute my experience in industrial design to this project, which has the potential to improve public health in Peru and eventually in other regions affected by dengue. I believe it would be beneficial for the success of the project if I could remain involved in the scaling process as well.

The current plan is to first test it within a community and approach it as a social project. The idea is to introduce between 100 and 500 Toads in the coming months in a community in Tumbes, which has a high incidence of dengue, and gather their feedback. Before considering large-scale production, this development needs to be validated in communities, and we need to perform cost calculations for scaling.

Francesca Masoero works as a curator, cultural organizer and researcher. She is part of LE 18, a cultural space in Marrakech, where she initiated QANAT. With a background in critical theory and political economy, she explores notions resistances in multiple forms, including testing collective-making processes within and beyond the art field, and researching the politics and poetics linked to watery worlds and to forms of being together otherwise. Since 2019, she also collaborates extensively with the Dar Bellarj Foundation (Marrakech, Morocco).

Collective edible garden at Dar Bellarj. Photo: Louisa Aarrass

6.4 QANAT Collective

BY FRANCESCA MASOERO

The qanat (or **khettara** in Moroccan darija) is the millennia-old subterranean water-harvesting system that allowed Marrakech to become a (once) archetype of a garden city. An extremely ingenious infrastructure, the qanats exploit only the force of gravity to connect rich water tables often at the feet of mountains (for Marrakech the Atlas Mountains), moving these waters towards semi-arid or arid planes. Through that, the qanats did not only supply a semidesert city in sustainable ways, but also embodied and shaped a specific social contract and a water culture grounded on principles of ecological care and solidarity operating within and beyond the city borders.

This system materially now belongs to Marrakech's past, as it was dismantled piece by piece with the 'modernization' of water management imposed by the French colonial system first, and the independent state after.

Marrakech's present is one of a touristic capital sold as the fantasized quintessence of the orient, where many of its inhabitants are forced to leave the historic neighbourhoods of the medina to move to its poorly built new peripheral margins. The latter are now surrounded by eighteen luxurious golf resorts—a bizarre number for a city in a semidesert region. While the water table has dropped from ten metres to more than two hundred metres below the surface in a few decades, the palm-grove, historically the city's green lung and one of the main communal spaces for its inhabitants, is today half dried out and half transformed in an exclusive, secluded residential area.

Originally, Laila Hida (founder and director of LE 18) and I conceived QANAT as a curatorial and research programme comprising artistic residencies, an exhibition, a series of conversations and walks involving cultural practitioners, architects, geographers, historians, and activists. The starting point in 2017 was to delve into the politics and poetics of water as embodied by, and embedded in, the material and symbolic legacies of the qanat in Marrakech and other parts of Morocco.

The increasing enclosure of communal spaces, combined with socio-economic uneven stratifications and spatial segregations, have produced dispossession and inequality, a critical lack of liveable space to get together and socialize, entangled with deep environmental degradation. It is by moving from this, and from a desire to commit to some of the fractures that traverse Marrakech, that the rhizomatic infrastructure of the qanat became our matrix to rethink the way we relate to each other and our environments.

Moreover, what became clear within one or two years from that first programme in 2017 is that this body of water and earth made of hundreds of intersecting lines and circles, translocally present across the Mediterranean, part of Asia and South America, would also become the aspirational or working image of a much richer and diverse project.

Indeed, if QANAT (the project) keeps thinking with water infrastructures as a way to engage with and learn from multiple manifestations of communal practices in Morocco, it has also become a collective and multi-local action-research platform that has been growing in many

Ours, participative artistic intervention by Nabil Himich in the city of Marrakech, Charita (2023 edition). Photo: Salah Bouadeh

directions and places, nourished through time by predispositions, reiterated invitations, and friendships.

Throughout its still short history, QANAT has been led at times by two, three or even ten people, and involved a much larger number of contributors, so that fluid as water and porous as the earthy walls of the qanats, the project has expanded through new channels dug by those that wished to join and share a specific research line, a particular methodology, adding a new word, a new world to our tentacular network.

In early 2019 I invited artist Jeanne van Heeswijk to facilitate the second of a series of gatherings inviting artists, activists, and researchers that I had launched to open up the possibility for QANAT to become a collaborative platform. In that moment, and most certainly partially thanks to Jeanne's capacity to hold space and enable listening, not simply collaborative, but actually collective sentiments from within a quite diverse group emerged and found articulation in a shared grammar. In an article Jeanne reflects on what she understands by 'preparing for the not yet', and there she asks: 'How can people feel that they can have agency as a collective body [...] when they don't know the collective body yet? How do you build a collective understanding of a territory when territories are fractured?'

What she argues is that preparing for such a Not Yet is a process of becoming, a balancing act between individual and collective needs where it is critical to be able to let go, even if only temporarily, part of yourself, and allow yourself to be affected by the ways in which others see the world.

What she describes is then a practice of collective making based on 'radicalizing the local': rooting oneself with others in newly defined ways by listening carefully and understanding different positions and capacities to take risks, and by offering each other time and space to learn and unlearn from each other.

Preparing for the Not Yet also encapsulates a call for contextuality: it asks to deeply and affectively engage with, and learn from the multiple flows forging the otherwise, as these latter are (re)producing radical imaginaries that already exist on the ground, often embodied by collective memories, vernacular knowledges, and by the many acts of resistance and solidarity, populating our cityscapes.

This relation to place is something very central to the work of QANAT, and I want to offer some examples in relation to QANAT's work in Marrakech first.

QANAT delineates a cartography that revolves around the notions of memory, desires and imaginaries. It tries to interlace the reactivation of contextual collective memories of water and land, cosmologies and political ecologies, with the creation of spaces to bring desires, knowledges, and practices triggered by them closer, as a way to make shared future imaginaries emerge.

We combine 'time-spaces' of collective learning and experimentation through workshops, public programmes, research sessions in various formats; we commission and sustain artistic research, and we develop frameworks of dissemination from exhibitions to podcasts, to interventions in public space and small publications, imagined as entangled carriers to let sprout what we have gathered.

In Marrakech we have been exploring the effects the decline of the qanat have had on our relationship with water, with the idea of the commons, of public spaces and of being in common. We map the existing legacies of this system while trying to identify which meaningful spaces we can occupy to sustain such traces.

This contextual learning process has increasingly involved different communities: starting with the cultural and research scene gathered around LE 18, to continue to the community of women and children primarily from the Medina, the old town, through the alliance with the foundation Dar Bellarj. Dar Bellarj neighbours LE 18 and works intensively with various inhabitants of the 'old city' to re-centre their voices, knowledge and practices as the heart of the cultural landscape and living heritage of the city.

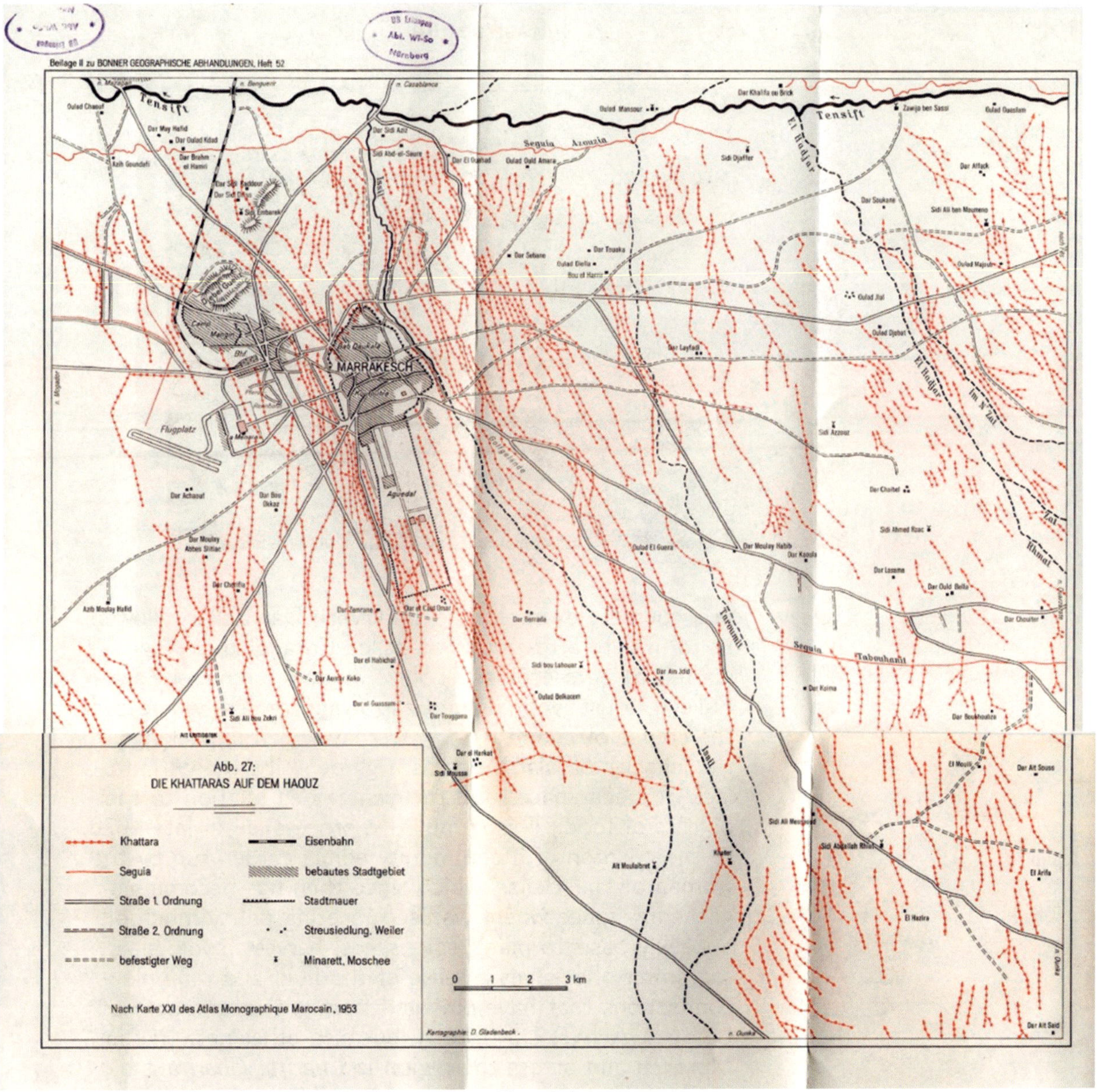

Map of the khettaras around Marrakech, 1974. Courtesy Thierry Ruf

Within this vision, in 2019, Dar Bellarj's director Maha el Madi invited curator Rim Mejdi and myself to design a yearly-running programme called the Collective Workshops. This was meant to enhance synergies and dialogue between our programmes and communities, by creating a learning space around the city's cultures, ecologies and transformations inhabited by participants and invited artists and researchers.

Much of the action-research QANAT has developed over the last five years in the city indeed owes to the rich and generous fields of knowledge, trust and exchange that Dar Bellarj has been sustaining for over twenty years.

Qanat system, aerial view, South Morocco

To give you a concrete example, the work with Dar Bellarj allowed us to initiate a research line called Entangled Ecologies, started by agro-ecologist Louisa Aarrass, architect Sara Frikech, artist Nouredinne Ezarraf and myself, where over the course of five years now we have been unpacking the legacies of Marrakech as a garden city, the current lack of green spaces and their entangled relation to the politics of food. This cumulative process led in late 2020 to the creation of a community edible garden run by the women of Dar Bellarj which since then has become a gathering space where we keep meeting pretty much on a weekly basis to plant, save seeds, harvest, cook, sing...

Another programme that gives a sense of the fluid and multiple connections that have emerged from within and across LE 18, QANAT, Dar Bellarj, but also other networks in Morocco and across the region is *Charita*. *Charita* is a project of artistic interventions in public and communal spaces in Marrakech, which draws its name from the world-wide children game of the hopscotch—in French la marelle. Just as the game does, *Charita* adapts to and engages with the street, creating ephemeral, otherwise inexistent playgrounds, focusing on some of the dynamics affecting Marrakech.

Started in 2018 by Rim Mejdi, Nouha Ben Yebdri (curator based in Tangier) and myself as part of a different collective called Madrassa, since 2020 Charita was 'adopted' by Dar Bellarj. This allowed us to sustain the programme and further combine artist-led interventions and citizen-led ones that formalised some of the action-research we have been developing through QANAT and the Collective

Workshops across different neighbourhoods of Marrakech.

Over the last years, we invited artists and researchers to explore the spaces of the palm-grove and different bodies of water in Marrakech and in other oasian regions in Morocco, bringing these in dialogue with practices emerging from other localities along and partly beyond the aridity line.

And so, while Mauritius-Italian artist Mati Jhurry critically looked at the making of the palmgrove as the ultimate dream-scape, a puzzling and selectively accessible image of paradisiac opulence; Palestinian farmer Yara Dowani has explored the possible connections or reciprocal learnings between drought-resistant farming practices in Southern Morocco and Palestine. While artist Jumana Emil Abboud offered to walk through the Marrakech palmgrove by listening to how water sites embed folktales and rites, expanding on her body of work around Palestine's water springs and spirits, artist Islam Shabana has been drawing connections between the history and speculative potentials of in-risk-of-drowning underground cisterns in Alexandria, Egypt and increas-ingly dried underground reservoirs in Morocco.

What I want to share with this is that beyond their existence in the time-space of the festival itself, not only the work, but primarily the relationships and friendships first enabled there, have partly already and will certainly continue to unfold into new channels, circles and pools, from Morocco to Scotland, from Palestine to Italy. This has especially been the case since 2021, when through different invitations Palestinian artist, researcher and curator Shayma Nader and myself in particular developed new collaborations in dialogue with organisations such as the Om Sleiman Farm in Palestine, the research platform SWAY in Wales or the cultural space CCA in Glasgow.

I hope these examples offer you a sense of the methodologies and organic proceedings that have emerged from within and throughout QANAT's channels.

Having said that, there are many shades to cast on the ways in which the project has unfolded over the years too.

To start, the very making of QANAT as a collectively owned or at least shared space of research and action has been a challenging, (still ongoing) and maybe partially failed process.

At various instances in our story, moments where a cohesive group had come together to collectively own and share the pillars of our infrastructure, were followed by its breakdown. Often this was triggered by a failure to pause, to listen carefully to each other, to avoid getting entrapped in a circle of production imposed by granting schemes or that we promised in exchange for some sparks of stability.

Yet, while failing and falling, there is also much to learn. As A. George Bajalia put it in a reflective piece we wrote together some years ago, instead of fixating on a failing cycle of production and consumption, we should pay attention to what we left behind, to those seeds casually thrown that are starting to root and germinate. We should learn to appreciate the habits we have been forming with one another, and how those habits have put our practices, ideas, and networks in common, rather than focusing on what we should claim (collectively or individually) ownership on.

There is an Arabic proverb that says: 'they planted and so we ate; we planted so they would eat'. These words encapsulate the wisdom of elders as they gift an unknown future. It speaks to the care towards a planted seed whose fruits they might never see growing, let alone taste.

This proverb encapsulates well the spirit our work aspires to be, or at least sustain.

As the *qanat* reminds us, nothing is set in stone. Rather, I think that communal (infra)structures could be thought as mud, an opaque and malleable matter made of water and earth, because, as Jeanne would put it, the collective should be thought of as an activity: not a pre-existing social structure, but as an acting towards, a coming together and dissolving and recomposing.

Forming collectives is then perhaps like sedimenting and flowing together, a never-ending process which nonetheless allows the formation of ripples, rhythms, and motions.

Are we ready to listen and adjust with the flow?

6.5 The Geomorphosis Cycle: The Wet Knowledge at Nabi Daniel & a Paradise in Walls of Nothingness

INTERVIEW WITH ISLAM SHABANA
BY HENRIËTTE WAAL & CLEMENS DRIESSEN

My project is not nostalgic nor involved in the exoticizing trends of 'indigenous origins of things'. It is discussing 'Cosmotechnics': a different trajectory based on ancient knowledges or even mythology that might eventually reach a resolution combined with modern technology.

HW What led you to work on the underground water system of Alexandria?

MAPS

IS It was pure coincidence. One day, I was having a break from teaching and went to nurture my love for maps at the Alexandria and Mediterranean Research Center.[1] I was examining the first non-colonial maps of Alexandria's infrastructure made by Mahmoud Bey El-Falaki. They show the irrigation systems and water canals around the city.

And then I saw a pattern of bright dots on one of these old maps. I asked the researcher who showed me the maps what these dots were, and she said: 'Oh yes, we used that to map the undocumented water infrastructure in the city'. And then she explained that basically there is another city under the city, there is another Alexandria underground. They documented over 200 cisterns, but based on historical records some people say there must be over 700 of these hollow structures, completely neglected.

1 CEAlex, which had invited me to do a piece for their annual 'Alexandria Heritage Days' festival.

HISTORICAL MYSTERY

The cisterns collected rainwater during the wet season, storing it for use during drier periods. An engineer, Stavroula Spanoudi, who did her Master's thesis on the cisterns, referred to it as a complete mystery how the building of this technical construction under Alexandria had started. Records suggest these systems existed even before Alexandria became an iconic city, even before the Pharaonic empire. Constructed using robust materials, the cisterns feature vaulted ceilings and sturdy pillars, which helped distribute weight and prevent collapse. Literary sources going back to ancient times tell us that the cleaning and repair of the system was a perpetual concern.[2] But how people at the time had built such an expansive and complex structure remains unknown.

2 See: *The Water System of Alexandria* – CEAlex – le Centre d'Études Alexandrines. www.cealex.org/research/ancient-studies/the-water-system-of-alexandria/
Muhammad Al-Awad & Yasmine Hussein, *Alexandria Past & Present*, 2020

GEOMORPHIC EVENT

A story goes that it happened due to a geomorphic event in the Mediterranean [the Crete earthquake in the year 365] that caused many tsunamis to hit Alexandria. Parts of the city disappeared underwater into the depths of the Mediterranean and people fled into the desert. When they came back and tried to dig new wells, they found the earthquakes had opened fault lines, which allowed saltwater to reach the freshwater aquifers. People had to depend on rainwater. Later, these systems were connected by canals to the Nile.

What caught my attention was that Alexandria had already been flooded by the Mediterranean once, but it hadn't disappeared forever, and then because of the water system, Alexandria carried on existing.

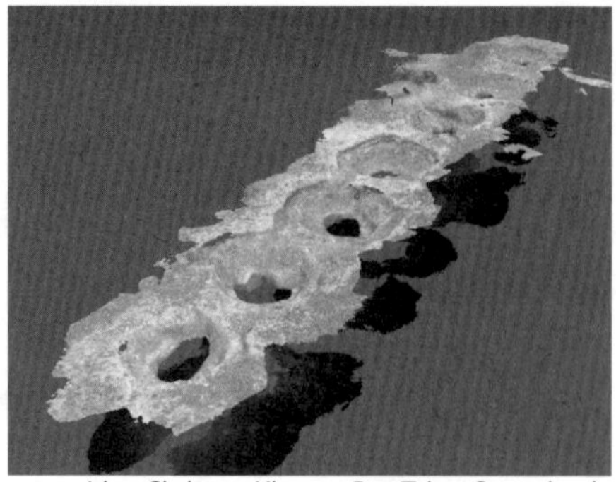

Islam Shabana, *Khettara Ben Tahar, Sanaoubar / Izdihar*, as part of 'The Geomorphosis Cycle', 2023, digital render

Later on, the system of cisterns disappeared, and people neglected them completely.

Alexandria is going to be flooded again by the Mediterranean in the future, in less than fifty years according to predictions of the rise of the sea level and the gradual sinking of the city, when it may disappear again. So, I started to connect this appearance/disappearance dichotomy. And I wonder what the cistern system could mean for the near future.

METHODS AND MYTHS

I usually go to collect mythologies and oral histories. I wanted to find a nonhuman element that can somehow carry the story that I want to talk about: about the city and disappearance in the future, and about the systems or the underground life of Alexandria.

The cistern system of Alexandria is extremely complex, and it's all interconnected somehow underground and partly overground, in order to transfer water from one system to another. And Alexandrians at the time even had the right to own their own cisterns. So, some were private, but most were public. But water would not be used excessively, you were not allowed to do that, unless you stored it privately for a month.

This appears in a book by al-Maqrizi, which is commonly translated into English as 'Al-Maqrizi's Plans' or

Interior view of the small Gharaba cistern, Alexandria, Egypt. © Centre d'Études Alexandrines

'Al-Maqrizi's Topography',[3] and also in the work of Leon The African.[4] Especially al-Maqrizi had a very poetic and fascinating description of the underground life of Alexandria, or how Alexandria has these massive passages between the systems that were used for maintenance, but were said to be big enough for a knight to go around the city of Alexandria, underground, through these arches.

3 The full title is often translated as *The Plans and Topography of Egypt by Al-Maqrizi*, reflecting its detailed description of Egyptian geography, historical monuments, and urban plans during the medieval period, p. 243.

4 Johannes Leo Africanus, *Cosmographia et Geographia de Affrica*, 1526.

HW Was it easy to find the cisterns? Did you take the map and try several of them, or how did that go?

IS Unfortunately, we don't have access to most of them, so I had to sneak into one of the systems.

SUFI SHRINE

Unable to obtain formal access, Shabana approached a fixer called Mina, a city guide who gives free tours in Alexandria. Mina gave no explanation but ordered him to follow. Finally, they entered the Nabi Daniel mosque, a functioning mosque, where daily prayers were going on. On the other end of the mosque was a door. Behind that door was a room with four Corinthian columns. As if through a portal, they stepped from Islamic Architecture into a Roman Greco style room. In the middle of that room was a hole in the ground with a wooden fence around it. A wooden staircase led them twenty-five steps into the ground. There they stood under an arch with four arches in different directions attached to it. Once leading to downtown Alexandria all the way to the Citadel, now all directions were blocked after a few meters (by a developer who wanted to build a department store building beside the mosque and filled the hollow underground with concrete, Shabana figured out later). But there was a better surprise than the concrete. Directly under the hole there was a shrine of Sufi saint Mohammed Al-Mousely. And there, in this insane coming together of Alexandrian cultures, Shabana knew he had a project. He wasn't sure where he was heading but he just started to 3D scan everything with Lidar 3D on his phone: the shrine, the blocked tunnel, the arches.

Mina pointed at an opening revealing the floor beneath them—he couldn't scan it; it was too dark—and told him that no one had been inside that hole underneath Prophet Daniel.

SINKHOLES

A lot of systems were discovered by sinkholes. A famous incident involved a construction worker falling into a deep sinkhole at a cathedral site. The worker survived with major injuries and the cathedral decided to turn the magnificent cistern he discovered into a museum. Another story is about a woman who disappeared into a sinkhole. When they tried to fetch her in the underground passages they couldn't find anything. The event occurred in 1973 on Prophet Daniel Street. And then of course, the story started to roll and roll and became bigger. The tale was reported by a journalist of a well-known magazine and linked to supernatural beliefs.[5]

5 Moufid Fawzy in *Rose al-Yūsuf* magazine.

HW What did you do with all the stories and beliefs you collected?

IS In Egypt there is a saying that the ground would open up and swallow you. It's a term you use to curse someone or to warn someone. And sometimes it's also used for embarrassment, like: I hope that the ground opens to swallow me, because I'm so embarrassed.

The story of the lady made it all come together for me. Most of the people I talked to think that it's related to Jinn. Jinn is an Islamic type of demon. So, it's like the genie but more serious, not the funny one of Aladdin. People fear Jinn. It overlaps with witchcraft and being possessed and stuff like that. So, people have this story because she was a newly married lady or was about to get married. Some people even call her the vanishing bride of Alexandria. And that there was a Jinn who was in love with her. Because she's human, he couldn't marry her, so he decided to make a sinkhole and kidnap her underground.

In the video works I made, one video uses Biblical verses from the prophet Daniel that I found, on which I projected my own interpretations. The piece is a kind of prophecy about the disappearance of Alexandria. The other video is situated in the future, where someone is sending a technical report from the area of Nabi Daniel that is now underwater. You don't get much detail. I speculated that people will need to flee again into the desert, because of what's going to happen with the sea level, and then they will need to dig cisterns again in the desert. But in the video the city completely disappeared and they need to know how to build cisterns. So they decide to send robots to scan the water systems, and then the robot disappears. And while trying to find the robot, they hear the voice of a lady singing mysteriously. That's the project in Alexandria, I called it The Geomorphosis Cycle, referring to the act of changing landscapes, either by a natural disaster, by climate change or by human interference. The first chapter is called The Wet Knowledge at Nabi Daniel, because it revolves around all these stories and myths on the underground area of Nabi Daniel Street in Alexandria, the area where the sinkhole appeared and swallowed this lady in the 1970s. And where sinkholes keep occurring every few years.

Islam Shabana, performative reading & soundwalk at *Khettara Ben Tahar, Sanaoubar/Izdihar*, as part of 'The Geomorphosis Cycle', 2023, Marrakech, Morocco

HW What did you aim for with the 3D scan?

CD And how do you envision your intervention in this 3D space?

IS I wanted to conjure a feeling of an operation room, a futuristic operation room. It looked cool but it's just failing technology. And this is where I question and critique modern technology. And the way modern technology is negating or ignoring the Global South. There are a lot of techno-fixes that are now speculated about how to save the Earth, but we don't have a say in this. Also, these are very accelerationist and pragmatic solutions meant to save the areas where most of the wealth and most of the technology is, meaning Europe and the West.

COSMOTECHNICS

My project is not nostalgic nor involved in the exoticizing trends of 'indigenous origins of things'. It is discussing 'Cosmotechnics': a different trajectory based on ancient knowledges or even mythology that might eventually reach a resolution combined with modern technology.

According to the Chinese philosopher Yuk Hui modern technology stopped progressing, in the sense that it killed what he refers to as techno-diversity.[6] And for him, going back to the origins of technology, which is thickening as a term and as a school of thinking and related to the word Cosmos as in universal understanding, that also is linked to the morality of this whole system of understanding.

6 Yuk Hui, *The Question Concerning Technology in China: An Essay in Cosmotechnics, Urbanomic Media*, 2016. Yuk references Latour, Stiegler, Lyotard and other influential thinkers of the West and introduces the history of modern Eastern philosophical thinking largely unfamiliar to Western thinkers, including Youlan, Zongsan and Nishitani, to elaborate an original reflection on the nature of technology. The central idea is that Chinese thought must question globalized technics and systemize its own cosmotechnics and the history thereof.

So for him Cosmotechnics is a techno-diverse way into technology and how to develop from different cosmologies rather than a singular modernism.

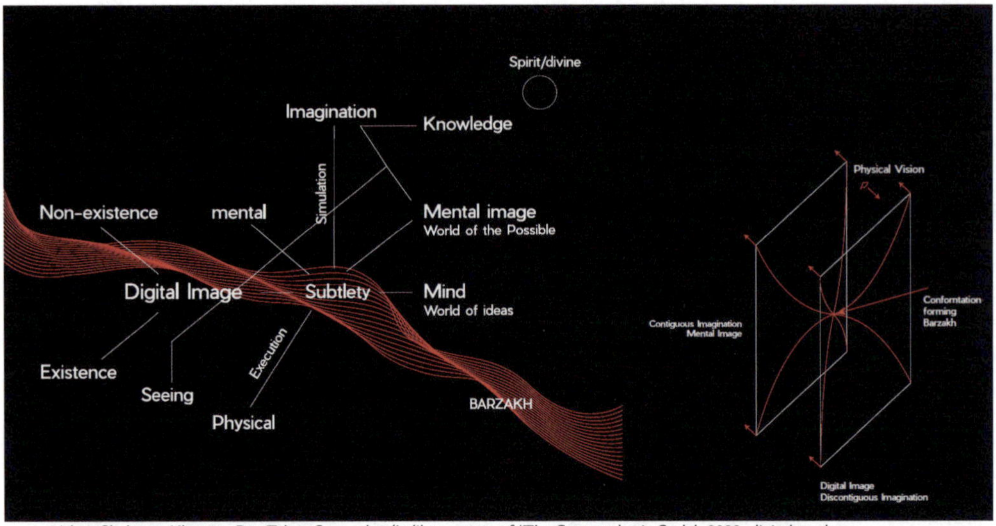

Islam Shabana, *Khettara Ben Tahar, Sanaoubar/Izdihar*, as part of 'The Geomorphosis Cycle', 2023, digital render

Yuk started by examining Chinese medicine as an example for CRISPR techniques, and how it's very unscientific to refer to such fields of alternative medicine or Chinese medicine, but at the same time, it's working. It's there. Of course, there are a lot of hoaxes, but also this classification is a problem. Because in our mental modernism, we cut off certain development of practices and ideas because modernism classified them as mythology, savage, witchcraft or whatever. In Cosmotechnics, these are all possible epistemic starting points to build knowledge and eventually transition to different technical approaches or even complete systems. In Cosmotechnics as Cosmopolitics he urges every culture to reflect on the question of cosmotechnics for a new cosmopolitics to come, since he believes that to overcome modernity without falling back into war and fascism, requires not refusing but reappropriating modern technology and the possibility of different technological futures.[7]

7 Yuk Hui, 'Cosmotechnics as Cosmopolitics', *E-flux Journal* issue #86, 2017.

In my work. I don't suggest certain technical solutions or make scientific claims, but I shed light on ancient, pre-modern technologies which thrived for centuries. The cistern systems of Alexandria were built by least more than 300 BCE when Alexandria wasn't even an important port, and they were used until the Ottoman Empire. So there was a humongous chunk of time that these technologies were functioning just fine. And we might need something like this in the future, especially when technology fails, when we don't have electricity to run machines, we might need to go back to mechanical structures that can live on their own with air or water. An example of cosmotechnics are the early automata from Islamic scholar Ismail al-Jazari's 'water machines', which use water as a source of power.[8]

8 Yuk Hui, 'Cosmotechnics as Cosmopolitics', *E-flux Journal* issue #86, 2017.

CD Is it just a weird form of superstition that leads us to certain forms of modern technology while ignoring others that are almost taboo. Is this Jinn, maybe a way to open up that world?

IS It is the question of nature, of living, rather than that of technics, that is at the centre of the story. The problematic thing is the human-nature divide, that happened philosophically. We negate nature or divide nature from us by centring

Islam Shabana, performative reading & soundwalk at *Khettara Ben Tahar, Sanaoubar/Izdihar*, as part of 'The Geomorphosis Cycle', 2023, Marrakech, Morocco

humans. I suggest that it is not just going back to nature but also going back to the mythology of nature, like the concept of Gaia or similar concepts in other cultures. Right now, planetary thinking is a trend, but it's been there in a lot of ancient civilizations like Pharaonic, Amazigh and Touareg (which I originally descend from). Or in mysticism like Sufism.

If you flee the flooding by the Mediterranean, you will have to seek shelter in the desert which is also unliveable because of severe droughts that are expected to worsen as well. So, we northern Africans are stuck in between. It's dangerous, I think.

In parallel to the Alexandria chapter of Geomorphosis, I also worked with QANAT Collective on neglected underground water channels in Marrakech called Khettaras. A lot of stories I collected there were also saying that Jinn would live in such places. And according to Leon The African, the king who first built Marrakech gathered wizards to cast strong spells to prevent people from discovering the source of waters that feed all the Khettara systems in Marrakech. My first intervention was two videos following the structure I built in Alexandria. One is based on a text from The Land and the Sea by Ibrahim Al-Koni, a Touargy-Libyan writer, and offers these excerpts:

They confirm that the water of the Aghmat River reaches Marrakech, taken from near Aghmat and flowing through underground channels. Several kings ordered investigations to discover the source of this water flowing to Marrakech. Some people entered the channel at the point where the water arrived, carrying torches for light. After traveling some distance in the channel, they felt a strong wind that extinguished their torches, with a force they had never experienced before [...] Historians say that the king who founded Marrakech, based on the predictions of some astrologers, foresaw that he would fight many battles. Therefore, he used magical arts to create all these obstacles in the canal, so that no enemy would know where the water to the city came from, and thus they would not be able to cut it off.[9]

9 Al-Hasan ibn Muhammad al-Wazzan al-Fasi, known as Leon The African, *The Description of Africa*, 1526, p.136.

While the other yet unfinished video is situated in the future where an engineer who reports on supernatural activities happening in one of the Khettaras, and it turns out Jinn might be reusing it to collect and save water during a strong drought hitting the area.

My second intervention was a collaborative video piece with artist Mati Jhurry where one of the Jinns talks about his existential crisis trying to cope with this vast modernity that's happening and modernization that is killing the environment and affecting the ecosystem.

Another intervention is dearest to me because it is my first performance in public. As an Egyptian it's dangerous to do public performances, you know.[10] It starts with a poem by my father, in which he talks about the rain that never comes. In the second part I talk about neocolonialist narratives of deforestation versus desertification. Then we walk as a group and we hear a female voice. We walk, we hear the voice, we stop. I speak, we walk again, we hear the voice. The voice was a manipulated voice of a mothers' collective who work in Dar Bellarj in Marrakech, reciting the rain ritual, an Islamic adapted version of an ancient Amazigh ritual singing for Anzar (the god of Rain) asking for rain during droughts.

CD You're bringing to life these underground spaces and you also emphasize this mental realm or this poetic imagination. It is almost like the material infrastructure underneath that prepares our minds to, for instance, wait for rain, and to relive these rituals. There appears this real connection between the physical infrastructure that prepares our kind of mental opening up.

IS For me the digital image is a kind of a transcendental medium using the framework of ethnography, drawing on the Sufi philosopher Ibn Arabi[11] and Ibrahim Al-Koni: I discovered that he wrote verses that look like Sufi poetry about the lands and the sea. And then somehow it felt like the computation of the project for me.

11 See: www.islamshabana.info/manifesto

HW Will there be a next chapter in this water work for you?

IS I come from the Kharga oasis with natural springs that are dying. They were the most beautiful things I've ever seen in my life. We call them in Arabic Eyen, which means the eye of water. We had different types of springs. Some of them were hot and those were good at healing certain bone diseases. Most of these eye springs disappeared. They are dying, because of the forced modernization and mechanization of irrigation systems introduced by Jamal Abdel-Nasser in his plan for the Sahara area of Egypt.

I'm thinking about going back to work in my hometown. One of my dreams is to establish a residency there to invite artists to work in the middle of the desert on this part of the land.

10 www.peopleanddancefloors.com/how-to-ban-working-class-music-politics-and-popular-culture-in-egypt. Under former Egyptian President Anwar Sadat, the Egyptian Federation of Artistic Unions (FAU) was established. Under Law 35 (1978) any artist or musician performing in public must be a licensed member of the appropriate syndicate for cinematography, music or acting. Artist syndicates in Egypt tend to deny the validity of newer and emerging cultural forms, which are often deemed to be artistically degenerative and even immoral by the (government appointed) artists and musicians who run them, and whose livelihoods (including cultural capital) depend on classical forms of music. A more recent addition to the law requires every public performance to be accompanied by at least six other syndicate licensed classical musicians, regardless of the main genre being played. In 2015 the government gave the FAU powers to shut down gigs at the discretion of whichever official is in charge. Recent actions of the union suggest that its main role is to police culture and cultural workers rather than support them.

6.6

Sjoerd Knibbeler, *Reservoir*, 2023, photogramme on baryta paper, various sizes

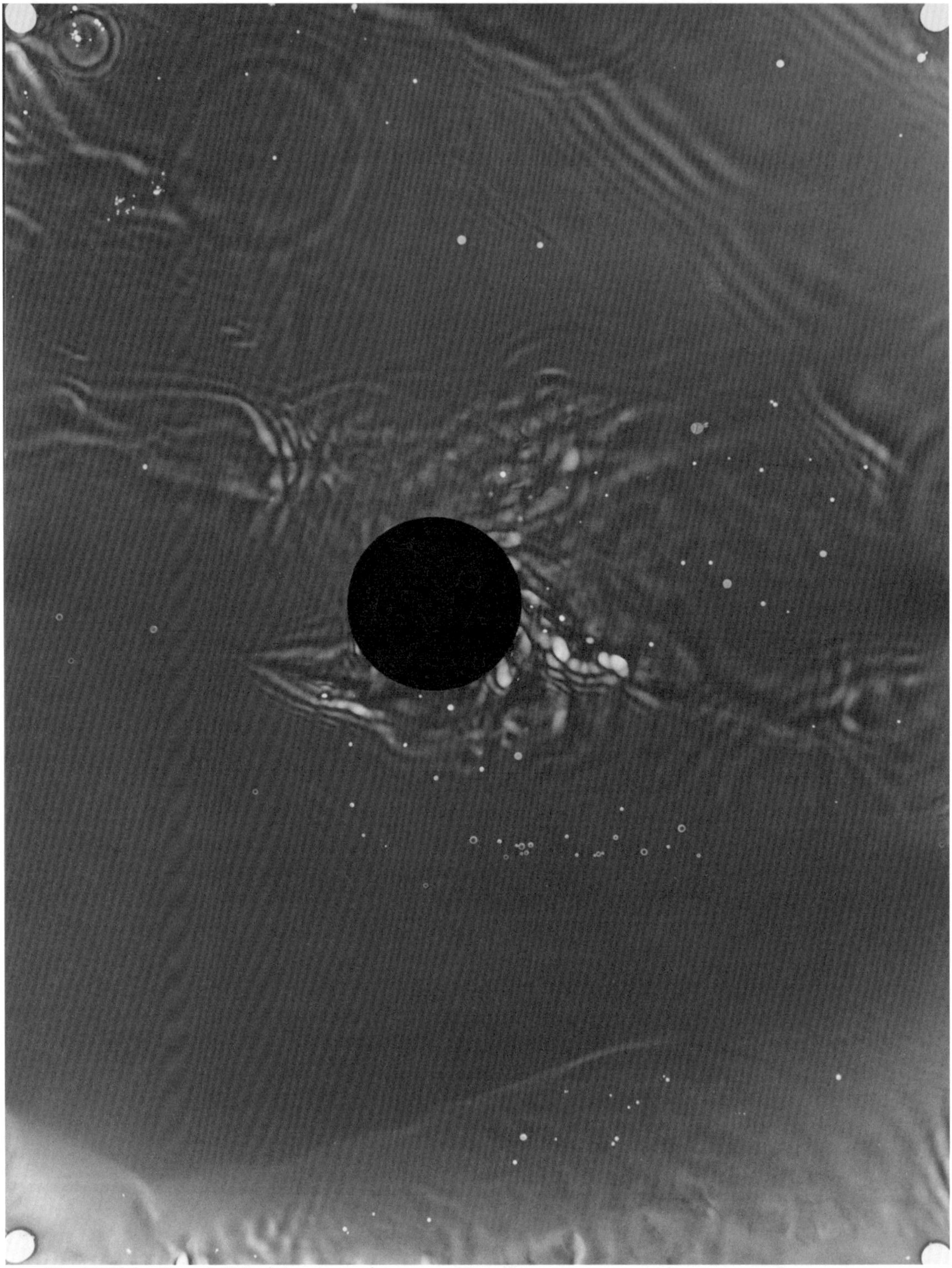

Sjoerd Knibbeler, *Reservoir*, 2023, photogramme on baryta paper, various sizes

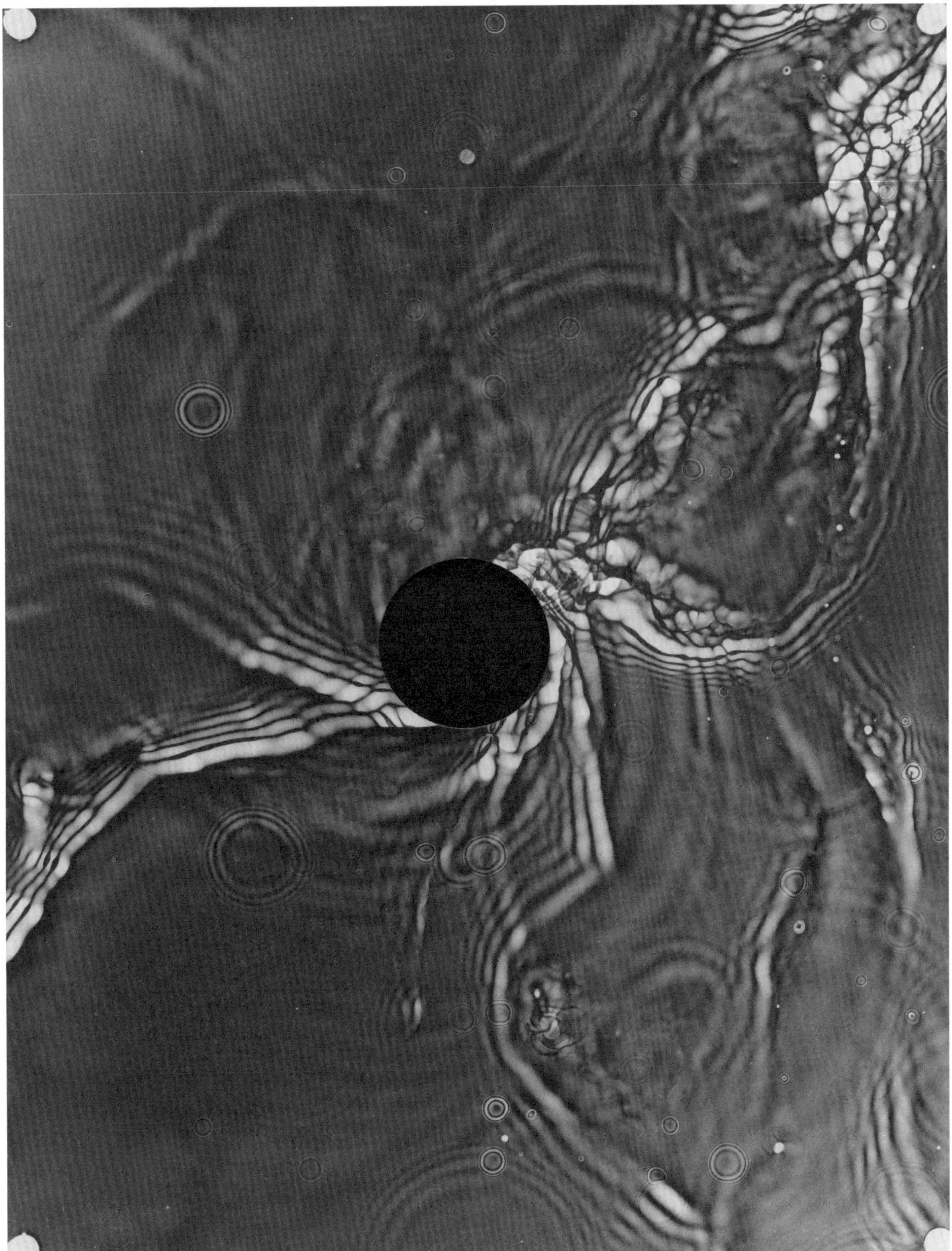

Sjoerd Knibbeler, *Reservoir*, 2023, photogramme on baryta paper, various sizes

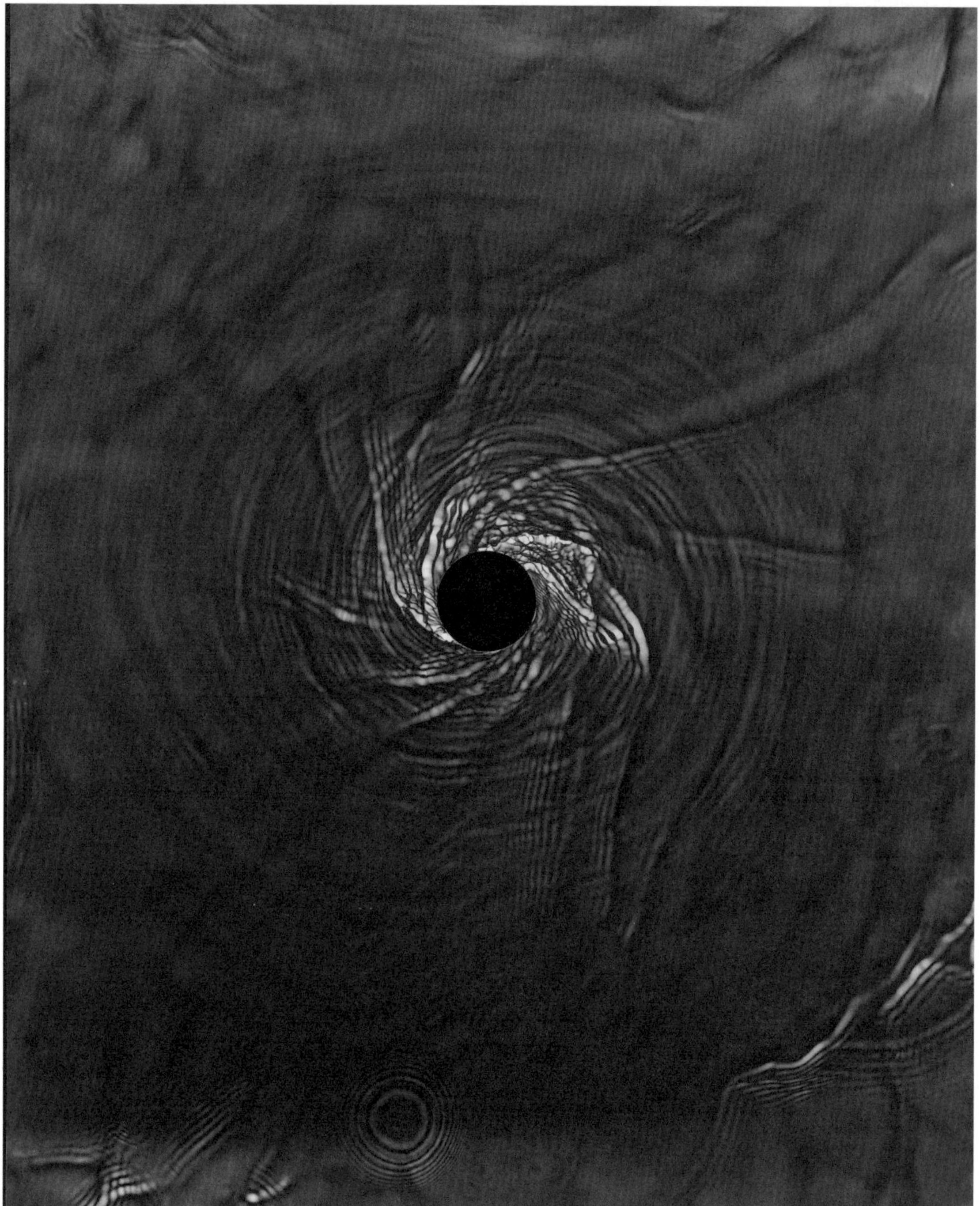

Sjoerd Knibbeler, *Reservoir*, 2023, photogramme on baryta paper, various sizes

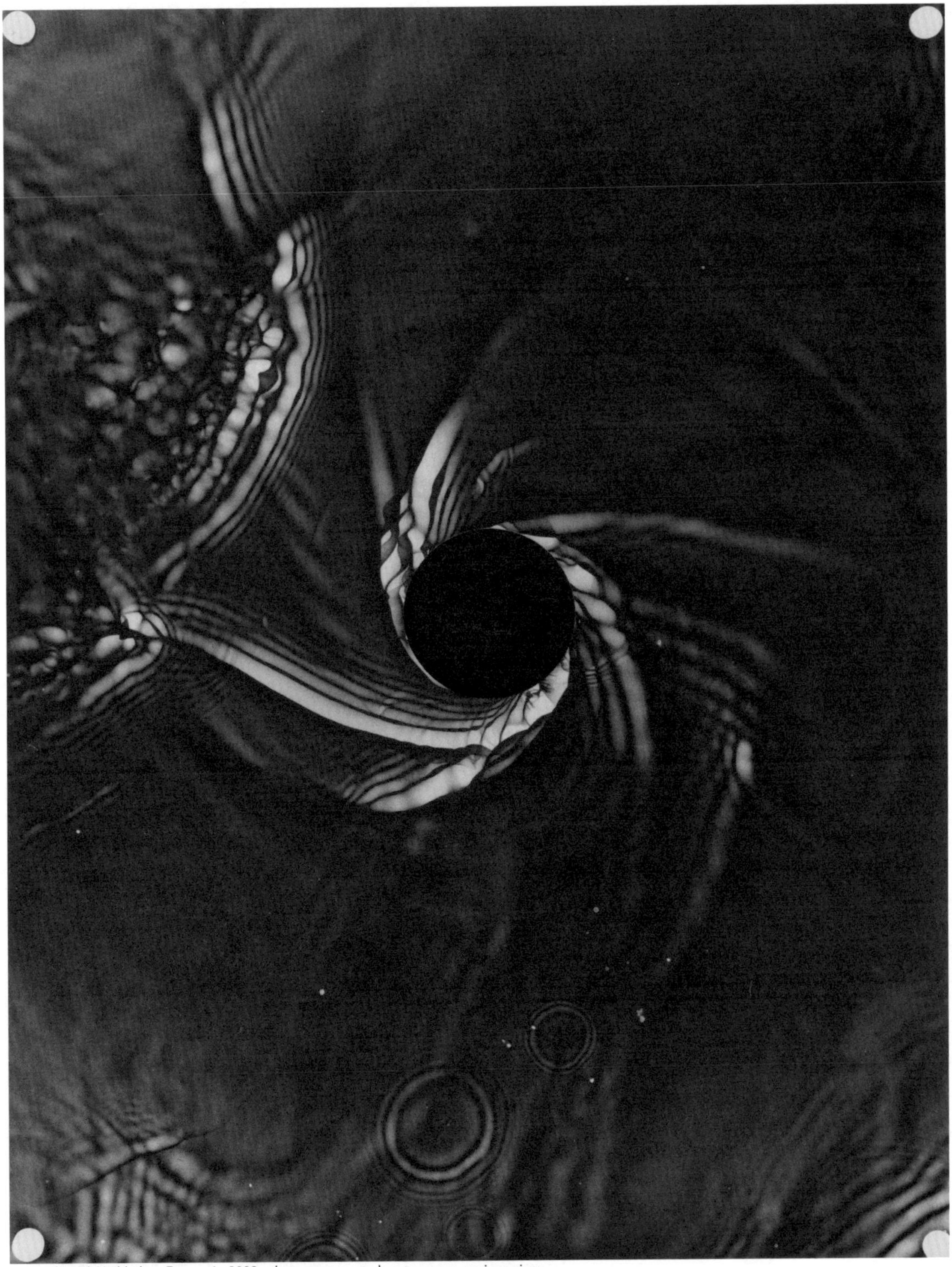

Sjoerd Knibbeler, *Reservoir*, 2023, photogramme on baryta paper, various sizes

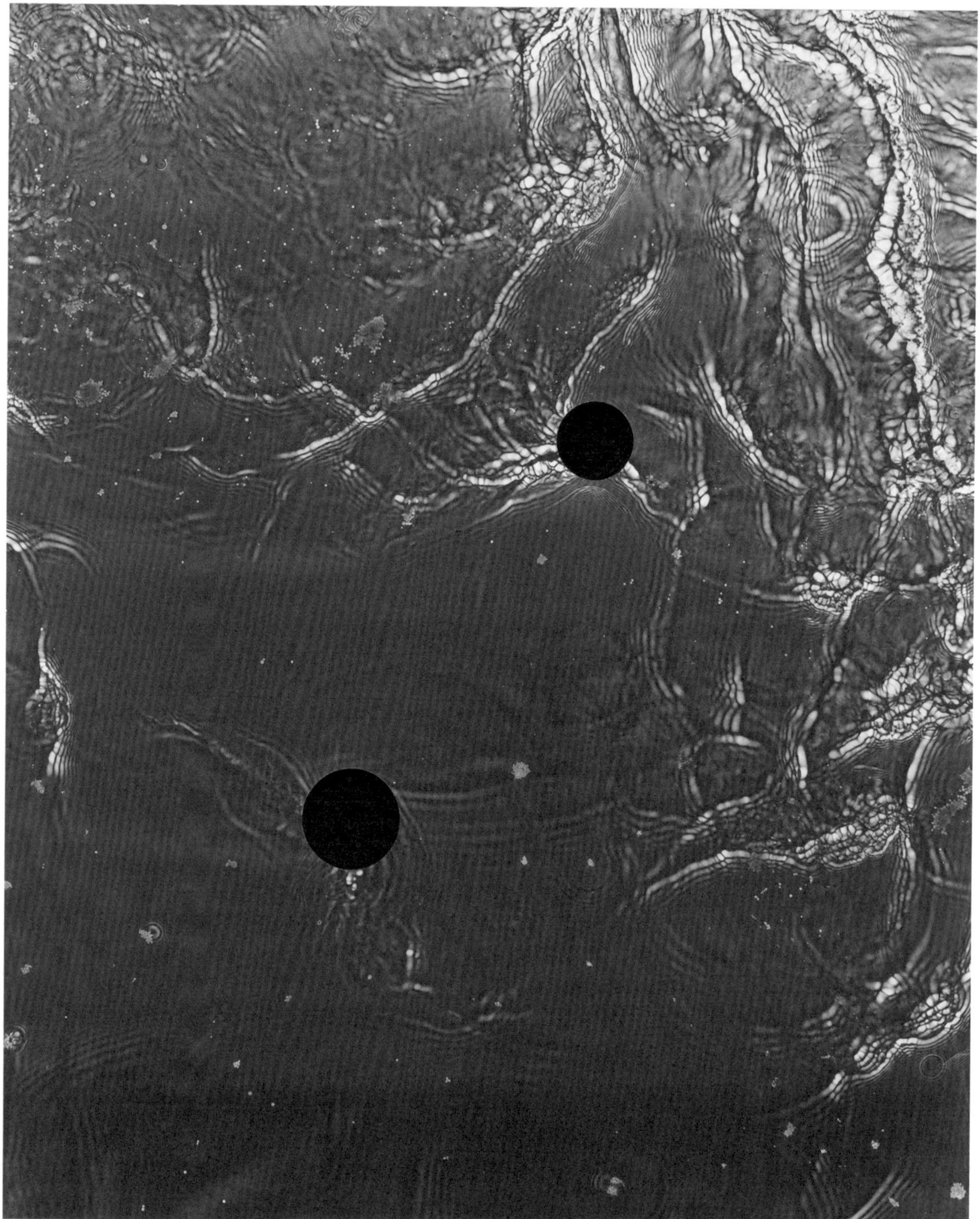

Sjoerd Knibbeler, *Reservoir*, 2023, photogramme on baryta paper, various sizes

6.7 Peat Regeneration in the Veenweide Atelier

BY HENRIËTTE WAAL

The term ‘carbon sink’ may bring to mind the Amazon rainforest rather than peatlands. However, the top 20 cm of peatland stores as much carbon as the vegetation of a tropical rainforest. Peat forms from plant material that doesn’t fully decompose due to the absence of oxygen and the acidic still water surrounding it. Peatlands are transitional landscapes, between water and land, life and death, holding personal and collective histories. If water is pumped out, memories of living with peat lose their bearing, the soil oxidates, and millennia of stored organic matter release their gases back into the atmosphere.

Woudagemaal, 2023, Lemmer, The Nederlands. © Omrop Fryslân, Onno Falkena

Across Europe, peaty soils were drained for agriculture, prioritizing production over greenhouse gas emissions. The 2022 European Nature Restoration Law aims to restore fifty per cent of drained peatlands by 2050, raising water levels to rewet landscapes and enhance biodiversity. It sounds simple, but it isn’t. The process involves different types of peat, where each has different water permeability and plant composition, while holding unknown pollutants. Additionally, families have farmed these lands for generations, and wet peatlands are often seen as economically unviable and associated with vivid memories of poverty.

In the early 2000s 'De Wilnisse Bovenlanden' and 'Groot Mijdrecht' in low lying west of the Netherlands were designated as 'test polders' to experiment with re-wetting. Being 6.7 metres below sea level, these are one of the deepest drained polders in the Netherlands. These are the polders where I grew up.

Linked to this rewetting transition, artist Sophie Krier and myself won a pitch to make a public art work for the area. *Hunnie* ('them' in local slang) was an experimental approach to recreation, routes and infrastructural elements re-imagining our relationship with nature in anticipation of freshwater fishes, meadow birds and a diverse assemblage of fungi, bacteria, and wet vegetation that were about to arrive. *Hunnie* involved farmers, community and city dwellers together in site-specific, participatory actions like shared philosophical conversations with fish and commemoration of Lakenvelder cattle,[1] fictional hunts, ditch wading, and a greenhouse symposium. These activities led to the creation of a field workstation, a field classroom and several bridges making the landscape accessible and a place for new encounters.

Wet crop experimentation, Hegewarren, Friesland, The Netherlands. Photo: Eva Pfannes

1 Originates from the Dutch words 'laken', meaning 'blanket' or 'sheet', and 'velder', meaning 'field' or 'meadow'. This cattle breed has a striking colour pattern with a black or brown body and white patches, resembling a blanket draped over the animal. Lakenvelders are known for their gentle temperament and historical significance in the Netherlands and have been traditionally used for milk and meat production. The high-yielding Holstein-Friesian breed, developed in the United States, has largely replaced the Frisian-Holland breed over the past fifty years.

Although the project was a direct critique of their somewhat dogmatic top-down approach, the Province was particularly pleased that *Hunnie* generated excitement and enthusiasm, easing opposition to rewetting plans and the concerns for mosquitoes. Sophie, who wasn't born in this polder, often remarked how fascinating it was that polder residents were so deeply attached to these monocultures of green meadows and neatly arranged ditches—an industrial 'nature' less than a century old. If trees or other plants sprout spontaneously in the meadows, residents remove that 'weed' themselves. I suddenly learned why the water in this polder sometimes has a brownish hue—it's the carbon from the dissolved peat staining it. *Hunnie* was a formative project for both of us. The wooden cross-section table, depicting the process of reclamation, excavation, and drainage of the peat—capturing the tragedy of this enigmatic landscape—left a strong imprint once completed.

Henriëtte Waal & Sophie Krier, *Cross Section Table* (part of 'Hunnie'). The formation of the Ronde Venen told through cross-sections. Periods: 1000 (peat mound), 1400–1600 (land reclamation), 1600–1800 (peat extraction), present (new nature). 390 x 100 x 100 cm, 2013. Domestic oak beams, resin, bronze sculptures

Later, when I started working on Atelier LUMA in the south of France, delving deep into the resourceful nature-cultures of the Camargue wetlands, I again collaborated with farmers, ecologists and diverse teams to explore bioregional practices and materials. Seven years later in collaboration with BC Architects and Assemble, the work of Atelier LUMA culminated in a bold showcase

Campaign image for the exhibition 'Veenweide Atelier: Design for a Sunken Marsh', Fries Museum, 2025. Photo: Tryntske Nauta

of bioregional architecture, highlighting the potential of a symbiotic relationship with nature.

When Arcadia came to visit Atelier LUMA, they asked how my approach could address the Feangreide (Frisian for Peatlands), my emotional connection to these carbon-rich wetlands resurfaced.[2]

2 The legacy organization of Leeuwarden-Fryslân European Capital of Culture.

At that time, Dutch farmer protests made the news. Peat meadows were (and still are) marked with upside-down flags, signalling resistance to governmental policies. The province appears to be in a burnout. There is pain and stagnation all over. Amid farmer protests and polarized debates, how do we shift from exploitation to a respectful, symbiotic relationship with peatlands? Beyond compensation, regulation and policy, how do we inspire, mobilize, and transform?

Innovation in peat meadow areas could take many forms: Planting wetland crops using horsepower, or switching to smaller cows that sink less deeply into wetter soil but produce higher-quality milk, fostering a healthier milk and cheese culture. To create sustainable and economically viable reed and cattail landscapes, we need well-designed products such as prefab panels, insulation walls and more, but also designs to maintain manual harvesting (mechanical cutting damages rhizomes and wildlife), as well as facilities for seasonal workers. It's not just about robots and new technology—although those could be part of it as well.

Farmer Gerbrand showing how he transports his heifers (young cows) by boat to let them graze on an island, 2023, Fryslân, The Netherlands. Photo: Eva Pfannes

Relying on scales and economic models quickly leads us back to the old economy. We want to move away from monocultures like ryegrass, but also avoid transitioning to monocultures of wetland exotics like miscanthus. Must we inevitably resort to scaling up with XL farms and building industrial factories for wet crops? Or are these very business models lacking imagination? How can we renew the concepts and frameworks for researching and managing peatlands? What are embodied ways for farmers to experience the health of their water and soil? These are the questions the Veenweide Atelier aims to explore, without being elitist, hobbyist, or nostalgic, but by going where it hurts—ecologically and socially—to create a physical and mental space to unlearn and transform.

Henriëtte Waal looking at sphagnum moss. This tiny plant can store twenty times its own weight in water, 2023, Fryslân, The Netherlands. Photo: Eva Pfannes

In 2023, the Atelier started in the development areas of Friesland's peat meadow region, where thick layers of peat exist and water levels are set to rise, protect and buffer in the coming years. Through four research

Veenweide Atelier i.c.w. Faber Futures, project 'Fit-to-Farm Models for Biofacturing', colour-generating micro-organism from Daniel Rozen/ Leiden University, 2025. Photo: Jonathan Doornenbal

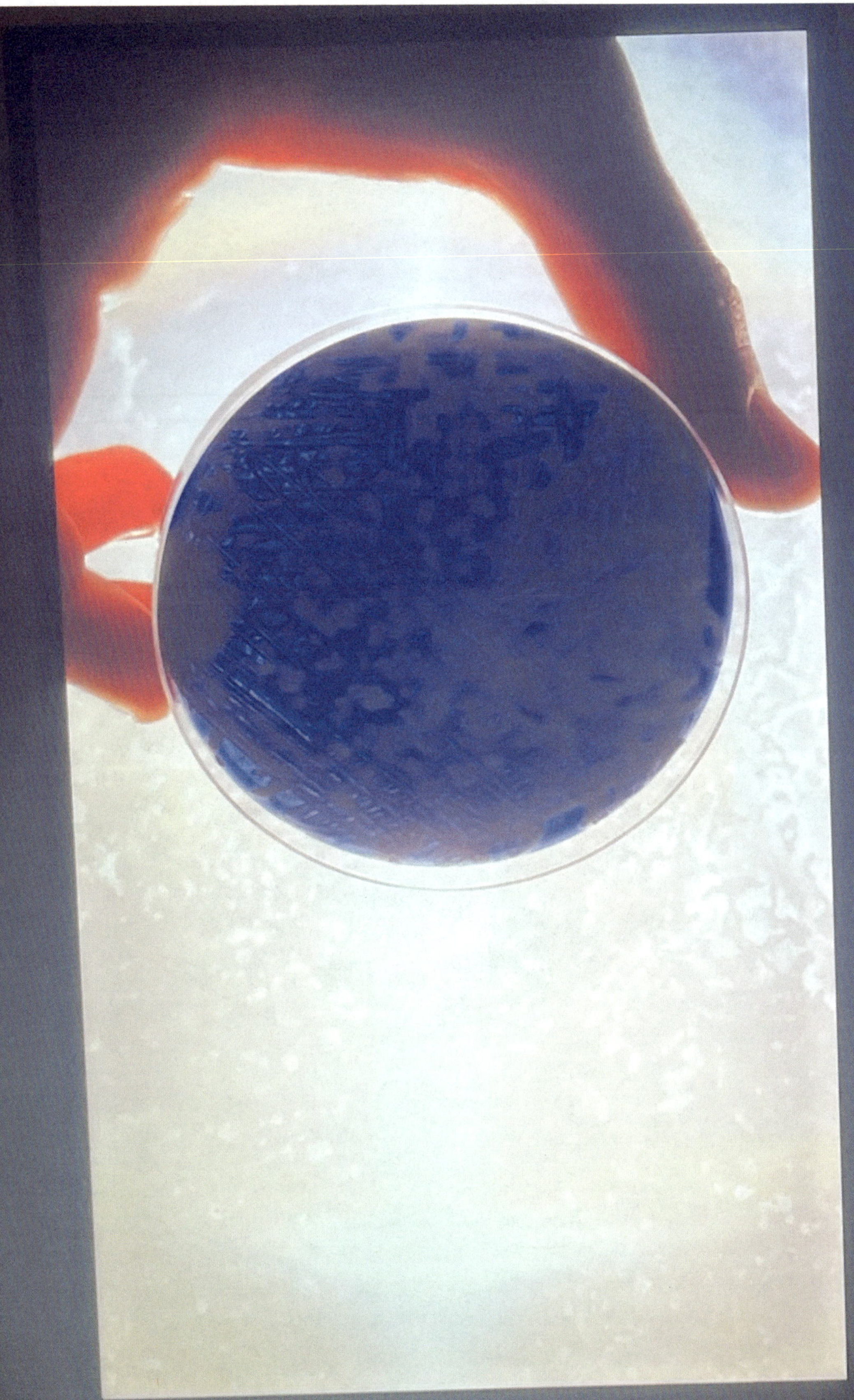

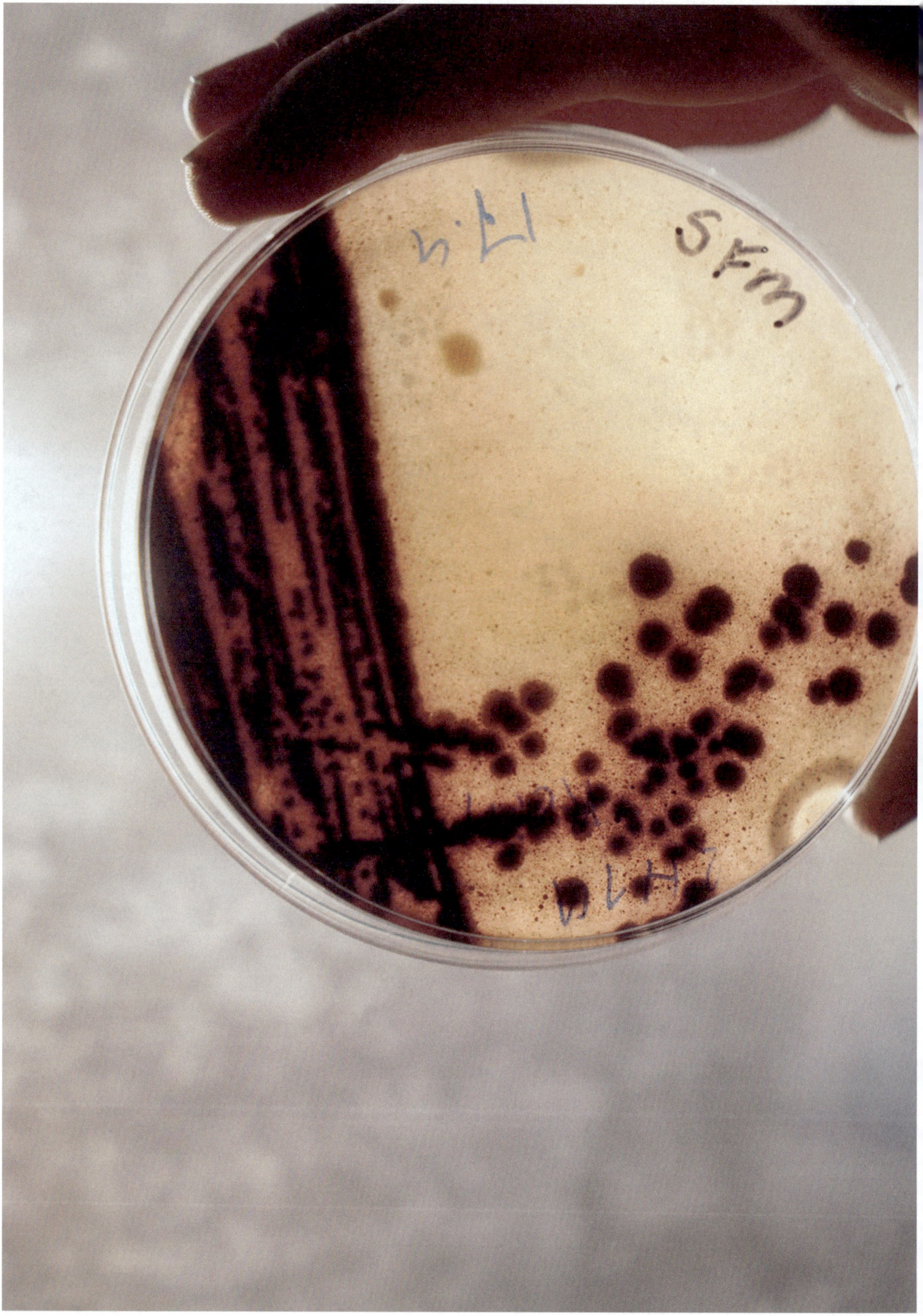

Farmers Sjoerd Miedema and Janna van der Meer and others planting cranberries in wet grassland, 2024. Drone camera. Courtesy Akkrumer Goedland, Fryslân, The Netherlands.

projects conducted by designers in collaboration with farmers, scientists, and governments, we focused on agroecology, wetland crops, future models, and a new peat narrative. The first phase concluded with the Feangreide Beraad, that brought everyone together to collectively imagine possible futures. In 2025, new initiatives will follow, building on previous research. Together with the Peat Meadow Innovation Farm, the Atelier plans to organize 'design sprints' to define experimentation and imagination needs, focusing on the themes of Plant, Animal, Soil, Water, and Business Models. The goal is to bring together expertise and define concrete collaborations that result in demonstration sites and moments where the new peat landscape can be experienced.

There are still very few initiatives in the design field that explicitly focus on rural areas and bring together different scales (landscape, building, product, material). By visualizing and storying new peat narratives, both in the areas themselves and in urban contexts, the Veenweide Atelier hopes to inspire many to love and care for peat meadows—where the unavoidable raising water tables is not merely a harbinger of poverty and the end of farming.

It should not be about a single business model but about working together on several different models and ways, making a new culture. 'It is not about "expanding the range of choices" (liberal freedom) but is intended to transform the kinds of beings we desire to be.' (Arthuro Escobar)

Or as Yoshiharu Tsukamoto from Atelier Bow-Wow recently mentioned in a lecture on bioregionalism: 'The task for the designer or architect today is to focus less on architecture and more on a network of relationships that together make regeneration happen'.

Together with Eddy Wymenga, Jeroen Wiersma, Jitze Peenstra, Lenora Ditzler, Janna Bystrykh, Jasper van Belle, Tjeerd Veenhoven, Janna van der Meer, Sjoerd Miedema, Gisanne Hendriks, Pieter van der Valk, Kees Boonstra, Stefien Smeding, Aen Age Jongbloed, Coen Verweij, Eva Pfannes, Sylvain Hartenberg, Takuma Johnson, Soujanya Krishnaprasad, Stephane Barbier Bouvet, Friso Wiersma, Erik Slor, Clemens Driessen and others we gathered a series of dos and don'ts in rewetting the peat landscapes, that aim to guide ongoing collective efforts:

Veenweide Atelier i.c.w. Faber Futures, project 'Fit-to-Farm Models for Biofacturing', colour-generating micro-organism from soil, 2025. Photo: Jonathan Doornenbal

DOS

Polyculture
Poetry
Being sensible
Appear in unconventional ways within ordinary settings
Take time to thoroughly assess current water- and ecosystems
Focus on processes
Focus on questions
Focus on verbs
Focus on embodied ways of knowing instead of cognitively constructed knowledge
Find practical ways of relating to species, ecosystems and their functions
Build community around your project
Build multidisciplinary and diverse collaborations
Consider wildlife needs
Include local, contextual knowledge and cultural and temporal value dimensions of landscapes in your design process
Do the maths of your ideas
Set up monitoring systems
Adjust based on monitoring results
Develop strategies for preventing re-drainage, such as community education
Provide alternatives for affected livelihoods
Tailor rewetting methods to each specific peatland
Create peat enthusiasts
Observe as a participant, not as an outsider
Talk about grass
Learn from cattail landscapes in the Senegal River Basin
Bridge urban-rural divides
Forget conventional business models
Recalibrate 'heritage'

DONTS

Monoculture
Being dry
Being idealistic
Being too comfortable or predictable
Rush the process
Focus on outcomes
Focus on answers
Focus on nouns
Thinking only from an 'agricultural perspective'
Hyping crops instead of ecosystems
Talking bad about plants ('weeds')
Forget grass as a wet crop
Focus solely on hydrology
Assume one-size-fits-all solutions
Neglect local input
Not being ambitious enough
Driving things top-down
Patenting
Participating as a designer in the bureaucratic meeting culture
Believing ugly things work just because they have a low footprint
Forgetting beauty
Staying comfortable within your own group
Disregard cultural landmarks
Permit over-access to tourists
Overlook methane emissions of rewetted peatlands
Fail to acknowledge the symbolic significance of peat meadows
Ignore cultural and historical significance of peat extraction
Let negative framing dominate
Resisting change without seeing futures

6.8 The Soviet Water Computer and the Death of the Aral Sea

BY CLEMENS DRIESSEN & JENNIFER VEILLEUX

Before—or parallel to—the development of electronic circuits, computers ran on water. Some have even argued that the Baroque gardens that powered complex mechanical fountains were the first programmable calculators, 'fluid circuits etched into the earth'.[1]

1. Alexander Trevi, 'Gardens as crypto-water-computers redux', *Pruned*, 2013.

In the 1930s, Soviet analogue computer scientist Vladimir Lukyanov built a room-sized device, a machine made up of a series of glass tubes and metal switches: the hydraulic integrator.

The device could solve partial differential equations with water running through valves, indicating the rate of flow between chambers. In the 1940s, this machine was used to design and calculate the effects of the Karakum canal: the world's largest hydraulic engineering effort.[2] The canal diverted the water of the Amu Darya river, one of the two main rivers that fill the Aral Sea, for navigation and irrigation. This canal made shipping possible through Uzbekistan and Turkmenistan, and turned approximately one million hectares of the Karakum desert into irrigated land used for monoculture cotton production.

2. Igor Zonn, *Karakum Canal: Artificial River in a Desert*, 2012.

With the new water system that the canal created, from the 1960s onwards the Aral Sea started to dwindle and desiccate. By the year 2000, what was left of the Aral Sea had become two separate highly saline waterbodies. Around 2010, the once third-largest freshwater lake in the world had largely become desert salt flats and marshes, its fishing villages and ports abandoned. By 2020, the Aral Sea had lost approximately ninety per cent of the water it once held.[3] Dust storms, a result from the environmental changes, cause 75 million tons of toxic pollution to spread annually.[4]

3. Alexander Izhitskiy, Georgiy Kirillin, Igor Goncharenko, Abilgazi Kurbaniyazov, Peter Zavialov, 'The world's largest heliothermal lake newly formed in the Aral Sea basin', *Environmental Research Letters*. 16. 2021.

4. Greenaralsea.org, This is Aral Sea.

One could see this as the Soviet version of modernist hubris; science harnessed by the state ordering the radical transformation of landscapes to increase prosperity by exponentially increasing cotton production. Or was it a miscalculation, perhaps exacerbated by the idea that the medium of computation operated as a model; the water-based knowledge infrastructure directly reflected the thing to be optimized and controlled, where the valves operating watery equations didn't translate well to the scale of the landscape in the end.

Hydraulic integrator designed by Vladimir Lukyanov, 1936. Courtesy Polytechnic Museum, Moscow.

Others have argued it was not the canal diversion itself but the lack of lining in the canal on porous ground, making it leak and lose excessive amounts of water. So not communist (or state-capitalist) extractivism and grandiose hydro-engineering is to blame per se: it was the half-baked execution, not making the real-life water system resemble the water computer enough that made the calculations turn out differently on the ground.

Still another version of the ongoing Aral Sea disaster stresses how the canal and river system is no longer managed as in Soviet times, to mutual satisfaction of Kyrgyzstan and Tajikistan sharing the river water with Kazakhstan, Turkmenistan, and Uzbekistan. Seen in this way, it's a post-Soviet case of 'river basin bullies', or 'hydro-hegemons',[5] where river sheds are the medium and staging ground for geopolitical power play, when countries upstream can retain and divert river water as they see fit.

5. Mark Zeitoun & Jeroen Warner, 'Hydro-hegemony—A Framework for Analysis of Trans-Boundary Water Conflicts', *Water Policy*, 8(5), 435–460, 2006.

In recent years, some of the last remaining bits of the Aral Sea have stabilized. But in debates over restoring the now largely dried up lake—one can question the extent to which a sense of urgency is felt, of reverting the river system to its former state and with that perhaps restoring the lake.

Whatever the motives, most people have now forgotten what the landscape looked like before the canal and the Karakalpakstan region has become the poorest of the area. The remaining population is suffering from economic marginalization, population decline, diseases and lack of resources, the first of which is water.

6.9
Westland

By Clemens Driessen

COVERING THE LANDSCAPE IN PLASTIC

The making of this water basin is a communal event, a weekend outing, when the entire village participates in covering swaths of land in plastic in the service of the Westland greenhouses that surround it. Is this a key cultural moment, expressing the aspirations and shared ideals of the community? Rendering a landscape into a state that leaves no space for growth of anything other than the planned indoor kind.

This seems to be what in Latin America is called a mink'a, a work-party, the occasional gathering of a community for a particular purpose. Drawing on Inca traditions, it occurs outside of capitalist wage relations. Participants do not get paid to work but join to help out someone and thereby strengthen local bonds.

What understanding of their landscape, and how to make it productive, makes this a communal project? How did these people collectively come to experience the aesthetics of this space in a way that rigidly shaped earthen basins covered in dark grey plastic as an actively welcomed meaningful intervention?

The whole village, it seems, realizes the importance of having sufficient water to feed the plants in the greenhouses, as well as the necessity of retaining rainwater to prevent flooding, while all available land has been turned into impermeable glass structures.

Historisch Archief Westland No T-000898 and No T-000900. Photos: A. de Bok

Commerce

Commerce

7
Commerce

BY HENRIËTTE WAAL & CLEMENS DRIESSEN

Water: too much, too little, too polluted. A string of problems, are they somehow connected? Each of these are occasions to earn money. Capitalism noticed the water crisis, and this isn't necessarily good for water, or for those of us exposed to the water crisis.[1] Money is not an issue. It's not that there is no money. Never before has there been so much financial interest in water. But markets won't improve access; turning public goods into commodities may be a response to scarcity, but it is also a cause of that same scarcity. Privatization and its fresh-faced twin, charity, threaten a wholesale takeover of the global water crisis.[2] Water is the sentinel of the limits of corporate capitalism on this planet. Not that we needed a sentinel. All this makes commerce even dirtier than before.

1. Filippo Menga et al., *Thirst*, RGS-IBG panel session, August 2024, London.

2. Filippo Menga, *Thirst: the Global Quest to Solve the Water Crisis*, Verso Books, 2025.

But people do like to get paid. Where to send the invoice for an ecosocial intervention? We were not able to pay all who contributed their time and insights to this book, signalling our inability to make ecosocial design pay off. What models of exchange, sharing, and valuing are there that somehow can work across commercial interests, private/public ownership, academic partners?

Opening image: Balenciaga storefront featuring one of the Spring Cooler sculptures by Moreno Schweikle, 2023. Photo: Moreno Schweikle

3. B. Elzenbaumer, F. Franz, K. Krois & A. Mattozzi, 'Eco-Social Design: Who Pays for It?', *The Design Journal*, (20:sup1), 2017, S4678–S4682, www.doi.org/10.1080/14606925.2017.1352965.

How to fund ecosocial design?[3] What could be a benign innovation ecosystem for water solutions? Who earns anything when rivers are drinkable? Well, farmers, for instance, and the producers of textiles and beverages—all benefit and they could work in such a way that they wouldn't cause pollution downstream. How to design for a world that does not seek to optimize profit, banking on cheap nature? Ecosocial design for commoning, exchanging affect, not just meeting individual demand of capitalist consumer desire? While trickle-down economics is drying up; and governments are drip-feeding the arts at most. Ecosocial redesign must get a place within infrastructure budgets, claiming the biobased economy, or socially embedded nature conservation, moving beyond the restrictive labels of 'culture', or 'aesthetics'.

Ecosocial design is moving away from feeding on corporate dreams of market shares—thinking in terms of the need to make a profit, to monetize, commodify. It crafts new ways of storying of what it means to design, to succeed, or even to fail. It challenges the conventional views of knowledge (exclusive), ownership (individualistic), and markets (purely efficient).

Ecosocial design has no obvious business model and sits uneasily with commercial exploitation. But then again, bottling water can be a beautiful scene to behold, as Henriëtte discovered in a small, family-owned beer and soda factory in the Black Forest. Beholding a production line where an idea is turned into an object at full speed has a certain magic. And interestingly, in some areas, local breweries have successfully blocked fracking initiatives by aligning commercial and communal interests. Can there be appropriately scaled commerce in the bioregion? Making water that connects, rather than global brands that taste the same everywhere and erode cultures of working with water?

When tracing the costs and benefits of water works, we can get inspired by diverse economies that acknowledge all the unmonetized modes of valuing and exchange, that build on relations and valuation of reproductive work, or maintenance.[4] How can we brand this type of worlding and consuming? Ecosocial artistic expression seems to resonate more easily with this curiosity for all the work that goes into keeping things running.

4. J.K. Gibson-Graham, 'Diverse Economies: Performative Practices for Other Worlds', *Progress in Human Geography*, (32:5), 2008, pp. 613–632.

Many water works are not rocket science nor high-tech, ecosocial design requires time and dedication, love of place, wonder and connecting. How to find space in budgets for what is actually fairly cheap? Getting paid for not polluting, rather than budgeting the endless cleanup?

7.1 Selling from the Source—Von der Quelle ins Regal

BY HENRIËTTE WAAL

As mineral water or deep-source water traverses vast distances through various geological formations, lingering beneath the Earth's surface for centuries, it emerges not as simple water but as intricate blends. This journey imbues it with salts, magnesium, calcium, sulphur, and trace elements.

Source in the Northern Black Forest, Germany, 2024. Photo: Henriëtte Waal

Depending on the presence of natural carbon dioxide, these waters may be still or sparkling. Electrolyte water contains minerals that carry an electrical charge. Alkaline water—with a higher pH level—can neutralize acidity in your bloodstream and enhance hydration. However, these waters can also contain heavy metals and radioactive elements, making strict quality regulations essential. Meanwhile, many drinking water companies tend to remove not just pollutants but also beneficial minerals in the name of guaranteeing uniform quality.

In the Northern Black Forest of Germany, known for its thermal springs, Mineralbrunnen Teinach has been producing mineral water since the Middle Ages—official records date back to the year 1300. Today, the Teinacher area and the landscape are protected, and its underground flows of water are subject to intensive quality management. Beneath this forest lies a careful and multidimensional subterranean interplay of ecology and technology.

Teinacher mineral water has an age structure between 40 and 10,000 years old, lives—yes it is alive, according to those

Water hadde dudde da, a honesty bar presenting Knötti, a newly developed knotweed drink, made in the Northern Black Forest, 2024, Egon Eiermannkirche, Pforzheim, Germany. By Henriëtte Waal, Amalia Shem Tov, Moreno Schweikle, Anthea Oestreicher, Jan Schulz and Ketterer Brewery. Courtesy ORNAMENTA 2024. Photo: Karolina Sobel

Water storage Stadwerke Pforzheim, Germany, 2024. Photo: Henriëtte Waal

Betriebswasser source (Source of the operating water), Mineralbrunnen Teinach, Germany, 2024. Photo: Henriëtte Waal (top to bottom)

working with it—between 80 and 140 metres below the surface and flows with a speed of 50 to 80 centimetres per day. It travels through sandstone and when it comes to the surface it has a temperature of more or less eleven degrees. The Teinacher Hirschquelle, a precious healing water, moves slowly through granite at depths of 400 to 500 metres.

Maintaining the springs to ensure their well-being is the primary responsibility of one man, Jürgen Kappler. His tasks are meticulous, from establishing new wells to ensuring operational efficiency. Heavy rains, he explains, can bring fine particles of coloured sandstone into the water. When this happens, they have to make sure to remove that from the water.

Teinacher's facilities, surrounded by pine forests, include a lab where Markus Blaich, Head of Quality Assurance, tests water samples for microbes, nitrates, and hydrocarbons. Automated conveyor belts in the packaging facility transport bottles, which are rigorously cleaned and filled. The facility processes 25,000 litres per hour, totalling 600,000 litres per day. Still, Teinacher sells only regionally, to ensure the water extraction does not exhaust the geologic layers that contain water. The taste of this water stays in the landscape.

Technological advancements allow for the maintenance and regeneration of wells, extending their productivity. Kappler and his team continuously seek new boreholes. 'When drilling for new water sources', says Kappler, 'we have to be careful not to connect the groundwater with the source. Sometimes the difference between success and failure is only two metres'.

Besides the drilling for mineral water, Matthias Hinsch from the Office for Environmental Protection of the city of Pforzheim tells us that the drilling for a mineral used to liquefy metal for the global car industry is set to resume in the Northern Black Forest. He assures us that the extremely polluting part of the process will take place outside the region.

Testing the color of the knotweed drink at Ketterer brewery, Pforzheim, Germany, 2024. Photo: Amalia Shem Tov

Environmental concerns are paramount. The threat of fracking looms, with its potential to contaminate water sources. This destructive extraction method stands in sharp contrast with the careful stewardship of mineral water. Seventy-seven municipalities in the region have spoken out against fracking in their territory, recognizing the importance of preserving the environment for sustainable water commerce.

In the Black Forest, Teinacher's regional approach to mineral water production promotes environmental responsibility. Unlike global giants like Nestlé, Teinacher and other bioregional companies negotiate to establish joint production facilities, akin to wine cooperatives, to share costs and resources. This model supports sustainability, ensuring the lively waters of the Black Forest continue to flow for generations to come.

Knotwood juicing at Mediamatic, Amsterdam, The Netherlands, 2024

KNÖTTI

It was only during the late 1800s that bacteria (including waterborne from the sewage contamination) were recognized as agents of disease. Beer, however, was usually safe, because the water it was made from came from deep wells, and was boiled as part of the pre-fermentation process.

Ketterer, a family-owned brewery in Pforzheim was founded in the late 1800s by my great-great grandfather Wilhelm Ketterer. I'm Michael Ketterer, the fourth-generation brewer. Like my ancestors, I continue to focus on quality and heart. In our times, the term 'heart' for me takes on the extra dimension of responsibility. Responsibility for the region, for our resources, and for future generations. Our beer is brewed according to the German beer purity law, with four ingredients: water, hops, malt and yeast. We brew enjoyable traditional styles. Our Edelpils has won German beer awards.

Bottling day of the knotweed drink at Ketterer Bottling Station, Pforzheim, Germany, 2024. Photo: Ray Formili Fendi

Running a brewery, maintaining quality and tradition is dedicated and hard work, especially as the world changes. Pforzheim grew and therefore our brewery got disconnected from its source of soft water that the region is famous for and that my great-great-grandfather once brewed with. Now, we use tap water—a mix of pure H^2O and limestone water. Not only has our water changed; the drinking habits have evolved as well, influenced by the young generation and newcomer communities who are not into drinking beer.

When ORNAMENTA paired me and our brewery with artist and brewer Henriëtte Waal,[1] I asked her: 'How do I ensure that my Turkish neighbour, when he returns to Turkey, takes my product with him and shows it with pride to his family there?' She said, 'I'm not sure but let's talk' and she came over with her mobile outdoor brewery.

Creating a non-alcoholic drink is not where my brewing heart lies, but it was a challenge I embraced through this collaboration. Besides beer, we also make cola mix and orange

1 First held in 1989, ORNAMENTA returned in 2024 as an exhibition for contemporary art and design in Pforzheim and neighbouring municipalities and communities in the Black Forest. Curated by Willem Schenk, Jules van den Langenberg and Katharina Wahl.

Betriebswasser source (source of the operating water), Mineralbrunnen Teinach, Germany, 2024. Photo: Henriëtte Waal

River Nagold, 2024.
Photo: Henriëtte Waal

lemonade with the name Ketti. The Ketti label was designed in the 1980s, has never changed since, and almost has a cult status now. Apart from the label and the name, there is little local, authentic, or artisanal about Ketti but it is a rapidly growing market for us.

The next time Henriëtte came over to the region I met her in a smoky local bar. She brought her colleagues Moreno Schweikle, Jan Schulz, Anthea Oestreicher, and Amalia Shem Tov and said: 'We would like to work with you as a collective and move away from traditional beers. We'll start from understanding the water and it will not be what you expect'. I accepted the adventure.

They explored local waters and plants, shifting their focus from hops to neophytes; plants that newly arrived in the forest ecosystem. New species like weeds from Japan, whom no one really wants to acknowledge, let alone knows what to do with them. The team found similarities to the disconnection with the changing community I was talking about.

I had never heard of Knöterich—the German name for Japanese knotweed—before. Now I know it was once brought here by beekeepers, to boost the bee population. But its dense clusters can overshadow other species and seriously hinder the forest. Once you know what it looks like, you'll see it everywhere, especially along the waterfront, flourishing in watery soils. One morning, on her way to harvest knotweed together with herbalist Roswitha Hild, by the river Nagold, Henriëtte encountered two men with brush cutters. Roswitha explained to

Henriëtte Waal, Ray Formilli Fendi & Andrea Groppi harvest knotweed by hand, close to the Nagold river in the Northern Black Forest, 2024, Germany. Photos 1 & 2: Henriëtte Waal. Photo 3: Andrea Groppi (pp. 353–356)

Knötti Fridge with Knötti bottles. The label is inspired by the Ketti labels, Ketterer's soda line and comes from the 1980s. On the side is a cash box for donations to support the second brew of the project. Photo: Sander van Wettum

the man, 'Knotweed tastes similar to rhubarb but more lemony, and its young shoots resemble asparagus'. The men looked somewhat bewildered. One of them turned toward some uncut knotweed next to him and tenderly stroked the plant—knotweed began to gain recognition.

Back at the brewery, a sugarcane crusher served as an effective knotweed juicer, producing the necessary quantity for our kettles. Some youngsters jumped in, bringing mad energy and diversity to the scene. Henriëtte and her team experimented with various aroma hops in combination with knotweed, finally crafting a refreshing flavour. Imagine: if you were to put this prototype into production it could seriously do something for the forest!

The team made it all come together in an honesty bar: a self-service drinking experience that re-roots our region with itself: a funky mishmash of industry, working-class mentality, precious waters and brewing traditions, nostalgia, herbal knowledge, smoky bars, and newly arrived living beings. Sour Knötti Boba pearls crafted from local waters, knotweed and algae make the experience complete.

It is only the beginning of a Knötti ecosystem.

Coltrane McDowell is an artist and designer known for his poetic visual narratives and community-driven projects. A cum laude graduate of the Social Design master's program at Design Academy Eindhoven, Coltrane has since delivered meaningful, collaborative initiatives across France, Egypt, and Kenya, blending creativity with social engagement to inspire change.

7.2 The Cow that Sits in the River

BY COLTRANE MCDOWELL

SHOFCO Aerial Water Distribution System, 2023, Mathare, Nairobi. Photo: Coltrane McDowell

To look forward sometimes you have to look back. There was a time in the not too distant memory of many Nairobians when the rivers that ran through the city were clean. Nairobi,

derived from the Maasai 'Enkare Nairobi' means the place of cool waters. While they may not have been drinkable, twenty years ago these waters were at least clear, they did not run in the myriads of colours that can be seen today: dark black, browns, red and blue.

Depending on the neighbourhood and industries along their banks, the rivers absorb colour into their smelly sluggish bodies as they snake through the city.

In the last twenty years Nairobi's official population has tripled. Rapid urbanization—a sign of growing economic wealth also sheds light on the problems nations such as Kenya have with water and sanitation within their colonial urban centres. 'You'll find that an area's zoning changes', Mshila Sio, said to me. 'So, where there was one house on one acre, you may have an apartment block with a hundred apartments, but it's going to the same sewer line, the same water has to be supplied'. Mshila runs Omiflo, a floating wetlands wastewater treatment system. The concept is based on traditional constructed wetlands, yet a floating wetland uses less space and is easier to set up. For Mshila, wastewater treatment is ecosocial, 'there's a lot of dignity that comes with dealing with sanitation properly. If you can clean water for people, everything else cascades from that'. He sees the potential of his systems in a decentralized wastewater treatment network because even the cost of transportation of mobile wastewater, with every kilometre, is enormous.

While Omiflo currently supplies to private clients, they are hoping to partner with the government to aid in the clean-up of the Nairobi rivers. There are many challenges, one being the government itself, slow to take up new ideas and with a history of corruption it's hard to even gain an audience. Still, Mshila has hope given the new Nairobi Rivers Commission (NRC) set up by the current president William Ruto. Ruto in an impassioned speech at the launch of the commission cited combatting climate change as a driver for the rivers clean-up among other widespread problems related to rapid urbanization.

Few have more to gain from the city-wide river clean-up than the hundreds of thousands if not millions of Kenyans living

Water distributing service, 2023, Mathare, Nairobi.
Photo: Coltrane McDowell

NA USAFI
MTAANI
DONATED BY
KARIUKI
K1
MP
DuraCoat
superfast

along the riparian zones of the Nairobi rivers. Many live in small, corrugated metal houses, neighbourhoods of informal to formal settlements. In the rainy season, homes closest to the rivers may flood, some are even washed away, while the small unpaved paths leading between homes become rivers of mud with streams of untreated sewage. The rivers in their sluggish and polluted state still serve peoples' uses. Most informal settlements exist along the riparian zones of the city, this land is technically illegal to build on, yet many who live here have done so for generations.

In Mathare, one of Nairobi's oldest settlements, the river is both a lifeline and a source of major problems. Pauline, a community mobilizer and friend of Mathare, began our interview with a sigh, the meeting she had come from still weighing on her. She began, 'I was telling them that the government wants to reclaim the river because the river wants to be reclaimed and it wants to be clean. Now for this community of us, that was like sad news. Although they're saying it is because of climate change. But for us when we told them it was a shock.' Evictions are nothing new in Mathare, however there is a sense that the evictions taking place through the NRC will bring an end to some neighbourhoods all together. The message to clean the rivers of Nairobi may sound noble, how this is done, and who is involved will determine the outcomes. Many homes in Mathare are built illegally along the river, yet there are examples of luxury high-rises and town houses doing the same and they are not met with the same treatment. In citing climate change, Ruto is evading the question of who in Nairobi will get to enjoy cleaner rivers.

Clean water access was not always a challenge within Mathare. Government-funded water infrastructure was readily available for the communities it served. An example being the public toilets in settlements with no street names or house numbers, these toilets became orientation routes. Pauline told me how people would gather at the toilets to talk and discuss things, and formed friendships. This changed in the last twenty years when public water infrastructure such as toilets and water pumps became privatized or controlled by gang cartels. In the wake of these changes, numerous water solutions sprouted.

Sitting above the corrugated metal roofs, blue poles hold blue pipes, and within them water flows from a blue building with large white letters printed on it: SHOFCO. SHOFCO is but one among many actors in Mathare providing water distribution services. Their model of 'aerial piping' is intended to circumvent the cartels who are responsible for blocking pipes or redirecting water elsewhere if people do not pay for their services. Speaking with Margaret, a community mobilizer from SHOFCO, she told me how people could trust the water because they could see it. Walking in Mathare, you will notice the numerous NGOs and governmental organizations funding new water projects for Mathare citizens. Each vendor prices their water differently, while the water sources vary in cleanliness or taste. In the absence of strong governmental infrastructure or institutions, a flurry of smaller-scale actors has come to meet the needs of people—each vying for a neighbourhood to exert their presence in.

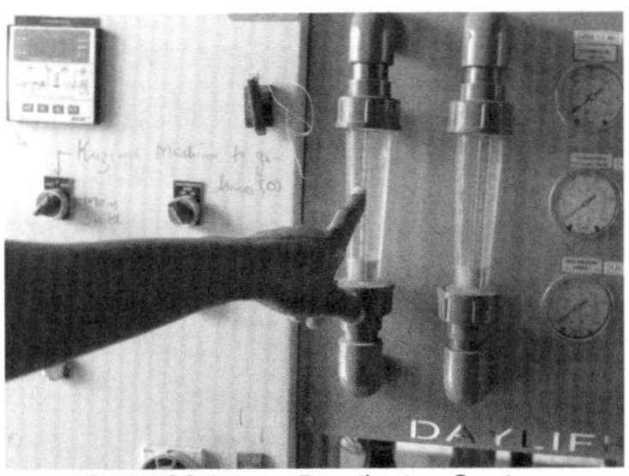

SHOFCO Water Distributing Service, 2023, Mathare, Nairobi. Photo: Coltrane McDowell

Pauline and her friend Esther had grown up remembering a time when you would go swimming in the river. While swimming no longer happens, teeming along the river clouds of smoke can be seen and the unmistakable smell of fermented sugar. Chang'aa the name given to the home-made alcohol produced across Kenya is made in abundance here. Mathare is seen as the place with the 'conkest' in both strength and taste, Esther describing the Mathare varietal as something akin to whisky. She told me how 'even the president, our first president of the Republic of Kenya, knew that there's a cow there that is in that river and people are milking it and you cannot finish that milk'. The river has a long history in Mathare of being used for distillation. The water used to cool the pipes, often a garden hose and copper coil connected to an oil drum, bringing the murky ferments into a clear vapour form. Generations of Mathare citizens have diligently milked the cow that sits in the river.

Chang'aa is criminalized but is a major source of income for many in Mathare. For both Pauline and Esther, daughters of distillers, and who had become distillers and alcohol sellers themselves, chang'aa is a good thing for the local economy. Distillers work day and night along the river in small community circles called 'Jikoni's' (the Kitchen). The term relates to both the act of distilling, which is often described as cooking, and the communal aspect the Kitchen serves. This community is born out of necessity, a small fee is given for every newcomer who wants to distil, this is used as a communal savings when needed to pay bribes to the police. The Jikoni will also help individuals who are arrested, offering bribes or putting pressure to release them. 'I was not cooking', Esther said, 'but then at the end of the day, I came to

SHOFCO Water Distributing Service, 2023, Mathare, Nairobi. Photo: Coltrane McDowell

know that the job that was in front of me that is the work that I depend on, so I started with one drum, I went to two, and now I have five drums'. While the challenges to distil may be immense, the rewards have put many through school, provided jobs where there are none, and given people a sense of pride and dignity.
that sits in the river is sick, its belly is gorged on plastics, its milk runs black, and it belches a stench of sewage and cleaning products. These are the Nairobi rivers. While it may be easy to think that nothing can be done to reverse the motions of progress and development, it was only a few years ago that the rivers ran clear. Now the Nairobi rivers are at the forefront of a debate to save them. How this is done will determine the success of such actions in the future. As can be seen from the previous examples, there are numerous ways in which community structures and the social architecture of informal settlements becomes a far more powerful tool for collaboration than top-down governance. With weak government institutions, many Kenyans, for better or for worse, have had to adapt.

To clean the city's rivers will be no small feat, given the continual urban growth. Omiflo's proposal for decentralized floating wetlands systems could work. The net result of many smaller systems would bypass logistical challenges of wastewater transportation, and larger and more expensive water-cleaning infrastructure to provide an overall cleaner river for all. One of the main challenges for wastewater recycling is the revaluation of wastewater itself. Currently in Kenya there are few incentives for recycling wastewater and associations of smells with existing sanitation systems mean most would not want these systems in their backyards or neighbourhoods. However, if like the production of chang'aa wastewater recycling could become a new chain of economic value, people along the rivers' sides could become the custodians for keeping the rivers clean. The systems are already there, new technologies are not needed but the harnessing of communal will. Community-lead floating wetlands, small enough for people to manage, and localized to provide for the direct needs of people are one way forward. Large-scale forced evictions along the rivers' banks will not put an end to the challenges of cleaning the rivers, it will only move these problems further downstream.

7.3 Biotech Dyeing —or Creating Ecosystems for Microbial Life to Change the Fashion Industry

INTERVIEW WITH NATSAI CHIEZA (FABER FUTURES)
BY HENRIËTTE WAAL & CLEMENS DRIESSEN

Natsai Chieza is an ecosocial designer making waves in the fashion scene with her groundbreaking work with pigment-producing bacteria on garments as an alternative to polluting chemical dye processes.

With synthetic biologist Christina Agapakis from Ginkgo Bioworks (an American biotech company) she launched Normal Phenomena of Life, a brand in the realm of biotechnology and sustainable fashion.

CD Does your practice focus on water, or is improving water use more of a side effect of trying to change the fashion industry?

NC Retroactively it is easy to say 'yes of course', but my story is maybe more generative than that. I confront design practice with a biophilic lens. One of the key revelations working with Streptomyces (bacteria) was the minimal water use of these organisms, compared to traditional textile dyeing processes. For this organism to live, and to live well, it only consumes the water necessary for its existence. It requires only organic nutrients, as opposed to toxic chemicals and solvents to make it work.

Chieza began working with the micro-organism Streptomyces bicolour by coincidence, while sort of moonlighting as a designer-in-residence at University College London's Department of Biochemical Engineering. Initially she was introduced to the organism by her collaborator, Professor John Ward, because of its colour-producing properties. She soon realized that the organism had more to offer than just pigment. Instead of following a traditional industrial logic

Lab collaboration between Natsai Chieza and Veenweide Atelier

of extraction, Chieza questioned the need to extract pigment, realizing that the organism was motivated to produce pigment because of its biological context.

HW If your starting point is how the organism lives, instead of what it produces, what are your ideas about growing and scaling your work?

NC Working with living systems as I have been doing teaches you very quickly that the industrial lens limits the discovery of more than anticipated outcomes and standard models of scale-up and distribution. Everything is built around industrial expectations. But what does biological exploitation look like in the bio-industrial revolution? How could we not do that?

Biology can be seen as the most distributed technology on this planet, and it scales by finding its niche and fitting its context. Think of a large rainforest. People ask: can we scale biotechnology? Well yes, and especially if we lean into what motivates biology to do so in nature every day. A thriving ecosystem is a function of resilient relations across many local and global flows of value. Scale-up in a commercial sense and therefore from an investor's perspective is a story of individual flourishing: the unicorn monopoly and singular metrics of value in terms of return on investment. We will need to overcome this singular story to reveal the expansive possibilities of working with biology. We will need to create new models and effective deployment. We can learn from Nature's ability to scale the systems we design and make it commercially viable, it's not a zero-sum game.

HW What is it that you would prefer to avoid?

NC The dominant industrial logic for developing and scaling biomaterials is the idea of the drop-in replacement. Find a biological molecule that can be commodified through cash value per metreage or weight. To add value, you're dropping it into existing machinery, into existing supply chains, into existing power structures. The same financial incentives remain. A culture of overconsumption persists. It is marketed as change, but I'm sceptical of this because in my view you change nothing, apart from divesting from fossil-fuel based materials. The latter is a good thing, by the way, so herein lies the

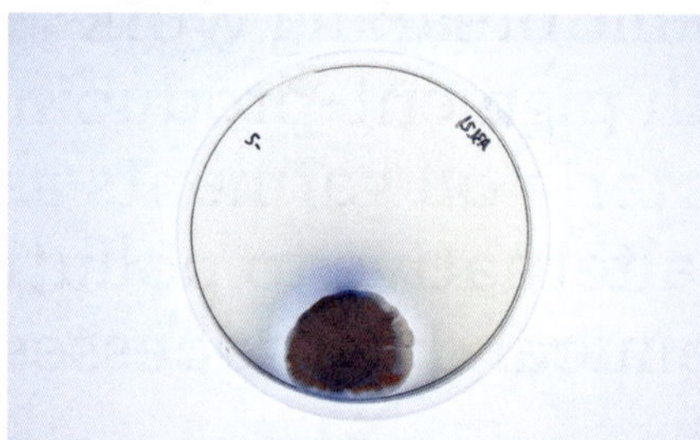

Faber Futures, Single Organism Portrait, Project Coelicolor

paradox. Without a systems-change that goes beyond the drop-in replacement, where we develop these technologies to create meaningful value and infrastructure across the value chain and for more equitable outcomes, I just see all of these new innovations ending up in landfill in spite of it all.

Chieza advocates for a shift from industrialized biology to biologically-informed industry. The former perpetuates existing production structures, supply chains, and power dynamics, potentially causing more harm in the long run. Instead, she proposes a holistic approach that integrates biology into various aspects of society, bringing in the humanities and fostering a more sustainable and equitable future.

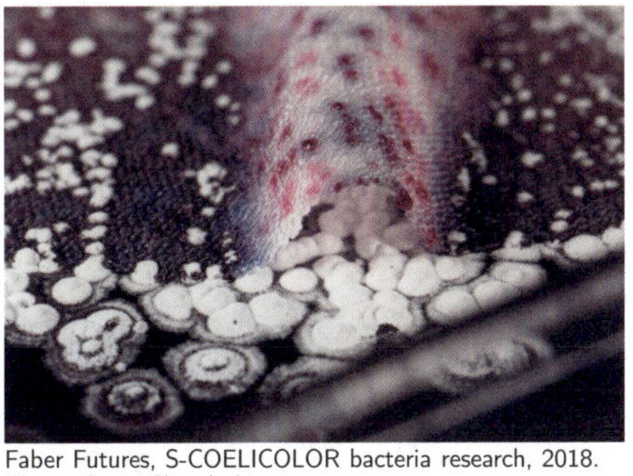

Faber Futures, S-COELICOLOR bacteria research, 2018. Photo: Scott Chimileski

The form Chieza chose is Normal Phenomena of Life, a product innovation pipeline and consumer-facing brand that brings design-led products made with biotechnology to market. The brand functions as a research and development lab, offering transparency into its processes and materials. By retailing products and storytelling, Chieza aims to create an accessible interface that broadens the understanding of biotech tools and materials.

Normal Phenomena of Life is committed to taking the consumer on a journey of empowered discovery, illuminating how biodesign encompasses design with wild-type organisms and engineered living systems. Chieza also envisions a shift towards distributed biotechnologies, where carefully calibrated licensing models generate equitable outcomes for all. The goal is to enable more people globally to engage with and have access to the tools of biotechnology, fostering a distributed biotech model with diverse stakeholders.

CD What does it look like to bridge the cultural realm with the technological? What is that like in terms of materials or working with these organisms, or having these conversations?

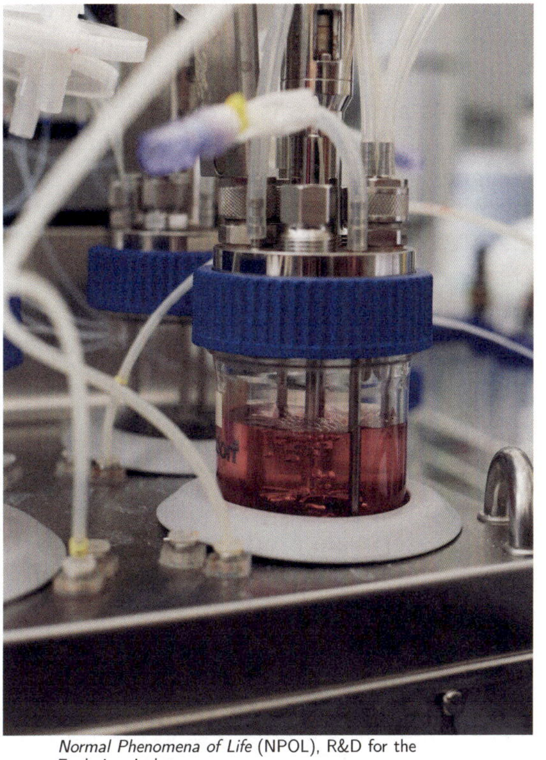

Normal Phenomena of Life (NPOL), R&D for the Exploring Jacket

NC Bridging the cultural realm with the technological, particularly at Normal Phenomena of Life (NPOL), is about creating harmony between human innovation and the natural world. At NPOL, we see nature as the original technology, and our work is focused on leveraging biotechnologies that are climate-positive and rooted in sustainability. This approach is not just about creating products; it's about reshaping entire value chains to honour nature and foster industries that thrive without harming the planet.

In terms of materials, we work with living, dynamic processes—like garments dyed with pigment-producing bacteria or lighting made from bioconcrete grown by microbes. Each creation embodies both human craftsmanship and the ingenuity of nature, showcasing how biotechnologies can redefine everyday objects in fashion, beauty, interiors, and industrial design. These aren't just technological advancements; they represent a cultural shift in how we understand and interact with the materials and systems in our lives.

Through collaboration between designers, scientists, engineers, and creatives, we create ecosystems that elevate biodesign beyond products, shaping industries and providing a blueprint for a more sustainable future. Our work is a modern expression of a new Bauhaus—a marriage of culture, art, and technology that offers a vision of what is possible when human innovation aligns with nature's wisdom. In this fusion of culture and technology, we are creating not only desirable products but also new systems and narratives that challenge and transform the marketplace. At NPOL, we're proving that biodesign can offer solutions that nurture both people and the planet.

HW What are the places and structures that allow the biorevolution you envision to happen?

NC We need to establish dedicated infrastructures that sustain these endeavours. But the expectations and results demanded by the market may at times prevent the deeper, more meaningful work required for significant advancements. The places and structures that allow the biorevolution we envision to happen are those that foster interconnected, distributed bio-economies across diverse geographies. Biodesign, as a practice, is deeply rooted in nature and indigenous ways of living, and its development should not be limited to the West. It requires global collaboration

Normal Phenomena of Life (NPOL), Exploring Jacket. 100% GTOS-certified organic silk, with a toxin-free, water-efficient microbial dye process. Fermented and produced in the UK

and an equitable distribution of resources to truly thrive. This means establishing interoperable infrastructures that can span different regions and support the growth of bio-economies in various contexts, regardless of existing patenting regimes or infrastructure barriers. When it comes to building infrastructure, there are practitioners who are developing bioregional and culturally specific solutions in other regions of the world including South America and parts of Southern Africa. Biodesign and its interaction with ecology and society is not confined to Western Europe. Certain material systems, like those leveraging fungi and algae, offer more accessibility for broad, distributed deployment due to their lower barriers to entry. These natural materials can be utilized more easily in various geographies, helping build bio-economies that don't depend on highly centralized or resource-heavy systems. However, creating interoperable systems that allow for equity across different regions is one of the major challenges we face in scaling this biorevolution.

Perhaps most crucially, bridging disciplines is necessary to address the complex questions the biorevolution brings up. We need places where biologists, designers, engineers, economists, and creatives can come together to collaborate. Only by integrating these varied perspectives can we achieve the high-resolution understanding required to build the bio-economy of the future.

HW To your knowledge, do major brands explore ecosocial design production?

They can do, but success levels vary of course, as there is a lot of greenwashing. If your sustainability budget is coming from the marketing budget that's a conflict of interest for obvious reasons.

Faber Futures, S-COELICOLOR coloured samples

It's really important to think about what finances change and how this is part of a structurally integrated business strategy.

There are some notable initiatives that are demonstrating the willingness of brands to 'learn in public' how to develop new production models that are rooted in positive ecological and social impact. What they do with those learnings is more difficult to assess from the outside, however Atelier100—funded as a joint initiative between H&M and Ingka Group, the largest IKEA franchisee—was an interesting example of global brands partnering to prototype new models that support local cultural production and distribution.[1] IKEA is accustomed to this kind of experimental institution building with Space 10,[2] so despite Atelier100 no longer being operational, it will be interesting to see what it catalyses next and beyond that context.

1. Atelier100 was an incubator established by H&M and IKEA encouraging hyperlocal production and emerging and discovering London-based creatives from a greater variety of cultural and economic backgrounds. First products were successfully realized throughout the programme.

2. Space 10 is IKEA's research and design lab dedicated to creating better and more sustainable ways of living.

HW How do you want to involve consumers; how do they affect what's going on upstream?

To involve consumers meaningfully in the biotech design process, it's essential to engage them from the outset with tangible products that bring life to the abstract ideas or the speculative imaginaries. Products embody so much more than the material. They bring the human and more-than-human relations of supply chains to life. This is a powerful vehicle for engagement if a brand wants to tell that story, and it allows consumers to interact directly with the materials and processes behind the innovation. We want to create a feedback loop between the consumer and the upstream design and technology processes, enabling more rapid progress towards a bio-economy and more meaningful contributions that inform its culture.

7.4 Spring Coolers

INTERVIEW WITH MORENO SCHWEIKLE
BY HENRIËTTE WAAL & CLEMENS DRIESSEN

From bulky blocks of ice that were difficult to move to the sleek, modern-day office dispensers, the evolution of the business of cooling water has been remarkable.[1]

1 www.susan-schuppli.com/MOVING-ICE

Balenciaga storefront featuring one of the Spring Cooler sculptures by Moreno Schweikle, 2023. Photo: Moreno Schweikle

In 1938 American plumber Luther Haws invented the first electrical self-contained water cooler—the kind very similar to what we use today—although these original water coolers used heavy glass bottles, preserving the 'difficult to move and transport' part. In the 1980s, plastic vessels replaced the original glass ones and water coolers then spread over the globe in the 1990s. As office environments became ever more commonplace with the advent of computers, water coolers developed as an integral part of office culture, providing sparkling, hot, chilled, and ambient water; just press the button.

Artist Moreno Schweikle, a product of the 1990s himself, grew up in the German Black Forest region known for its natural springs and wells, and a thriving mineral water industry. In this environment where water design is omnipresent, he noticed that although water coolers hold a humble yet significant place in office culture, sparking conversations and serving as meeting points, this function had not been taken up design-wise. 'Despite their obvious function and underlaying social dimension the objects often don't embody desire and look a bit sad', according to Schweikle.

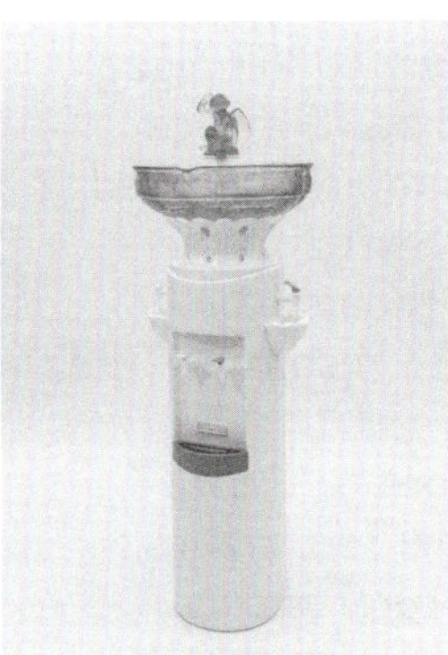

Moreno Schweikle, *Spring Cooler*, 2023

He started looking into neoclassical drinking fountains on ancient city squares once designed as gathering spots for those travelling with horses and as status symbols because of their complex technologies. By fusing water-circulation techniques, decorative neoclassical elements, and the perhaps postmodern industrial design language of the water coolers, Schweikle creates sculptures that draw immediate responses from viewers.

His work caught the eye of Balenciaga's creative director Demna Gvasalia, who acquired multiple Spring Coolers for Balenciaga's 'Art in Stores' project, showcasing art in the brand's flagship stores. 'SPRING S-700' fountain was first displayed in Milan and then travelled to other boutiques throughout Europe. The fountains 'SPRING A-100' and 'SPRING G-300' travelled through America and Asia.

MS Knowing Demna's work, I've always been intrigued by how their collections reflect the amalgamation of different aesthetics, through the infusion of digital culture in all aspects of daily life. I sympathize with how the garments but also the stores and runway shows capture this tension between tech-enthusiasm and nostalgia.

One of the store's visual strengths is the absence of nature. Each cooler functions as a tiny oasis in this sterile atmosphere. It's almost like the running water lures you in to then actually notice the sculpture. I'm fascinated by these 'levels of primal attraction' as it happens on a subconscious level beyond style, personal taste or cultural imprint.

The sleek look of the Spring Coolers turns out to be the result of a messy, intuitive and playful process that started online. Schweikle found water cooler models on 3D platforms and began downloading and manipulating them digitally. He also sourced physical models from eBay and other marketplaces in order to 3D scan them and use the digital clones to construct the fountains. 'The coolers mark an interesting moment in the history of product design. Most of them have been created on the brink of digital fabrication and often hint towards historic architectural elements. Sometimes a shape is reminiscent of a Greek Ionic pillar or Roman arch. At the same time, these objects are designed to relate to the nature of water, conveying purity and sensuality. This mix of influences results in quirky combinations that have a bodily quality. The 3D-modelling softwares are my tool to study these shapes.' To materialize modified versions, Schweikle works with a mix of contemporary production methods such as 3D printing, silicon moulding and resin casting, and more traditional crafts such as silver plating and Trompe-l'œil painting.

MS I want the coolers to work a bit like a double exposed photograph, where both the original cooler and my interventions are visible at the same time. I hope that each fountain, just like their ancestors, incite interaction.

Schweikle's Spring Coolers, while keeping up a wonderfully clean appearance, are a playful critique on the sterile, commodified world of water. Maintained by conveyer belts and automated water supply systems, water coolers present a simplified image of water. 'The Balenciaga sculptures need maintenance. Water coolers live from that. They have to look sterile, so you need to add these powders to kill algae growth and check pump performance regularly.' Combining plumbing knowledge with artistic expression, Spring Coolers are both functional and thought-provoking, inviting viewers to reconsider their relationship with water.

In the future, he hopes to create work that embraces 'dirtier' aesthetics, Schweikle says, and a natural ageing process, like historical fountains that gain beauty over time. By incorporating elements that allow for gradual change and decay, he aims to celebrate the dynamics, renewal and deep wisdom of water.

7.5 Meilinck: Allegory of Nature and Human Nature

By Clemens Driessen

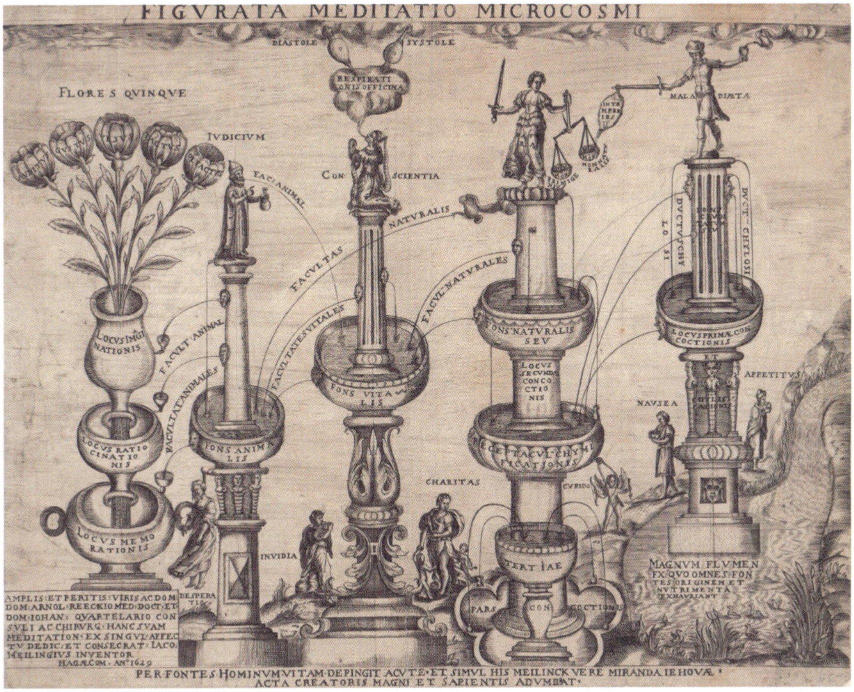

Allegory of nature and human nature (microcosmos). Anonymous, after Jacob Meilinck, 1629. Rijksmuseum Amsterdam

A model showing the interlinkages of elements of the fountain of nature and life in relation to Christian virtues and human bodily experiences.

The image emerges from the Renaissance tradition seeking to depict the nature of reality in a single image, integrating micro- and macrocosmos as a series of interconnected fountains. This type of schematic image making was considered to be itself a source of knowledge, here revealing ways in which different phenomena—bodily experiences, moral virtues, psychological conditions, and environmental factors—are part of a single integrated process. The fluid model resonated with this form of knowledge being associated with alchemy and the practice of distilling: the material procedures for gaining forbidden knowledge and pure substances through evaporation and condensation. The genre of diagrammatic drawing came under heavy critique from astronomer and mathematician Kepler for its mysticism and symbolic riddles, leading to a heated debate about the role of images in the development of the sciences.[1]

1 S. K. Heninger Jr., *The Cosmographical Glass: Renaissance Diagrams of the Universe*, Huntington Library Press, 1979.

7.6 De-Extractivist Poetics in the Semi-Periphery

INTERVIEW WITH MIRKO NIKOLIĆ
BY HENRIËTTE WAAL & CLEMENS DRIESSEN

I knew the water was not only polluted with heavy metals, but also that the location was not far from a sewer exhaust pipe. Sewage water is still largely untreated in the whole country unfortunately. I had cut myself and I was scared in the murky water, which added some tension to the whole thing.

We reflect with artist, researcher, and activist mirko nikolić on the impacts of current 'mining booms' in north and southeast Europe, the exploitation of the Balkans and nikolić's belief in de-extractivist poetics, which is how they call the work to create space to reground a common life with communities who opt-out of accelerated, large-scale and unsustainable extraction.

MN Today there is a mass demonstration in Belgrade. British-Australian mining giant Rio Tinto wants to mine lithium and other critical raw materials in the lush Jadar Valley in the west of Serbia to meet the EU's current lithium needs. This would significantly damage the valley ecosystem and pollute the waterways. Last week, protests against the mine already took place in several cities across the country. Scientists indicate it will certainly destroy one of the key water-bearing areas in Serbia and argue that it is critical to prioritize environmental protection and public health over economic benefits to ensure a sustainable future.[1]

1. D. Đorđević, J.M. Tadić, B. Grgur, R. Ristić, S. Sakan, J. Brezjanović, & B. Šolaja, 'The influence of exploration activities of a potential lithium mine to the environment in Western Serbia', *Scientific Reports*, (14:1), 2024, 17090.

Elsewhere in the country, near the eastern Serbian town of Bor, women block a bridge in shifts to protest against the expanding mining operations by the Chinese Zijin Copper company, which is producing copper while polluting waterbodies and land with mine tailings, making their village unhealthy and unsafe to live in. Even if their centuries-old village is full of memories, they have seen that further coexistence with the mine is impossible and demand a new settlement, with a piece of peaceful land, a school, a church and a cemetery.[2]

2. www.aljazeera.com/gallery/2024/4/20/in-serbian-village-women-fight-to-escape-encroaching-mine

More than 8,000 years ago at this same spot in the river Danube watershed: imagine a group of six hunter-gatherers sitting in a circle around a hole-in-ground smelting pit made from river clay. The people work together in coordinated roles, some wait outside the circle as a replacement for the blowers and there is one drummer to provide the blowing rhythm.

These hunter-gatherers living in the Danube watershed around 6000 BCE are considered the oldest European civilization. They were smelting ore, made pottery and remarkable sculptures of women and fish or water fairies. Along with the position these sculptures had in the house shrine, they suggest a connection with river gods.[3]

By revisiting the deep history of the metal mines in Serbia and the Balkans, mirko nikolić aims to generate a more complex understanding of the ecological changes affecting the living conditions of humans and more-than-humans alike. Their environmental and eco-focused work is a fusion of research, activism and artistic performance, featuring a striking combination of dense theory, visuals, and queering practice.

3. M. Radivojević & B.W. Roberts, 'Early Balkan metallurgy: origins, evolution and society, 6200–3700 BC', *Journal of World Prehistory*, (34:2), 2021, pp. 195–278. www.link.springer.com/article/10.1007/s10963-021-09155-7

HW Moving to Finland and Sweden made you intervene as an artist in this condition you grew up in and the theme of extraction. How did that work?

MN Through Frontiers in Retreat, a network of art residencies coordinated by the Helsinki International Artist Program I had the chance to stay in Finland twice, in 2015 and 2016, and learn about the 'mining boom' in Finland and northern Europe. I was particularly struck when I learned about the massive Talvivaara disaster, multiple tailing dam breaches in 2012/2013, and the fallout of the disaster and the social movement opposing the mine.[4] And I started to learn about the exploitation of the indigenous Sámi lands here in the North.

Through the same network I also got a chance to do a residency back in Serbia. I was interested in metallurgical mining as the pillar of modernity and capitalism—and how through water these mines keep leaking from antiquity until today. Especially the last 100-120 years, when industrial mining in Timočka Krajina, in the mountains of what is today the east of Serbia, developed. The Balkans have colonial histories. Having in mind these existing legacies of pollution,[5] we now don't have any reason to believe that our countries will be able to adequately monitor and manage mining waste because of the influence the investors and the big companies have, as well as the lack of transparency and democracy in the current political system.

At the time, I started doing lots of different kinds of visualizations, cartographies of power networks and hand drawings of financial flows. I became more aware that I do have some sort of embodied trauma. I'm still in the process of sorting this out. I grew up in an extremely air-polluted environment, but it wasn't like a big deal somehow in the 1980s and 1990s. But now it's a huge issue and gathering momentum in Serbia.

4. www.ejatlas.org/conflict/talvivaara-mining-company

5. N. Atanacković, V. Dragišić, J. Stojković, et al., 'Hydrochemical Characteristics of Mine Waters from Abandoned Mining Sites in Serbia and Their Impact on Surface Water Quality', (20), 2013, pp. 7615–7626. doi.org/10.1007/s11356-013-1959-4
N. Atanacković, V. Dragišić, V. Živanović, S. Gardijan & S. Magazinović, 'Regional-Scale Screening of Groundwater Pollution Risk Induced by Historical Mining Activities in Serbia', *Environmental Earth Sciences*, (75), 2016, pp. 1–8. www.link.springer.com/article/10.1007/s12665-016-5983-9

HW Can you explain what is happening currently in Serbia with new environmental alliances and social groups?

mirko nikolić performing at the Borska river, December 2022

View on the unprecedented mining expansion in Timočka Krajina, which is devastating the Veliki Timok River, and the destruction of Starica Hill above Majdanpek, July 2022. Photo: mirko nikolić

MN Environmental politics have gained significant traction in the Balkans. And some wins, shorter- or longer-lasting, have been made. For example, Rio Tinto's project was cancelled in January 2021 (even though it was subsequently 'revived' in the summer of 2024). Members of green groups have successfully entered parliament and also started to have influence on the municipal level in some areas, which is remarkable. There is a lot of work to be done. I have been following, documenting, and telling stories about the evolution of the movements through art and writing, and I am personally involved in some of the work as a citizen or an activist. Since I am diasporic, I am trying to support on the international level. International solidarity and coordination is crucial for environmental and climate justice.

HW In the video of your performance at the confluence of rivers Sava and Danube, we see a constellation of buckets floating in the water. Dressed in a metallic bodysuit and big rubber boots you are carrying water around, transferring it from one bucket to the other. What did you aim for with the performance?

MN It was part of an exhibition 'Overview Effect' at the Museum of Contemporary Arts in Belgrade.[6] Since the Museum is near the river, I decided to also hold a performance situated on a small beach where the Sava meets the Danube.

6. www.msub.org.rs/exhibition/overview-effect/?lang=en

It started at sunset in autumn. I knew the water was not only polluted with heavy metals, but also that the location was not far from a sewer exhaust pipe. Sewage water is still largely untreated in the whole country unfortunately. I had cut myself and I was scared in the murky water, which added some tension to the whole thing.

The performance was putting together the genealogy of mining-related struggles in the Balkans and beyond with the women's labour of carrying water historically. So, it's a feminist critique of male-coded extractive industries from the standpoint of water carrying and reproductive labour, carrying but also cleaning. The buckets contained aquatic plants known for their metal-absorbing properties, but in this instance, they were more symbolic than effective since there should have been many, many more of them.

It is an important job for the future: manual cleaning in a kind of very affirmative sense, and as such it should be paid properly and actually should be centred by society. I draw here on Mierle Laderman-Ukele's maintenance art works, and generally on the feminist tradition in arts and beyond, so also following and understanding the work of women in my family, today and through history. To some extent, this is basically a reperformance, a reproduction in a different historical moment and from a different body marked by patriarchy as well as toxicity in a different way.

nikolić and their students also visited and stayed at former mining sites in Sweden and Finland. This activity, they explained, was meant to reflect on the past, understand the present, and contemplate the future.

MN I see what I'm doing as my own voice in this long chain of people doing different things. I am trying to highlight my positionality as being in-between Northern Europe and the Balkans, the semi-periphery of the EU,[7] with its own cores and peripheries, and this dialectic always firmly caught up in the 'ecologically unequal exchange' between the global North and South. The exact place of the non-EU part of the Balkans is particularly tricky here, since it is hard to establish if it is over-consuming or poor, and who exactly is doing that. It requires a very detailed class analysis and cartography.

7. www.blogs.mediapart.fr/berliner-gazette/blog/140622/no-cheaps-here-not-anywhere-after-extractivism-semi-periphery

This place-based performance, *water is (non)life* (2020-2022) was a series of small-scale actions and they were essentially individual, just me with the plants. I just find it increasingly complicated to invite people, especially of course thinking about the local environmental justice organizers in terms of their safety and security, and autonomy. Academia and art often involve extractivist protocols, it is very hard to get out from these asymmetrical relations.

Yet, the rapid expansion of extractivism in the western Balkans since 2020, during the pandemic and afterwards, challenged my approach and transformed into a more attentive documentaristic and collaborative approach. In the following cycle of performances, *To friends, upstream and downstream, downwind and upwind* (2023-present), I tried to weave a narrative between personal and collective, between North and South, following my diasporic subjectivity, and inviting more hands in the performances themselves. An upcoming publication and a film seek to tell this larger, yet also very personal, story.

HW What can we achieve with cultural production?

MN On one hand you want to use poetics to stimulate or celebrate collective action and the other aim is to deconstruct and dismantle undemocratic power. The powers that be can afford not to listen to other knowledges, not to learn. Interactions between corporate science and other types of knowledge ('local', 'traditional' et cetera) are on unequal terms, and if they take place, it is under a form of 'consultation' or 'dialogue', but the decisions have largely been taken. The case in question, for example: Rio Tinto commissioned several key faculties and public institutions to perform specific studies, which also meant that many academics were either obliged or felt obliged not to take a public stance on the project. Luckily, there has been a number of scientists who did speak out and support the movement, such as the article you cite in the introduction.

We often don't have a grounded understanding of what is possible in and through culture and what is not because of power structures and infrastructures. Same with academia. Beyond knowledge and poetry, we do other types of work, and that's where actually everything else happens, where reproductive work happens, where productive work happens, where all other types of work happen. A lot of this work does not necessarily happen through art and academia, it does not wait. The communities go down and clean this river. There is a point where we can't just talk about it anymore, it's about doing.

We need to acknowledge the work that frontline water and air defenders do—it is environmental justice work, cognitive, informational, affective, and manual. It's just absolutely insane amounts of work; it's not only mostly unpaid, but it's also criminalized and oppressed and suppressed. Recently, there has been a series of arrests of activists protesting the lithium projects of Rio Tinto and others across Serbia, and they are facing heavy charges.

HW In your writing you also point at early twentieth-century protests where peasants and miners joined forces.[8] There seems to be a potential to create new alliances around this river or around the idea of pollution?

8. mirko nikolić, 'Shutting Down Monocultures of Capital: Reinventing Labor Struggles by Forging the Connections Between Rural and Urban Communities', 17 July 2023, *Balkans Unmastered* www.berlinergazette.de/author/mirko-nikolic/

MN The experience and knowledge of a rural worker is essential. They have desires closer to the land they inhabit than an engineer or a manager. There arguably still are gaps between environmental and labour movements. And thinking around both, there's so much work in between. We cannot really say that there is a clear common vision yet, but there are many bridges being built. It takes time and work, again.

HW Was the performance also a way of forging these kind of alliances?

MN Yes, I aim to create a window to actually do all this work, which is a totally different type of work, it's much slower, it is about total reinvention. In this other space, we can imagine forms of 'mixed' urban-rural labour organizations, especially through cooperatives and commons. But these alliances are also of a completely different scale than art and research, it takes collective effort.

I recently discovered that my state-run pension fund in Sweden invests in the companies that extract and destroy places, the communities and their living environments, in the Balkans and across the continents, such as Rio Tinto and many others. It is not very hard to learn about the controversies and contestations around these companies. Lots of knowledge is produced and made available by independent sources and environmental organizations. One wonders then, in this context, what is ignorance and what is indifference, and how to push against them. The workers, the citizens, are most likely not happy that their savings undermine the existence of others. So, the question is what kind of meetings and understanding can be established across these vast differences created by the powers that be? What are the justice-based alternatives for rearranging our collective lives in solidarity across the hemispheres?

7.7 Liquidity

INTERVIEW WITH JULIEN FARGETTON
BY HENRIËTTE WAAL & CLEMENS DRIESSEN

Bhishti, traditional water carrier, Delhi, India. © Sahapedia 2024

The city of Vevey is the home of the Nestlé company headquarters. Nestlé is the world's largest producer of bottled water. In 1878, Henry Nestlé, the founder of the company, gave Vevey a public fountain to offer the water from the natural springs to the population.

When Julien Fargetton was asked to contribute to Food Culture Days, an art festival in Vevey, he came across a newspaper article about the multinational company privatizing water across the globe.

JF My original idea was to deliver water in Vevey using this animal leather bag that I saw someone carry on the street to distribute water in India, in an image next to the article in Le Monde about Nestlé privatizing drinking water and depleting the groundwater, which impacts water availability and food production in the region. And I found someone to send a Mashq, the traditional water bag, from India to Vevey. But then Covid happened, and the festival

Bhishti, traditional water carrier, Delhi, India.
© Sahapedia 2024

was postponed, and there was no one to pick up the package—so it was sent back to India and got destroyed. Two weeks before the festival I heard this, so I went to the flea market in Lisbon, and I got these two large bottles used for olive oil and wine. And I did the performance using these, delivering water from Henry Nestlé's public fountain in a village near Vevey and offering it for free to the cooks in the festival.

I often carry things in my works. I went with my two big bottles and had to walk five kilometres. It was around a thousand steps to go up to the fountain. I went there and I filled the bottles with water. And then I walked back to deliver this water. It was so heavy and it was physically hard. I felt pain on the way back.

Walking and carrying around twenty kilos of water also made me think about the fact that we are sweating. The sweat then evaporates, goes into a cloud and will turn into water again, which we may drink. It's this cycle, in which we take part.

HW/CD How to take on this massive issue of privatization and these giant corporations. Would you say your approach is about making things very local and personal?

JF I walk a lot and work with what I find on the streets. Doing that just makes me think about our situation, the context, and from this, I try to go further, to be something strange in the landscape.

I try to find details which you can open up to bigger things. I like these kinds of tiny actions. I don't like to protest or occupy and take up space. I'm someone who's a bit shy and so it's not my thing to go on the big stage and claim something. I'm just trying to do something, to express myself in a tiny way and to find the right way to talk about something. I want to talk.

To be very simple is not always easier. We always try to be complex, but to stay simple, it's a lot of work just to find the right idea and the right action. It's often easy to hide behind complexity. Also for large companies.

New Delhi is 10,000 kilometres from Switzerland, but in a way they are linked together through this town, through this company, which is part of the life for people from India, because they have their brand name written on this water bottle: 'Pure Life'.

I would have been happy to talk about this with people in Switzerland. It would have probably worked better if I would have had this leather bag, as a way that people come and talk to you: 'What are you doing? What is that?' Those types of simple questions. To then just interact socially.

Curators tend to invite Julien for his combination of humour and poetry—but he doesn't see it as trying to be funny. He works by honestly responding to a particular situation.

HW/CD Could humour be a way to confront these big powers? Or how did you conceive the project?

JF Humour. Yeah. I think it would have been strange to have me walking with this big leather bag. I thought it was the best way also to interact with people on the streets, going

Julien Fargetton, 'Liquidity', tapping water from the Nestlé source in Glion, 2022

from one fountain to another with this big bag and doing this back-and-forth. I suppose that a lot of people in the town work for the Nestlé Company, so it would have been a way to question them straight away. But I'm not from the area, so it's also just me passing by with an idea that I read in a newspaper and my own opinion about this. It's not questioning so much, but finding a way just to talk with people. And to do so in this public place surrounded by mountains where the water is running constantly; it's just very peaceful.

I formed my opinion in a way, but I'm always open to change my opinion, or just to diversify or complexify my opinion because also I don't know all of the points of view. So, it's always interesting to discuss and gather information or stories from people from a place and to mix it with your own feelings.

7.8
Mining Water/Mien Bléëch

By Henriëtte Waal

2015, Limburg, Belgium

Henriëtte Waal, *Mining Water/Mien Bléëch* water bottles, 2015, Limburg, Belgium. Courtesy Z33 Hasselt, Belgium. Photo: Sjoerd Knibbeler

Mining in the Limburg region of Belgium has caused the land to subside. The village of Eisden has dropped in elevation by as much as twelve metres. As a result, a lot of groundwater flows to this village, and it has to be pumped out constantly. Twelve Olympic swimming pools of top-quality potable water have to be discharged daily into the canals and the river Meuse, even though this sometimes causes flooding. That is a huge quantity of potable water in a world where half the people have to cope with a scarcity of water. Meanwhile, Belgians prefer to drink bottled water, consuming on average six times more of it than Dutch people do. Clever marketing by mineral water companies in Belgium has created the myth that water from bottles is cleaner, more natural and better tasting. As a result, Belgians are among the world's biggest drinkers of bottled water.

With the **Mining Water/Mien Bléëch** project, I illuminated this economic and cultural paradox, and designed a series of distinctive and refillable bottles for 'tapped water'. The bottle design reinterprets the refillable aluminium drinking bottle that mineworkers used below ground: 'mien Bléëch', in mineworkers jargon means 'my water bottle', from the German word for 'tin'. In contrast to their pressed aluminium predecessors, the new bottles are made of stainless steel and formed by means of hydroforming technology, using water pressure. This technique requires no moulds, making the process ideal for the production of small batches. Available in three sizes, the bottle comes with a film in which former mineworkers and other residents from Eisden talk about mine water, pipe water and drinking underground, where water and one's personal drinking bottle were a much-cherished good.

Henriëtte Waal, Ruben Castro and Roel van Tour, *Mining Water/Mien Bléëch*, Testimonials from Eijsden, Belgian Limburg. About mine water, pipe water and under ground drinking, 2015, film still, 13 min

Hydroforming by Expansor Nederland.
Developed in the context of exhibition 'Beyond Food and Design', 2015.
Supported by Kunstencentrum Z33 Hasselt, Belgium.

Conclusion

Conclusion

'Much more than an anticipation of or a simple desire for a coming good, hope is an effort to care for that possibility in committed, practical, and situated ways. From this perspective, hope does not pre-exist acts of hopefulness. Hope grows and expands as we actively care for the future. It is an iterative, cumulative, self-reinforcing work, an effort to do whatever one can—within the context of the many, always unequal constraints that shape our possibilities—to nurture a better future into being.'[1]

1. Thom van Dooren, *The Wake of Crows: Living and Dying in Shared Worlds*, Columbia University Press, 2019, p. 209.

CONCLUSION

BY HENRIËTTE WAAL & CLEMENS DRIESSEN

When people start to ask for hope, you know the situation is getting dire. This was not a book full of solutions. No happy marriage between economy and ecology. No quick fixes. No best practices. No clear culprits, nor saviours. These water works offer an invitation to do something, to come together around water, to imagine and make, even—especially—when faced with the overwhelming scale of what is at stake and unknown but troubled futures.

Making can seem futile against massive powers that seek to control, isolate, monopolize; while facing melting glaciers, saline water creeping into the soil, flash floods and multinational corporations polluting and bottling.

Is it vested interests that with all their might hold on to the idea that 'water is just water'? Or could it also be a lack of imagination? Moving across times and places, we have

Opening image: *Water Fairy*, from Lepenski Vir site located on the right bank of the Danube in eastern Serbia, within the Iron Gates gorge, Serbia, 7000–6500 BC. Courtesy Lepenski Vir Museum

seen sixty odd ways in which water is never just water. These designers, artists, hands-on makers, composers, stubborn engineers and curious anthropologists (and vice versa), all bring a certain openness to these issues—makers will make, not giving up in the face of ecological and social disruptions, but imagining different ways of responding, caring, turning upside down, relating.

These water works are about the joy, the poetry, the sorrow and the frustration, the beauty and the dread that honest attempts to remake watery worlds may lead to.

The projects discussed are never mere projects—but often reflect lifelong dedication to a material, a process, a place, the making of curious collectives, the search for some tangible form of knowledge. The book aimed to gather a broad range of ways of making and talking about these works. In many cases, key outcomes were not so much the immediate impact of the intervention but what these attempts towards changing ecosocial relations revealed.

The upshot is that water itself works—or can work, in an endless variety of ways, as long as it is approached with curiosity and wonder, as a means to build ecosocial relations, but never only a means.

This is not a systematic book, but a personal one. Driven by shared fascinations and the people we found inspiring and happened to come across or know well. One thing led to another; things got out of hand—that is how things can be made. At crucial moments we got help, unexpected enthusiasm, solid support, belief in an outcome that is not clearly defined but will be worth the trouble.

This may be key to finding liveable futures, in which design, art, engineering, composing and researching is about building relations, shaping possibilities, not giving up on making just and poetic ecosocial collectives.

Can We Distil a 'How To' of (Watery) Ecosocial Design?

START

You bring yourself in a certain place or situation. There is something in the water—or there's water in there?—in any case, you have a hunch that there is an ecosocial design potential, and then how to start? Starting means sensing, to be there and observe, and experience. Key therefore is getting access to a place and the practices that shape it. Organizing a 'meeting.' Taking time, hanging around, trying different things to create an interest from both sides.

PEOPLE

Where are the experts-enthusiasts and what are their experiences—how to understand my window, what open-ended questions to ask, what amount of time do we have, and what can we do with that? You may have a vague direction of where you want to get to with them. Exercise expectation management: explain what you are interested in, what part of the process this is. How will it gain meaning for a range of different beings—human and nonhuman. How to engage with scientists you would like to explore different ways of doing things together with, whereby the interest in learning from each other goes beyond the immediate project. It is about being interested in the people or entities you want to work with in a way that goes beyond the project. About developing a project 'lens'. About getting obsessed. You can work with microbial life and aim to squeeze out everything to make them productive, or to wonder how to digest together productively, learning how to collaborate. Communicate complex things in a simple and engaging way, taking everyone along.

SURPRISE

Don't go somewhere thinking you know what the problem is. Be open to be affected, to feel, to see depth, to get a taste for surprise. Maybe it's about how formal political institutions work, or informal political relations. Maybe it's about bringing a government engineer in charge of river management in a position where she can talk with a lady whose house was flooded fifteen times. It is only when you go places, when you immerse yourself, that you realize the microbiologists you have gathered for your project all turn out to play the same musical instruments.

SCALING

Immersing also involves zooming out, drawing connections between an object and the landscape, between people and the regional scale—to see geopolitics in a drop of water. Test things on different scales at the same time. See how logics radically change when moving across scales. Then you can enter into a series of messy loops, where ideas and materialization shape each other. Making is often all about the details, this means adjusting, learning. Make a connection between the microlevel and the planetary, play with language and materials. Use practice-based knowledge on both infrastructural and policy levels. The challenge is to see systems: being open to complexity and interconnectedness, but without seeking to only calculate and optimize simplified models and forgetting about places and those that do the work.

FLOW

Build a focus that is so intense that both ego and self-doubt dissolve and worries about failure fade away, bring focus in the present moment. Be prepared to accept that things won't go the way you imagine but keep the fire going, hold space. Gain momentum: celebrate milestones and use enthusiasm and impetus to make the next steps. Create from an understanding of deep connection, work from a sense of reverence. Be comfortable with the ambiguity of not knowing what you're going to produce.

POSITION

Ecosocial design is hard work, an open process instead of a closed brief takes lots of time; it is about figuring out your role every time again—how to relate, to work with people, governments, companies, activist groups. Don't be critical all the time but do build in moments of reflection. How to be independent and embedded in collectives; is ecosocial design activist or institutional? Two options often appear: going along and risk being used—or taking an outsider position, offering critique as a commentator, responding more strongly, going over the top and thereby making real issues visible. Building collectives and working with established makers in ecosocial settings—those that already know how to weave, grow, harvest, and build in a particular place.

FUNDING

In either case—how to deal with 'NDAs' (Non-Disclosure Agreements)? Contracts? Getting paid? At least realize there is a commercial side to these water works—but also that there's a range of ways of conceiving economies, getting valued, getting your work recognized as relevant, perhaps booked under 'participation' or 'research', or ecosocial maintenance.

MOTIVATION

Do not (always) stick to purity—of water, of moral positions, of ideology. Think of the local German bottled water company, the way it is hunting for sources: always a blend of modernity and craft, tech and tradition, true love of water and profit, community and competition. Imagining possible futures on a regional scale and then walking into a local bar where everyone is drinking knotweed sodas.

AFTERLIFE

Ecosocial design generates particular challenges—how to (re)present these projects, also a decade later? How to archive the process; how to make the work understandable and valuable; how to claim success—and failure—how to promise impact; how to allow others to learn and improve, and not make the same mistakes.

LOOPS

Revisit places and sense how situations have changed; when you go to new places realize these demand something new. Approach things as ever-changing in cycles. Make stories like music, not chronological. Take some time to adjust. Realize the value of long-term commitment.

Water Fairy, from Lepenski Vir site located on the right bank of the Danube in eastern Serbia, within the Iron Gates gorge, Serbia, 7000–6500 BC. Courtesy Lepenski Vir Museum

Water Fairy

By Henriëtte Waal

Water Fairy, from Lepenski Vir, a Meso-Neolithic site on the Danube River in eastern Serbia, 6300–5900 BCE, quartz sandstone, 23.1 kg. On display at the Lepenski Vir Museum.

The salience and social efficacy of boulder artworks from Lepenski Vir might have lain in their expressiveness of revealing, exposing and opening up—sketching the territory of animality. They embody complex beliefs about the connection between animals and humans, that share the same substance of being, which changes form in the perpetual movement of matter.[1]

1. Dušan Borić, 'Body Metamorphosis and Animality: Volatile Bodies and Boulder Artworks from Lepenski Vir', *Cambridge Archaeological Journal*, April 2005, 65.

The Water Fairy appears as a fish-human hybrid. She looks startled, in pain. Her most fish-like feature is her thick-lipped open mouth, with corners drooping downward. Her simple face is depicted on the boulder's surface without a neck, which aligns with the representation of a fish-being, enhanced by her rounded form, while her wide-open eyes take a more human-like position.

The Lepenski Vir is seen as the birthplace of art in Europe. An art of a relational and spiritual kind. The former inhabitants of Lepenski Vir were found buried parallel to the Danube, their heads downstream—facing the direction from which the migratory fish would arrive and depart. Archeologists suggest this positioning reflects the community's belief that the souls of the departed were carried downstream by the migratory fish, specifically the beluga sturgeon, the largest freshwater fish in the world, and returned in the spring.[2]

2. Ivana Radovanović, *The Lepenski Vir Culture: A Contribution to Interpretation of Its Ideological Aspects* 1997, 85–93.

The bond between humans and fish may have carried a mythical dimension, with the rock sculptures serving as the primary medium to express bodily metamorphosis and ancestral connections, bridging the realms of the living and the dead.

Lepenski Vir was submerged under the waters with the construction of the Iron Gates, the largest hydropower dams and reservoir system along the entire Danube. The unique vortex of the river and many spawning sites of the sturgeons were lost. And there, a white UFO landed in the Danube gorge. In this 'white cube' perched a hundred metres higher on the same riverbank, a 1:1 reconstruction of the sunken site can be visited. Here, one can come face-to-face with the haunting boulders—though they seem ready to break free from their perspex pedestals.

Mirroring inner mental worlds and our dependence on the river—where we can learn things from magnificent fish about letting go.

BIOGRAPHIES

Henriëtte Waal is an artist and designer known for her interdisciplinary work in artistic research, environmental design, and community engagement. She is the co-founder of Atelier LUMA, a bio design lab in the south of France, focused on bioregional practices and material engineering. Henriëtte has led fieldwork in Mediterranean wetlands, integrating southern European, African, and Near Eastern remote wetland communities. Her work spans co-fabrication systems for beverages, footwear, colors, crafts and spaces, as part of a search of new eco-social balance. Her projects have been exhibited internationally, highlighting her dedication to social and environmental resilience. Since 2013, Waal has guided Social Design master programs at Design Academy Eindhoven.

Clemens Driessen is a cultural geographer at Wageningen University, where he teaches in programmes ranging from Landscape Architecture to Molecular Biology. He is interested in the ways in which animals, plants, humans, and other organisms shape each other's lives in contemporary and historical landscapes. He believes in making and other forms of intervening to learn new ways in which shared worlds can be reimagined. Clemens has published in scientific and public facing outlets on both wild and domesticated cows, pigs, ducks, high rise farms, milking robots, rewilding, cultured meat, bumblebees, greenhouses, Cartesian landscapes, the plantationocene and beavers.

Bart de Baets is based in Amsterdam. His design for the Sandberg Institute's temporary master programme The Radical Cut Up was nominated for a Dutch Design Award. He designed *On the Necessity of Gardening: An ABC of Art, Botany and Cultivation* (Valiz, 2021), which was awarded with the highest prize in the category Best Book Design from all over the World by Stiftung Buchkunst. He holds a part-time teaching position at the Graphic Design department of the Royal Academy of Art in The Hague, and he has taught at the Gerrit Rietveld Academie in Amsterdam for fifteen years.

Valiz is an independent international publisher and addresses contemporary developments in art, design, urban affairs and visual culture. Their books offer critical reflection and interdisciplinary inspiration in a broad-based and imaginative way, often establishing a connection between cultural disciplines and socio-economic questions. Valiz is based in Amsterdam and connects to authors, artists, designers, institutes, bookshops, distributors and readers worldwide. www.valiz.nl

INDEX OF PERSONS AND ORGANIZATIONS

INDEX OF PLACES AND WATERS

COLOPHON

EDITORS
Henriëtte Waal
Clemens Driessen

AUTHORS
Mari Bastashevski, Janna Bystrykh, Ștefan Constantinescu, Carolina Domínguez-Guzmán, Clemens Driessen, Shahnoor Hasan, Clara Sika Helbo, Ils Huygens, Francesca Masoero, Coltrane McDowell, Joana Pestana, Mariana Pestana, Benedetta Pompili, Esha Shah, Mihnea Tănăsescu, Serina Tarkhanian, John Thackara, Daniela Tokashiki, Jennifer Veilleux, Henriëtte Waal, Ria Waal

PROJECT EDITOR
Eli Witteman/Valiz

COPY-EDITING
Eugenie Tee

NOTES AND INDICES
Elke Stevens

PROOFREADING AND INDICES
Jesse Muller

BOOK DESIGN
Bart de Baets

DESIGN ASSISTANCE
Haoran Zhi
Fabian Reichle

TYPEFACES
CMU Sans Serif
Resonanz (Out of the Dark)

PAPER INSIDE
Arena White Smooth

PAPER COVER
Sulphate carton

LITHOGRAPHY
Wilco Art Books, Amersfoort

PRINTING AND BINDING
Wilco Art Books, Amersfoort

PUBLISHER
Valiz, Amsterdam, 2025
Astrid Vorstermans

This publication has been printed on FSC-certified paper by an FSC-certified printer. The FSC, Forest Stewardship Council promotes environmentally appropriate, socially beneficial, and economically viable management of the world's forests. fsc.org

All essays in this book are peer reviewed.

INTERNATIONAL DISTRIBUTION
NL/LU: Centraal Boekhuis, www.centraal.boekhuis.nl
BE: EPO, www.epo.be
Europe/Asia (except GB/IE): Idea Books, www.ideabooks.nl
GB/IE: Central Books, www.centralbooks.com
USA/Canada/Latin America: D.A.P., www.artbook.com
Australia: Perimeter Books, www.perimeterbooks.com
Individual orders: www.valiz.nl; info@valiz.nl

THIS BOOK WAS GENEROUSLY SUPPORTED BY
Creative Industries Fund NL
EFL-Stichting
Cultuurfonds
Wageningen University & Research

ISBN 978-94-93246-45-4
Printed and bound in the EU, 2025

This publication is also available as an open-access PDF:
ISBN 978-94-93246-57-7